Autonomous Systems and Embedded Intelligence

Autonomous Systems and Embedded Intelligence

Lokesh Pandey

ARCLER
PRESS

www.arclerpress.com

Autonomous Systems and Embedded Intelligence

Lokesh Pandey

Arcler Press
4164 Lakeshore Road
Burlington, ON L7L 1A4
Canada
www.arclerpress.com
Email: orders@arclereducation.com

ISBN: 978-1-77956-972-1 (Hardcover)

ABOUT THE AUTHOR

Lokesh Pandey is currently pursuing his PhD in Mechanical Engineeing from Uttarakhand Technical University, India, where he also completed his M.Tech in Thermal Engineeing. He has more than 3 years teaching experience as a Faculty for Mechanical Engineering. He has published articles in reputed Journals and has also chaired many conferences and workshops.

TABLE OF CONTENTS

List of Figures .... *xi*

Preface .... *xvii*

Chapter 1 **Introduction to Autonomous Systems** **1**

Introduction 1

1.1. Historical Development 6

1.2. Core Technologies 11

1.3. Applications of Autonomous Systems 16

1.4. Benefits of Autonomous Systems 22

1.5. Challenges and Limitations 26

1.6. Future Trends 31

Summary 36

References 38

Chapter 2 **Embedded Intelligence** **43**

Introduction 43

2.1. Concept of Embedded Intelligence 47

2.2. Embedded Processors 53

2.3. Microcontrollers 59

2.4. Role of AI in Embedded Systems 63

2.5. Case Studies of Embedded Intelligence 68

2.6. Future Trends in Embedded Intelligence 73

Summary 78

References 80

Chapter 3 **Sensors, Actuators, and Perception Systems** **83**

Introduction 83

3.1. Types of Sensors 87

3.2. Data Acquisition 93
3.3. Signal Processing 97
3.4. Sensor Fusion Techniques 102
3.5. Actuators and Their Role 107
3.6. Applications of Perception Systems 112
Summary 117
References 119

Chapter 4 Control Architectures for Autonomous Systems 123
Introduction 123
4.1. Open-Loop Control Systems 127
4.2. Closed-Loop Control Systems 131
4.3. Hierarchical Control Architectures 136
4.4. Distributed Control Architectures 141
4.5. Feedback Mechanisms in Intelligent Systems 147
4.6. Comparison of Control Architectures 152
Summary 158
References 160

Chapter 5 Machine Learning for Embedded Intelligence 163
Introduction 163
5.1. Historical Context 169
5.2. Key Concepts in Machine Learning 174
5.3. Applications of Machine Learning in Embedded Systems 178
5.4. Challenges in Implementing Machine Learning for Embedded Intelligence 183
5.5. Tools and Frameworks 189
5.6. Future Trends in Machine Learning for Embedded Intelligence 194
Summary 198
References 201

Chapter 6 Navigation and Path Planning 205
Introduction 205
6.1. Historical Development of Navigation 210
6.2. Types of Navigation 215
6.3. Path Planning Algorithms 221

6.4. Applications of Navigation and Path Planning 226
6.5. Challenges in Navigation and Path Planning 231
6.6. Future Trends in Navigation Technology 236
Summary 241
References 243

Chapter 7 Communication and Networking in Autonomous Systems 249
Introduction 249
7.1. Wireless Communication Protocols 252
7.2. Vehicle-to-Vehicle (V2V) Communication 257
7.3. Machine-to-Machine (M2M) Communication 262
7.4. Internet of Things (IoT) Integration 267
7.5. Challenges in Communication and Networking 272
7.6. Future Trends in Communication Technologies 277
Summary 283
References 286

Chapter 8 Energy Management and Efficiency 291
Introduction 291
8.1. Power Consumption in Embedded Devices 297
8.2. Battery Technologies 302
8.3. Energy Harvesting Techniques 307
8.4. Low-Power Design Strategies 312
8.5. The Role of IoT in Energy Management 318
8.6. Case Studies in Energy Efficiency 322
Summary 327
References 329

Chapter 9 Human-Machine Interaction (HMI) 333
Introduction 333
9.1. Historical Development of HMI 337
9.2. Types of Human-Machine Interfaces 342
9.3. The Role of Artificial Intelligence in HMI 346
9.4. Usability and User Experience in HMI 351
9.5. Challenges in Human-Machine Interaction 356
9.6. Future Trends in HMI 361

Summary ... 365
References ... 368

Chapter 10 Safety, Security, and Ethics of Autonomous Systems ... 373
Introduction ... 373
10.1. Safety Concerns in Autonomous Systems ... 377
10.2. Security Challenges in Autonomous Systems ... 383
10.3. Ethical Implications of Autonomous Systems ... 388
10.4. Regulatory Frameworks for Autonomous Systems ... 393
10.5. Public Perception and Acceptance ... 398
10.6. Technological Advancements and Innovations ... 403
Summary ... 408
References ... 410

Index ... 413

LIST OF FIGURES

Figure 1.1. Leading industry verticals in the autonomous systems market
Figure 1.2. Cyber-physical systems (CPS)
Figure 1.3. History of autonomous system
Figure 1.4. Autonomous technology
Figure 1.5. C-V2X (cellular vehicle-to-everything)
Figure 1.6. Multiple sensor perception system
Figure 1.7. Application of autonomous systems
Figure 1.8. Autonomous vehicles
Figure 1.9. Drones and aerial systems
Figure 1.10. Benefits of autonomous
Figure 1.11. Connected and automated vehicles
Figure 1.12. An intelligent medical data augmentation
Figure 1.13. Delivery as a service
Figure 1.14. IoT platform integration
Figure 2.1. Embedded intelligence
Figure 2.2. Real-life applications of embedded systems
Figure 2.3. Intelligent embedded software
Figure 2.4. Predictive maintenance and how it is transforming the manufacturing sector
Figure 2.5. Applications of embedded intelligence in various fields
Figure 2.6. Embedded processors
Figure 2.7. Digital signal processing (DSP)
Figure 2.8. Architecture of an embedded system
Figure 2.9. Microcontroller and its types
Figure 2.10. Deep neural network
Figure 2.11. Embedded artificial intelligence for business purposes
Figure 2.12. Benefits of embedded systems
Figure 2.13. Smart home devices

Figure 2.14. Industrial automation
Figure 2.15. The future of AI in embedded systems
Figure 3.1. IMU sensor orientations
Figure 3.2. Architecture of the perception system
Figure 3.3. Types of sensors
Figure 3.4. Vision sensors work
Figure 3.5. LiDAR and how does it work
Figure 3.6. Sensor technology
Figure 3.7. Challenges in data acquisition and storage
Figure 3.8. Signal processing
Figure 3.9. Classification of the signal processing algorithms with specific options
Figure 3.10. Sensor fusion
Figure 3.11. Kalman filter
Figure 3.12. Applications and benefits for electric actuators
Figure 3.13. The global positioning system
Figure 3.14. Advanced driver assistance systems (ADAS)
Figure 4.1. High-level diagram of the control system architecture of an autonomous vehicle
Figure 4.2. Open-loop control system
Figure 4.3. Open-loop control system for filling a vat with liquid via a pump
Figure 4.4. Closed-loop control system
Figure 4.5. Closed-loop control system, the one in which the output is fed back and compared with the desired output (reference input). If there is a deviation, the controller generates the necessary control signal to alter the conditions of the plant or process in order to achieve the desired output (controlled output)
Figure 4.6. Hierarchical control architecture
Figure 4.7. Perceptual control theory
Figure 4.8. Distributed control systems
Figure 4.9. Classification of resource-constrained project scheduling problems
Figure 4.10. Understanding DCS in industrial automation
Figure 4.11. Feedback in artificial intelligence
Figure 4.12. Perceptual control theory
Figure 4.13. Different types of control systems architecture
Figure 4.14. Hierarchical distributed control network

Figure 5.1. Embedded machine learning

Figure 5.2. Decomposition methods for machine learning with small, incomplete, or noisy datasets

Figure 5.3. Machine learning methods

Figure 5.4. Integrating data: AI and machine learning

Figure 5.5. Machine learning types

Figure 5.6. Supervised versus unsupervised learning

Figure 5.7. Machine learning applications

Figure 5.8. Wearable technology

Figure 5.9. Machine learning challenges and solutions

Figure 5.10. Data security versus data privacy

Figure 5.11. Convolutional neural networks (CNNs) and layer types

Figure 5.12. TensorFlow lite

Figure 5.13. Object classification with edge impulse using raspberry

Figure 5.14. Machine learning trends

Figure 6.1. Navigation and path planning

Figure 6.2. Path planning methods

Figure 6.3. The evolution of navigation

Figure 6.4. Ultimate guide to survival navigation tools

Figure 6.5. Network GPS navigation in a modern city with future technology

Figure 6.6. Celestial navigation

Figure 6.7. Terrestrial and coastal navigation

Figure 6.8. Path planning algorithms for mobile robots

Figure 6.9. A* Algorithm

Figure 6.10. Dijkstra's algorithm

Figure 6.11. Path planning for autonomous vehicles

Figure 6.12. Researchers developing drones that can build structures

Figure 6.13. Future of maritime navigation

Figure 6.14. Hierarchical reinforcement learning

Figure 6.15. High-speed civil transport

Figure 6.16. Advances in sensor technologies

Figure 7.1. Communication and networking in autonomous systems

Figure 7.2. Wireless communication protocols playing a pivotal role in advancing communication frameworks necessary for the effective operation of autonomous systems

Figure 7.3. Wireless communication protocols in autonomous systems
Figure 7.4. Vehicle-to-vehicle (V2V) communication
Figure 7.5. Example of safety applications based on V2V communications
Figure 7.6. Importance of machine-to-machine communication for manufacturing
Figure 7.7. Applications of M2M in industrial automation
Figure 7.8. Internet of things (IoT) integration
Figure 7.9. Data exchange interoperability challenge
Figure 7.10. Robot wireless sensor network (RWSN)
Figure 7.11. Latency issues in autonomous systems
Figure 7.12. Future communications fast technology
Figure 7.13. A simplified tier topology of edge computing-based connected autonomous vehicles
Figure 8.1. Energy management and efficiency
Figure 8.2. Low-power embedded systems design and power management
Figure 8.3. Key factors for optimizing power consumption in an embedded device
Figure 8.4. Low-power consumption as a critical factor in smart devices
Figure 8.5. Three battery technologies that could power the future
Figure 8.6. Advancements in battery technology and their implications
Figure 8.7. Energy harvesting sources
Figure 8.8. Wireless sensor network (WSN) with one desirable sensor
Figure 8.9. Energy harvesting techniques and trends in electronic applications
Figure 8.10. Integration of renewable energy in microgrids and smart grids in deregulated power systems
Figure 8.11. Proposed system model
Figure 8.12. IoT in energy management
Figure 8.13. A network model for wireless powered communication
Figure 8.14. Wireless power transfer circuit working
Figure 9.1. Human-machine interaction (HMI)
Figure 9.2. The evolution of human machine interfaces
Figure 9.3. Graphical user interface
Figure 9.4. Types of human-machine interfaces
Figure 9.5. Exploring voice user interface (VUIs)
Figure 9.6. The role of AI in connected automation platforms
Figure 9.7. Five key points on how AI and ML enhance business intelligence (BI)
Figure 9.8. User experience in autonomous cars

Figure 9.9. HMI software development
Figure 9.10. Challenges in human-robot interaction
Figure 9.11. HMI improving user interaction in control systems
Figure 9.12. HMI trends engineers
Figure 9.13. The future of augmented reality
Figure 10.1. The ethics of autonomous vehicles: Balancing safety and control
Figure 10.2. Safety and security in autonomous vehicles
Figure 10.3. Implementing redundancy and failover mechanisms
Figure 10.4. Autonomous weapons systems (AWS)
Figure 10.5. Autonomous vehicles be protected against cyber security threats
Figure 10.6. Vulnerabilities to cyber-attacks
Figure 10.7. Data privacy and protection issues
Figure 10.8. High performance work system
Figure 10.9. Ethical dilemmas and privacy issues in emerging technologies
Figure 10.10. Algorithmic bias and fairness
Figure 10.11. Work on the future of regulating autonomous systems
Figure 10.12. Emergence of lethal autonomous robotics
Figure 10.13. Mean ratings of blame on the target vehicle and post-incident trust in both the human driver and autonomous system conditions across all scenarios
Figure 10.14. Autonomous vehicle perception
Figure 10.15. Value sensitive design

PREFACE

Technological advancements in autonomous systems are revolutionizing industries, societies, and daily lives in deep and significant ways. Machines that previously were performing rigid orders can now sense, reason, and adapt in real time to changing circumstances. From autonomous vehicles navigating city roads to robot caregivers assisting healthcare, the ability of systems to independently make choices without continuous human intervention has enabled efficiency, safety, and innovation. The revolution is not limited to a single industry—transportation, defense, energy, agriculture, and communications all have an interest in the growing autonomy of machines. These technologies challenge concerns of trustworthiness, reliability, and the responsibility of getting such systems to behave in ways consistent with human values.

Embedded intelligence is the heart of the revolution. It gives machines the power to reason over sensor inputs, learn from experience, and tailor performance while running under size, power, and processing limitations. Unlike legacy computing systems, embedded intelligence must balance autonomy and responsiveness, reliably functioning in the uncertainty of the real world. As societies become increasingly interconnected, the demand for systems that are able to perceive, decide, and act with minimal supervision continues to grow, thus making it a priority to comprehend the principles, technologies, and implications of autonomous systems in great detail.

Autonomous Systems and Embedded Intelligence is a unified discussion of the ideas, methods, and uses that define this domain. The book combines theoretical bases with practical design requirements and offers a wide scope for understanding how autonomy and embedded intelligence intersect to make machines capable of performing significant action. The chapters are structured to move from fundamental ideas to sophisticated applications and moral issues, equipping readers with a complete course of study from start to applied expertise.

Chapter 1 presents the topic of autonomous systems, outlining their history, establishing basic requirements, and opening the embedded intelligence role in enabling autonomy. Chapter 1 sets the stage by outlining the general scope of autonomous systems and explaining why they are distinct from standard automated devices.

Chapter 2 treats the fundamentals of embedded intelligence. It discusses computational frameworks, embedded processors, and the design principles of resource-limited systems that are nevertheless capable of displaying adaptive and intelligent behavior. Hardware and software elements are provided to readers that come together to make decision-making in the embedded environment possible.

Chapter 3 discusses sensors, actuators, and perception systems. These are the components which allow machines to perceive their environment and react to it. It discusses how sensor data is processed, combined, and converted into actionable knowledge, and how actuators make decisions and implement them as motion and reaction in the physical world.

Chapter 4 treats control structures of autonomous systems. Hierarchical, distributed, and hybrid control structures that manage decision-making and behavior are treated. The strengths and limitations of some control strategies are contrasted with real-life examples of how they are used across industries.

Chapter 5 introduces machine learning for embedded intelligence. It highlights the use of algorithms that enable systems to learn by experience, get better over time, and refine performance. Special emphasis is given to methods suited for embedded contexts where memory and power are constrained.

Chapter 6 covers navigation and path planning. The chapter addresses how autonomous robots determine their position, predict motion routes, and construct efficient and secure paths through structured and unstructured environments. Mapping, localization, and avoidance of obstacles are integrated and explained through real-life examples.

Chapter 7 discusses communication and networking in autonomous systems. As autonomy is increasingly dependent on connectivity, the chapter investigates protocols, wireless technology, and methods for secure and efficient communication among machines, humans, and infrastructure.

Chapter 8 addresses energy management and energy efficiency. Since embedded systems are typically powered by finite sources of power, energy optimization is vital. The chapter addresses ways to manage energy consumption, increase the utilization of batteries, and design systems that offer long-duration autonomous operation.

Chapter 9 covers human–machine interaction (HMI). In this chapter, autonomous systems are dealt with as to how they are designed to communicate with humans in natural and trustworthy ways. Interface design, feedback mechanisms, and psychological aspects of man's reliance on autonomous technology are topics covered.

Chapter 10 addresses safety, security, and ethics of autonomous systems. The chapter addresses protection to prevent accidents, protection from malicious interference, and the ethical principles that guide development and deployment. Students are encouraged to think about accountability, trust, and societal consequences of embedding autonomy into daily life.

The book is written to be generally accessible. It will provide a compact and coherent overview for engineering, computer science, and robotics students to autonomous systems concepts. Researchers and professionals in related domains will appreciate the exposition of new technologies and design practices. Decision-makers in business and policy will discover insights into safety, ethics, and deployment fundamental to decision-making in an increasingly autonomous machine-driven world. For study, research, or implementation, the book is designed to offer both conceptual and practical understanding and insight into the rapidly expanding market for autonomous systems and embedded intelligence.

CHAPTER 1

INTRODUCTION TO AUTONOMOUS SYSTEMS

Contents

Introduction 1
1.1. Historical Development 6
1.2. Core Technologies 11
1.3. Applications of Autonomous Systems 16
1.4. Benefits of Autonomous Systems 22
1.5. Challenges and Limitations 26
1.6. Future Trends 31
Summary 36
References 38

INTRODUCTION

The definition of autonomous systems is in a wide range of technologies and methods that enable machines to behave independently, make choices based on their environment without direct control from humans. Autonomous systems are for perceiving, reasoning, and acting in real-time with the combination of sensors, artificial intelligence, and sophisticated algorithms. The potential of adaptation to diverse states and learning from experiences is what autonomous systems are all about. This critical aspect enables such systems not only to undertake pre-programmed tasks but also to develop their own capabilities through continuous interaction with their environment. An important part of this definition is the focus placed on automation, which has progressed incredibly from its initial applications in factory environments, paving the way for more sophisticated systems of today across many sectors, transportation, manufacturing, and medicine being a few examples. The technical context of autonomous systems is constantly being updated and renewed, and this presents opportunities as well as challenges. In an attempt to advance these systems further, various subdisciplines are being established such as robotics, artificial intelligence, and machine learning, each contributing various methodologies and conceptual frameworks.

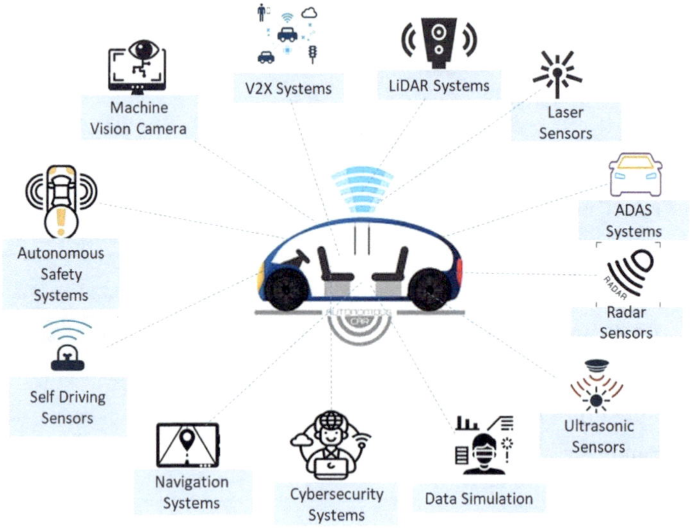

Figure 1.1. *Leading industry verticals in the autonomous systems market.*

Source: https://media.licdn.com/dms/image/v2/D5612AQEy-zDI4GLe4g/article-cover_image-shrink_720_1280/article-cover_image-shrink_720_1280/0/1689369974759?e=2147483647&v=beta&t=-rEVV2cqXd8e3SNpLG987YRcpRUg-rmZ4j1G5_gW2Ss.

The spectrum of autonomous systems ranges from simple automated appliances, such as robot cleaning devices, to advanced multi-agents systems that can dynamically coordinate and collaborate, such as complex drone groups or self-driving vehicles. Further, as abundantly established in joint activities and studies, such as those carried out by organizations such as NASA, the combined effect between hardware systems, software platforms, and functioning rules is a key driver to taking these technologies forward (N/A, 2025). Embodied evolution and other distributed learning methods have crucial roles to play in further advancing the efficiency and effectiveness of such systems, especially in relation to swarm-like collective behaviors (Alba et al., 2018). Examining the implications of autonomous systems reveals that their definition goes well beyond technological concepts to include ethical implications and societal ones. With autonomous systems more and more becoming a part of everyday life, issues related to safety, security, and responsibility must be considered. The potential of the reduction of errors along with the efficiencies generated in processes attests to the potential of such systems; however, this must be weighed against the responsibility that comes from letting such technologies loose. Therefore, understanding the definition of autonomous systems would not just be coupled with their technical ability but would also require a multidisciplinary understanding involving engineers, ethicists, and policymakers. This wide-angle vision is required to navigate the future path of these systems, promote their adoption while limiting potential risks, and address societal concerns from the public about their impact on society (Figure 1.1).

Explanation of Autonomy in Technology

The concept of technology autonomy is actually concerned with the capacity of systems to operate independently in their settings. Autonomy of this kind has its characteristic mark in the utilization of advanced algorithms and artificial intelligence, whereby machines are able to process sense data, make decisions, and act without direct human intervention. Autonomous technologies are increasingly common in most fields of use, including transportation, medicine, and manufacturing. For instance, within autonomous cars, the use of cyber-physical systems (CPSs) provides room for real-time interaction between hardware and software components to facilitate sophisticated decision-making processes. CPS development provides an example of how technology is able to migrate away from human-centric operations into more autonomous functions, demonstrating continuing advancements in autonomy that power the development of applications with capabilities for adapting and responding to dynamic conditions (Aghamohammadi et al., 2018). Moreover, the implications of autonomy carry over into areas of high importance such as healthcare, where robotic systems are being brought into operating rooms. Autonomous surgical robots are a significant step in health technology, with the potential to improve accuracy and efficiency in the provision of medical treatment. The robots can move with sub-millimeter precision, using real-time biosignals to control surgical deployment. Autonomy at this level enables interventions above human capacity or optimizing the overall outcome of surgical intervention by minimizing fatigue and human error. However, growing autonomy in surgery raises ethical and operational issues regarding the role of human surgeons alongside the implication of automatons being employed to carry out life-critical tasks. By surmounting such challenges, the healthcare sector is better able to exploit the benefits of autonomy in technology (Das et al., 2017). The study of autonomy in technology thus raises profound questions about the future interactivity between humans and machines. As autonomous systems develop, the boundaries of what may be achieved without human intervention are being pushed further. This shift needs to re-evaluate ethical standards, safety codes, and legal landscapes to ensure technology still serves society. For example, in autonomous vehicle landscapes, establishing stringent safety protocols is essential in a bid to endure the effects of safety hazards from non-controlled environments. Similarly, with medical uses, R&D needs to consider machine decision-making outcomes within life-and-death decision-making. Lastly, understanding and collaborating with the complexity of autonomy in technology is key to enabling innovations that are human-centered and contributing to overall well-being in numerous areas of life (Figure 1.2).

Figure 1.2. *Cyber-physical systems (CPS).*

Source: https://www.imm.dtu.dk/~jbjo/CPS_Diagram.png.

Types of Autonomous Systems

The classification of autonomous systems also largely hinges on their operating complexity and autonomy levels. The most basic level is fully autonomous systems, where they can function independently without any human interference. They can be autonomous cars or drones, which employ sophisticated sensors and algorithms to navigate and decide in real-time. Their design can also include aspects of machine learning and artificial intelligence so that they can develop based on varying circumstances and enhance their performance with time. The relevance of such systems is not only because of their ability to perform work with minimal human involvement but also due to the possibility of enhancing efficiency and safety in various areas, including transportation and logistics. Integration of assurance and safety models, as shown by Bloomfield et al. (2020), becomes necessary to offer reliability for such intricate systems. Semi-autonomous systems are yet another highly significant category with the requirement of human observation or intervention on particular tasks. Often applied where complete autonomy is not feasible, these systems achieve a balance of human control and automation. For instance, in autonomous driving technology, vehicles are able to run on freeways but require driver involvement on inner-city roads or difficult traffic situations. This two-mode operation can enhance trust and customer satisfaction as well as the benefits of automation. The synthesis of machine efficiency and human intuition is essential in such systems, with implications for the roles that human operators play in maintaining security and effectiveness. Thus, the relationship between human intervention and technology capability needs to be understood, with applicable information on assurance and resilience offered by Bloomfield et al. (2020). Lastly, decentralized autonomous systems exist, in which disparate agents are acting independently but together in order to achieve shared objectives. This arrangement is prevalent in many applications, including networked robotic systems and smart grid control. Unlike decentralized control strategies, decentralized systems depend on autonomous agent behaviors to deliver resilience and adaptability. Agent group formation is an inherent property of the system in which effective group performance arises through interdependent actors' knowledge-based interactions. As shown in, the kind of systems tends to stabilize and mature when agents concentrate on group-level learning instead of individualist approaches. An understanding of such interactions not only sheds light on the self-organizing potential of independent systems but also the importance of community dynamics in ensuring effective functioning and survivability within increasingly complex environments.

Importance of Defining Autonomous Systems

The significance of correctly defining autonomous systems cannot be overstated, especially as it disseminates in various sectors, including transportation, health care, and manufacturing. Correct definition ensures a common perception among researchers, engineers, policymakers, and the public about the characteristics and limitations of autonomous systems. This basis is required to establish effective regulations and standards that govern their use. Division between each other of various degrees of autonomy, from fully automated systems to others that require human input, provides clarity and enables stakeholders to effectively assess the risks and benefits of deployment. In the absence of common terms, the discussion on autonomous systems may become incoherent, leading to possible misuse and system breakdowns that can undermine public trust in these technologies. Another critical aspect of defining autonomous systems is safety and ethical concerns that surface in their design and implementation phases. As complex technologies, autonomous systems have the tendency to rely on algorithms to make decisions in real-time based on data inputs. A detailed definition provides the grounds for addressing problems in terms of accountability, especially in ethically charged conditions common in innovations like driverless vehicles. The integration of ethical principles into engineering and technological design necessitates standards for what constitutes autonomous operation. As our example with autonomous vehicles shows, the challenges to moral decision-making can be troublesome; thus, clear definitions can guide engineers to create solutions that encompass ethical principles. These systematic approaches of deployment not only ensure conformity with safety norms but also enhance social acceptance for these technologies (Dodig-Crnkovic et al., 2018). Moreover, the necessity for a definition of autonomous systems also extends to promoting interdisciplinary cooperation. Due to the multiaspect nature of autonomous technologies, input from different disciplines—computer science, engineering, ethics, and law—is necessary for their successful development. An established definition allows interdisciplinary experts to engage in beneficial discourse, suggesting concepts that combine technical possibilities with societal consequences. This cross-disciplinary synergy is most essential in settings like medicine, where autonomous systems may influence patient outcomes or operational efficiencies. In this setting, a precise definition establishes a shared terminology that transcends disciplines so that comprehensive evaluations of the potential advantages and disadvantages of autonomous interventions become possible. Lastly, a common understanding of what an autonomous system is paves the way for new means by which their potential can be responsibly drawn upon (Fagerberg et al., 1999).

1.1. HISTORICAL DEVELOPMENT

The evolution of autonomous systems through history can be traced back to the initial mechanical advancements of the 20th century, which paved the way for the modern sophisticated technologies of today. In the first place, the automation paradigm initially aimed at automating repetitive processes. These autonomous systems were mostly operated under human guidance and lacked advanced cognitive abilities. But with the evolution of technology, particularly computers and computer software, the boundaries of automating work expanded. Notably, advances in artificial intelligence in the 1950s introduced the concept of machines that learn and adapt, a key aspect of what we now understand as autonomous systems. It was this revolution that compelled researchers and engineers to study many complexities associated with autonomy, cognitive processes, and eventually, social interactions, leading to an increased understanding of how machines can operate independently within intricate environments, as explained in cognition and cognitive robotics research (Arsiwalla et al., 2017). Autonomous system creation is more than just technological innovation; it also reflects broader societal progress.

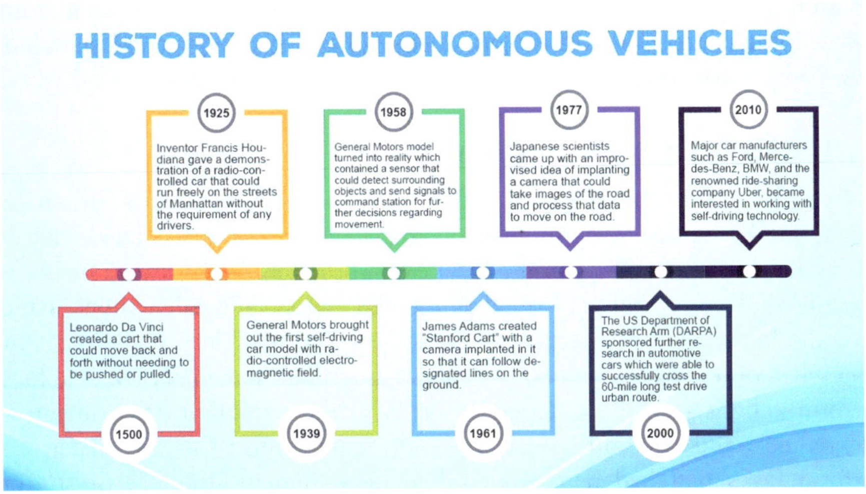

Figure 1.3. *History of autonomous system.*

Source: https://www.researchgate.net/publication/384613916/figure/fig2/AS:11431281281842810@1728026440507/History-of-Autonomous-System-2.png.

With intensified competition in the global market, nations have been striving to emulate effective business models, which have affected the manner in which they embark on automation and artificial intelligence. Emulation is often determined by peculiar national contexts, as uptake of external technology comes to be understood in relation to indigenous routines, practices, cultures, and education systems. For instance, robotics use in production is considerably disparate and is governed by historical training methods and labor relations for every country. The national specificity outlined above advances that copying high-tech technologies is subject to local learning patterns, which can lead to hybrid models of behavior specifically tailored to individual countries, thus shaping their autonomous system trajectories (Lorenz E). The convergence of subject domains in the fields of cognitive neuroscience, robotics, and artificial intelligence has also propelled even further the evolution of sophisticated autonomous systems. State-of-the-art developments have surpassed the scope of simple automation to encompassing consciousness and cognitive sophistication in machines. This has necessitated the need for a taxonomy system that classifies many types of consciousness in synthetic and biological systems and captures their similarities and dissimilarities. The generation of a complexity-based morphospace, which includes autonomous, cognitive, and social factors, provides an organized way of analyzing these systems. In addition to aiding comparative machine consciousness analysis, this framework also guides the design principles needed for successfully engineering conscious machines and hence driving the course of history toward the unified vision of mankind's future with autonomous systems (Arsiwalla et al., 2017) (Figure 1.3).

1.1.1. Early Examples of Automation

The origin of automation is referred back to pioneering practice in various fields, and the success was progress towards what we can observe nowadays in autonomous systems. Some of the early examples include clockwork machines in medieval Europe that performed automated actions and calculations, which gave the knowledge of mechanical precision. The machines had elementary principles of control systems, demonstrating that machinery was able to perform actions partially independently. With the evolution of civilizations, waterwheels and windmills were a glaring application of nature's energy to perform tasks that had heretofore relied upon human strength. Shift from human to machine processes not only lightened workloads but also yielded more consistent and uniform results, which are today cornerstones of stand-alone functions in all sectors, from industry to robotics. In the industrial sector, the 18th and 19th centuries saw the revolutionary use of automation in terms of machines such as the Jacquard loom, which utilized punched cards to control the weaving of complex patterns. This was a key step towards the capability of machines to execute programmed instructions without human intervention. The principles of the Jacquard loom later informed numerous systems, including the first computers, showing the interconnected evolution of automation. All these advances in technology paved the way for cross-industry production efficiency, enabling mass production and consistency that would revamp economies. The introduction of automated systems radically transformed labor patterns and established a configuration that was amenable to innovations to come, as is witnessed in the uses of automation in industries today, for example, autonomous surgical systems which employ robot accuracy within hospitals (Das et al., 2017). As automation evolved further, its application extended extensively across industries. In aviation, automation had already started early with the automation of navigation systems, which reduced pilots' workload and enhanced operating safety. Research by organizations like NASA indicates the evolution of these systems, with their importance in complex environments where better decision-making and timely response are crucial (Kopardekar et al., 2025). The development demonstrates how automation is used to address specific issues, such as reducing human error and enhancing operational effectiveness. Furthermore, contemporary applications like autonomous vehicles and intelligent transportation networks draw from these pioneering efforts, evidencing the long-lasting impact of early automation on autonomous systems today. In keeping with awareness of these innovations, we can more clearly understand how innovative technologies propel today's autonomy technologies as well as their corresponding social implications.

1.1.2. Evolution of Autonomous Technologies

The past of autonomous technology has been totally transformed since the early history of their creation, with exponential increases in computing power, data analysis, and learning algorithms. In the early years, autonomous systems were simple and employed for automating simple processes in quite limited repetitive and predictable situations. But with time, as the computing architecture has evolved, so have their capacities. The interoperation of intercommunicating systems is today a hallmark of this development, facilitating seamless communication and data exchange between various systems. This shift not only enhances operational efficiency but also expands the range of applications of autonomous technology to various sectors, from aerospace to medicine (Finkelstein et al., 1998). As researchers and practitioners produce such systems increasingly sophisticated, so has the focus fallen on addressing the complexities of real-world interaction, making demands for advanced frameworks of system architecture and management based on adaptability and resilience. No more important role is played by artificial intelligence in developing autonomous technology. As advanced algorithms were introduced, computers began to exhibit increasingly complex behaviors, such as making decisions based on vast amounts of data and predicting. This is also complementary with the evolution of cognitive robotics, where machines are not only trained to perform pre-programmed tasks but also learn and adapt through experience. Here social and emotional intelligence are being integrated into systems to facilitate more conciliatory relationships with human beings. This is consistent with newer paradigms that categorize systems based on types of complexities, like autonomous and cognitive. The investigation of the intersection of these categories contributes to a richer conceptual understanding of both synthetic and biological systems, toward a future where machines can possess varying degrees of consciousness and social engagement (Arsiwalla et al., 2017). Future developments of autonomous technologies will inevitably rely on interdisciplinarity research integrating conclusions from artificial intelligence, cognitive neuroscience, and social sciences. This alignment will enable the determination of new design principles needed for the creation of machines not only that are able to act autonomously but also that are able to meaningfully interact with societal environments. The possibility of conscious machines puts forward ethical considerations in terms of responsibilities together with their use and implications of their decision-making processes. Thus, their advancement requires thoughtful reflection on their implications for society and necessitates normative advice for their incorporation into daily life. The adaptive character of autonomous systems, underscored by their increasing sophistication and ability, implies that they not only transform businesses but also redefine human-machine interfaces (Arsiwalla et al., 2017) (Figure 1.4).

Figure 1.4. *Autonomous technology.*

Source: https://krakul.eu/wp-content/uploads/2020/10/krakul-1-scaled-1.jpg.

1.1.3. Key Milestones in Autonomous Systems

Autonomous systems' development is a landmark trajectory of technological advancement that is characterized by a chain of significant milestones that mark significant breakthroughs. The most important milestone was perhaps the development of some of the world's first autonomous vehicles in the late 20th century, which laid the basis for contemporary applications. Pioneering projects such as Carnegie Mellon University's Navlab and University of California, Berkeley's ALV brought forth the possibility of cars moving through spaces autonomously. These early experiments established limits for sensor fusion, navigation, and object recognition that would establish the basis for autonomous developments. By advancing these technologies, researchers began to raise interest in cars as well as robots and other autonomous systems, eventually spreading to more far-reaching implications between industries, from transportation to medicine. With ongoing advancement in research, even more sophisticated milestones were achieved, particularly in machine learning and artificial intelligence. The deep learning that emerged in the 2010s, in particular, sped up a paradigm shift in the way autonomous systems perceive and interact with the world. Advanced algorithms allowed improved data processing and decision-making capabilities, which enabled systems to adapt dynamically under different conditions. This is critical for applications where uncertainty dominates a world, such as city traffic or disaster relief environments. Moreover, the integration of cognitive robotics research, emphasis on embodied cognitive agents, and their sophisticated sensorimotor interaction has driven the advancements in social learning and autonomous cooperation. This technology does come with the demands for systems that not only can autonomously carry out operations but also become capable of learning and improving through their experience, thus increasing their functional applications across diverse contexts (Belpaeme et al., 2010). In the maritime industry, the implementation of Industry 4.0 fundamentals has revolutionized ship designing and operation. The focus on energy efficiency in autonomous ships has introduced considerable challenges while opening up opportunities for innovation in traditional practices in the sea industry. As advanced in more recent studies, the transition towards intelligent manufacturing and design techniques is a reflection of the need for functional and environmentally friendly ships with reduced operational and environmental impacts (Ang et al., 2017). Among the most significant milestones for the sector has been the development of autonomous ships that benefit from novel sensors and artificial intelligence technology to enjoy simplified navigation and reduced cases of human error. This not only endorses operational efficiency but also represents a crucial mile towards full autonomous maritime operations, reflecting the complementary relationship between technological innovation and ecological sustainability. The constant striving for energy-efficient, intelligent ships represents the expanding horizons of autonomous systems in delivering solutions to contemporary world issues.

1.2. CORE TECHNOLOGIES

In understanding the underlying technologies that enable autonomous systems, there is a need to value the central role played by Bayesian inference. This algorithmic approach gives autonomous systems basic capabilities to handle uncertainty and incomplete data. As these systems become increasingly embedded in numerous applications, from robotics to autonomous cars, the essential need for efficient processing ranks high. However, Bayesian inference is not simple in complexity but is a problem known to be NP-hard. Scientists are thus actively working on developing new methodologies to enhance the tractability of Bayesian inference without compromising on its precision. Computational paradigm breakthroughs such as parallel processing through multi-core processors and GPUs are central in shaping the way of future Bayesian computation. These developments not only provide more applicable Bayesian solutions but also promote the creation of more advanced autonomous systems that can learn to deal with practical problems (Alves et al., 2015). A second fundamental technology influencing autonomous systems is the development of networking solutions, particularly vehicular communication. Cellular-Vehicle to Everything (C-V2X) is a significant improvement over traditional Vehicle to Infrastructure (V2I) and Vehicle to Vehicle (V2V) communication by maximizing communications and incorporating robust cellular security mechanisms.

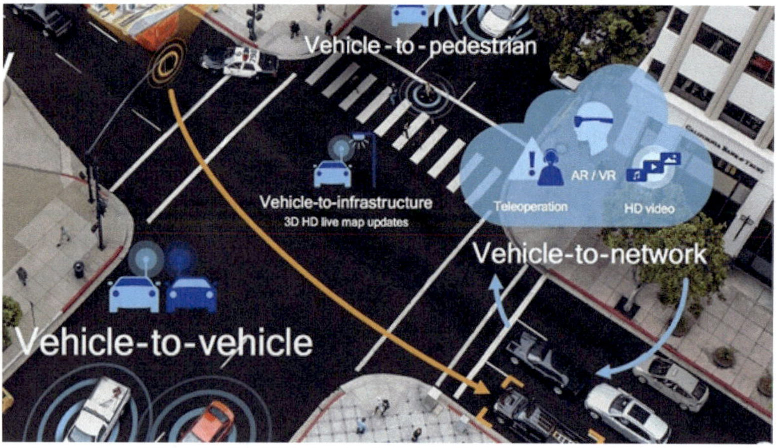

Figure 1.5. *C-V2X (cellular vehicle-to-everything).*

Source: https://cdn.everythingrf.com/live/1558545307929_636941421099334356.png.

The approach overcomes past shortcomings by reducing the number of entities required to support effective communication, hence maximizing operational efficiency. The transition from LTE-V2X to 5G-V2X is a groundbreaking development because the latter accommodates the demands for higher throughput, ultra-reliable links, and low latency. Whereas LTE-V2X set the groundwork, the prospects of 5G-V2X establish a new benchmark required for autonomous vehicle functioning smoothly as central elements of intelligent cities and advanced transportation systems (Guizani et al., 2019). Coupled with the advancement in Bayesian inference and networking, the contribution of data processing and machine learning algorithms cannot be overlooked in the autonomous system framework. These key technologies enable autonomous systems to learn from their environments and make good-informed choices. The use of machine learning algorithms for adaptive learning enables systems to improve with time based on experience. As autonomous systems encounter different situations, the ability to scan massive amounts of real-time data is crucial for ensuring safety and reliability. Besides, the ongoing progress in algorithm efficiency equips such systems to operate in advanced environments with minimal human intervention. Lastly, the combination of Bayesian inference, high-level networking by C-V2X, and advanced machine learning capability lays the foundation for the next generation of autonomous systems, triggering the floodgate of innovation across industries (Figure 1.5).

1.2.1. Artificial Intelligence and Machine Learning

The intersection of machine learning (ML) and artificial intelligence (AI) has been at the core of developing autonomous systems, providing the foundational algorithms to make machines function independently of human input. In autonomous systems, machine learning allows devices to process huge datasets, detect patterns, and improve their function without explicit programming. This capability is most groundbreaking in sectors such as transport, where self-driving vehicles apply advanced algorithms to process sensor data in real-time and make decisions that maximize safety and efficiency. As autonomous systems continue to advance, the engagement between AI and ML will drive revolutions in human interaction with technology, toward intelligent cities and more unobtrusive user interfaces across industries. The ethical implications of introducing AI and ML into autonomous systems cannot be overstated. Their construction and application raise very significant questions regarding privacy, security, and bias. For instance, reliance on data-driven algorithms guarantees that any bias in the data used to train them results in biased outputs, affecting vulnerable groups negatively. This highlights the necessity of proactive intervention during the design phase of autonomous systems. Value-sensitive design (VSD) is a strong framework that emphasizes embedding human values into AI design approaches. Involvement of stakeholders and ethical reflection at the outset can assist developers in ensuring that they work towards developing systems that are not just effective and efficient but also fair and equitable. This approach is particularly relevant in the case of rapidly evolving technologies that pose ethical dilemmas, such as those identified by Bellis et al. (2018). As machine learning and AI technologies evolve, the room for new solutions and issues will grow. Application of these technologies can drive numerous aspects of everyday life, from improving healthcare benefit enhancements by predictive analysis to optimizing resource allocation in urban development. However, the emerging scenario also calls for vigilant surveillance of the social implications of autonomous systems. For example, increased reliance on these systems raises issues of accountability and whether it is compromising jobs in certain areas, an especial risk to which vulnerable groups are susceptible and can be disproportionately affected. Also, the experience of the hybrid war in nations like Ukraine demonstrates the profound impact of technology on societal dynamics, emphasizing the roles youth—a group normally reflective of the risks and gains of digital technologies—play in shaping future interactions with autonomous systems, as discussed by Burov et al. (2020).

1.2.2. Sensors and Perception Systems

Sensors and perception systems integration is a critical factor in building autonomous systems that enable machines to engage with and comprehend their environment effectively. Sensors form the primary interface between such systems and the external world, sampling large amounts of data ranging from visual and auditory signals to spatial information. The data collected through these sensors is processed to build models of the environment in real time, which are critical to autonomous decision-making and navigation. Increased sophistication in sensor technology, such as LiDAR and infrared sensors, offers greater accuracy and dependability and minimizes the opportunity for failure in operation. This capability is particularly highlighted in the AWARE platform, which demonstrates the importance of integrated sensor capabilities to facilitate cooperation among aerial systems and ground sensors, as outlined by Andres et al. (2007). Furthermore, how sensor information correlates with cognitive processing within autonomous systems is emphasized by new theories of neural synchronization and information flow. It is important to understand how sensory inputs influence behavioral responses and adaptability of autonomy agents under varying conditions. Experiments have proved that, similar to biological systems, artificial perception systems can enhance performance with oscillatory dynamics. For instance, a model of coupled phase oscillators demonstrates that certain regimes of synchronization enhance the reconfigurability and adaptability of robotic agents and consequently maximize their operating efficiency in response to sensory input. Through the definition of such interactions, it is possible to understand how various sensory modalities interact and contribute to the overall functionality of an autonomous system, as discussed by Husbands et al. (2012). In brief, sensor and perception system development is central to the advancement of autonomous systems, ultimately determining their autonomous navigation capability within complex environments. As technology progresses further, a multi-dimensional approach involving advanced sensor technology and cognitive modeling will likely lead to more powerful and adaptable systems. The growing emphasis on collaborative frameworks, as outlined in the AWARE project, shows the potential for autonomous systems not only to exist independently but also to collaborate with other systems synergistically. This synergy would revolutionize applications from domains, such as disaster response and surveillance, to offer more efficient and responsive solutions to real-world challenges. Through ongoing research and development, the application of sensors and perception systems will continue to be at the core of the autonomous technology of the future (Figure 1.6).

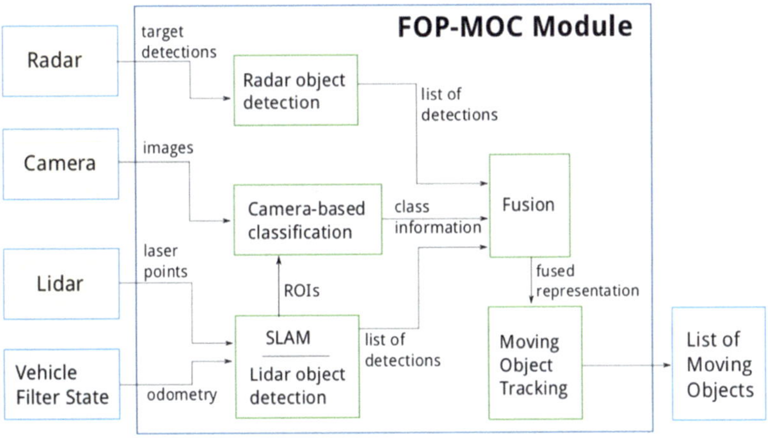

Figure 1.6. *Multiple sensor perception system.*

Source: https://www.researchgate.net/profile/Jelena-Kocic/publication/329153240/figure/fig2/AS:696154882842627@1542987666684/Multiple-sensor-perception-system-28.ppm.

1.2.3. Control Systems and Robotics

The combination of robotics and control systems is very critical in the creation of autonomous systems, and this influences their performance and application significantly. Control systems form the basis of robotic functionality and allow for accurate control of motions and actions in complex environments. This connection is critical to effective autonomous robot operation, as the robots must adjust to evolving conditions and user requirements. Moreover, advances in system design and control software have rendered robots more autonomous in function. With evolving advancements, the systems incorporate more machine learning techniques that enable them to adapt and optimize to accommodate their responses to changes in the environment and user demands. Thus, construction of effective control frameworks is paramount in the realization of intelligent robotic systems with autonomous functions in dynamic environments. The robotics' entry into daily life is also made possible by cutting-edge human-robot interaction (HRI) capabilities. Designing robots that are not only functional but socially compliant is a step towards development in this field. The literature emphasizes the necessity to take on a dual approach incorporating technology-led and human-centered design, as depicted in research studies. On the one hand, robots must be flexible and intelligent so that they can work autonomously; on the other hand, they must be able to be interactive and user-oriented so that they can be embraced by users. Both of these factors are crucial to the large-scale deployment of robots in environments such as Ambient Assisted Living (AAL) and home automation, where user experience becomes a dominant factor in determining the effectiveness of the technology (Dautenhahn et al., 2015). Comprehending and managing the nuances of HRI is therefore essential to creating autonomous systems that are reliable and trusted by users. The importance of robotics transcends individual applications, influencing broader technological innovations and economic development. Initially discovered in NASAs assessments of automation technology for space applications, the potential of leading-edge robotics has significant implications. The suggestion counsel to Congress underscored that investment in automation and robotics is not simply a technical undertaking but a strategic choice that has the capacity to significantly improve the U.S. economy (N/A, 2025). Now that robotics technology continues to evolve, it is those control systems that will ensure efficient performance across various sectors, from aerospace to healthcare. The planned merging of control systems with robotics not only enhances system efficiency but also ensures that such technologies are capable of meeting industries and society's evolving needs in general. Significantly, the merging of control mechanisms with robotics is the future for autonomous systems, a step that has far-reaching implications in several fields.

1.3. APPLICATIONS OF AUTONOMOUS SYSTEMS

Applications of autonomous systems are extensive and multifaceted, covering numerous fields such as robotics, transportation, and artificial intelligence. In robotics, for instance, autonomous systems have been developed to conquer tasks that incorporate efficiency, precision, and adaptability, particularly in complicated environments. Behavior Trees (BTs) have emerged as a fundamental structural technique of programming autonomous systems, enabling more intuitive handling of tasks and modularity. BTs allow the context switching among tasks with flexibility, a highly crucial ability in situations where circumstances can change rapidly, allowing robots to react accordingly and maintain operating effectiveness across a large variety of situations. As a result, BT applications in robotics thus emphasize their pivotal function in enhancing the performance and reliability of autonomous systems, according to their increasing involvement in mainstream industrial and consumer applications, thereby redetermining the manner in which activities are performed in repetitive and complex situations (Colledanchise et al., 2020). In decision-making, autonomous systems employ methods such as value-driven design to cope with the complexity of advanced planning procedures.

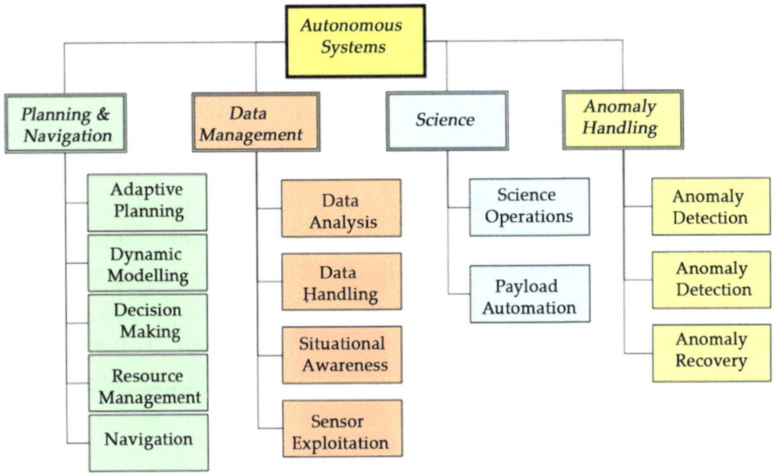

Figure 1.7. *Application of autonomous systems.*

Source: https://www.researchgate.net/profile/Elie-Allouis/publication/320585817/figure/fig4/AS:552940740251648@1508842753243/Application-of-Autonomous-Systems.png.

These systems are particularly beneficial within hierarchical decision environments, i.e., scheduling or resource allocation, where a preset set of rules might not be able to cope with the dynamics of real-world situations. Value-driven paradigm accommodates an adaptive style of decision-making automation, where system operations are synchronized with user priorities and requirements. Above all, the way it can perform in actual use such as TAC Brawler air-to-air combat simulation and autonomous underwater vehicle mission planning indicates the potential of the method to enhance the effectiveness as well as the efficiency of problem-solving procedures. These advancements demonstrate the capability of autonomous systems to not only support human decision-making but also execute intricate tasks that demand precision and flexibility (Burke et al., 2025). In addition, autonomous systems are applied in healthcare domains, ranging from robot surgery assistants to automated diagnostics. These systems enhance medical precision and have the potential to change patient care by providing accurate and consistent decisions. During surgery, robotic systems can perform intricate procedures in a minimally invasive fashion, thereby reducing patient recovery time and rates of complications. Further, the employment of autonomous systems in telemedicine also enables remote control and monitoring, thus ensuring greater access to patient care. Thus, as the technology that governs autonomous systems continues to progress, their applications not only enhance the efficiency of operations but also bring ethical and logistical problems that must be addressed in order to maximize their contributions to society while preserving safety and reliability (Figure 1.7).

1.3.1. Autonomous Vehicles

The coming of autonomous cars represents a transport technology milestone that is directly related to autonomous system advancements. Autonomous cars use a combination of sensors, AI, and advanced algorithms to travel without the need for human intervention. Focusing on efficiency and security, autonomous systems integrate various technologies such as Intelligent Transport Systems (ITS) that monitor and manage traffic patterns to reduce congestion and accidents. Within the set of ITS applications, independent systems operate in isolation from external communication with or reliance on other vehicles or infrastructure, with innovations such as blind spot detection and electronic stability control being pillars of their operating structure (Mol et al., 2008). Not only will this technology enhance user convenience but is also expected to dramatically decrease the incidence of road accidents, thus revolutionizing public safety in transportation. The role that radar systems play in the operation of autonomous cars cannot be overstated. The systems are core to the decision-making and perception of the vehicles, providing real-time data on surrounding environments. Utilizing principles of phased array technology, radar systems enable cars to detect obstacles, measure distances, and estimate speeds of objects surrounding them with high accuracy. Also, digital signal processing boosts the ability of radar to filter out noises and improve target identification (Rummel et al., 2019). With scientists continuing to innovate, integrating radar with other sensing modalities such as LiDAR and cameras will make autonomous systems more stable and robust. The capacity of radar systems to function under varying weather conditions further enhances their worth, enabling autonomous cars to travel safely and effectively under nearly any conditions. The progress to complete autonomy, however, depends not only on technological innovation but also upon a complex dance between regulatory, societal, and ethical concerns. Policymakers are faced with the dilemma of formulating all-encompassing rules through which such vehicles can safely be incorporated within existing traffic patterns. This calls for the development of regulations concerning liability in the event of an accident, user data protection, and ethics of decision-making algorithms employed by autonomous systems. Public acceptance is also of utmost concern; extensive outreach and education will likely be required to build trust and familiarity among current and potential drivers. Having ensured technological maturity and economic viability, collaborative effort amongst governments, industries, and researchers will make autonomous vehicles become widely popular, influencing the future of mobility and transport in a holistic way (Figure 1.8).

Figure 1.8. *Autonomous vehicles.*

Source: https://innovationatwork.ieee.org/wp-content/uploads/2018/07/iStock-829197466-1024x683.jpg.

1.3.2. Drones and Aerial Systems

Development of drones and aerial systems is a breakthrough in the overall context of autonomous systems that presents new possibilities that were previously unimagined. The technologies have evolved over time, and their applications have ranged from military to commercial use and recreation. For instance, drones are today used in various sectors like agriculture, where they assist in monitoring crops and precision farming to achieve more efficiency and yield. Besides, optical flow algorithm innovations such as the kernel cross-correlator (KCC) technique have made remarkable velocity and position estimation improvements, which are significant in autonomous navigation. The algorithms can effectively eliminate motion blur with good altitude and yaw rate estimations, thereby enabling robust management of trajectory under different operational conditions (Ji et al., 2018). The integration of sophisticated algorithms, thus, not only improves the functionality of drones but also their reliability in complex missions.

Figure 1.9. *Drones and aerial systems.*

Source: https://cdn.shopify.com/s/files/1/0033/6317/6560/files/dji_98ac945d-5ff8-4bdc-8e67-964a61d52c5a.png?v=1739391118.

The future of drones extends to urban air mobility, where they are considered delivery trucks and air taxis, revolutionizing transport in increasingly urbanized metropolitan areas. While there is the possibility of reducing traffic jams and accelerating deliveries, the socio-technical environment of drone application raises complicated challenges. There is proof to indicate that the current controversy surrounding civilian drones encompasses an abundance of hurdles, such as regulatory bodies and citizen fear regarding safety and environmental impacts (Biehle et al., 2020). Surprisingly, while there are significant economic forecasts linked to drone technology, these must be accompanied by a careful assessment of societal benefits. Actually, the correct incorporation of drones in the infrastructures of cities is not just reliant on technological progress but also on alleviating public fears and regulatory issues, hence ensuring the participation of citizens in debating these technologies. As technology and society continue with the further use of drones and aerial systems, they will continue to affect each other's path. The drone debate is characterized by polarization around innovative potential and social implications, which forces stakeholders to look at development as an inclusive process. Consensus-building methods and community involvement can play a significant role in the acceptance and perception of drones in communities. The literature highlights the imperative need for empirical studies to back up the expected benefits of drones in transport, such as traffic reduction and environmental amelioration (Biehle et al., 2020). It is through a critical analysis of both the pros and cons that society can maximize utilization of the potential of aerial autonomous systems while ensuring ethical and sustainable usage. This multi-disciplinary approach will be key to determining how drones become part of everyday life and contribute to larger autonomous systems (Figure 1.9).

1.3.3. Industrial Automation and Robotics

The merging of robotics and industrial automation into present manufacturing processes is a significant advancement in how work is executed throughout various sectors. Robotic devices traditionally were simple programmable units with fewer capabilities, primarily used for repetitive motion functions on the assembly line. More recent technological advancements have redirected the development of robots with greater autonomy, smartness, and adaptability. This advancement enables them to navigate across dynamic environments, respond to real-time data, and perform sophisticated maneuvers that were deemed impossible before. Therefore, industries are utilizing robotic systems not only as equipment but as a part of the whole that increases overall productivity and efficiency. The shift toward increased collaboration between human work and autonomous systems brings into focus an important trend in modern industrial landscapes, in which gaining access to the potential of technology is capable of optimizing activities and reducing the need for labor in increasingly competitive contexts (Gu et al., 2017). In light of such technological developments, the interface between robotics and wider digital transformation is reshaping conventional practice across multiple sectors. The innovation of new applications, with sophisticated sensors and artificial intelligence, allows unprecedented control and monitoring in the industrial environment. Furthermore, as robotic systems extend to numerous fields, including medicine, security, and micro-devices, they bring in complications that need consideration to governance in R&D. The collective work in developing theories of technology assessment shows a growing sensitivity to the need to shift policy-making models in order to deal with these revolutionary changes. The models must take into account the social, ethical, as well as economic consequences of bringing autonomous systems into the previously human-controlled domains (António B Moniz). Finally, industrial automation in the future will rely on the symbiotic interaction of human operators and autonomous systems. Technological advances in interface and user experience design have begun bridging the gap, and it is increasingly feasible for humans to work harmoniously with robots. Training and education programs are adapting to prepare the workforce of the future to work with advanced robotic systems. While organizations work towards attaining digital maturity, the onus of effectively leveraging automation falls not only on technological advancements but also on human responses to these advancements. Finding a balance between technological advancements and human-based methodologies will ultimately decide the success of industrial robotics and automation in mitigating modern challenges, thus ensuring that the advantages of autonomy are equitably achieved in a socially conscious manner (Gu et al., 2017).

1.4. BENEFITS OF AUTONOMOUS SYSTEMS

The intersection of autonomous technologies across industries marks the beginning of numerous advantages that significantly enhance operational efficiencies. Among them is their ability to improve safety through precise and consistent execution of tasks. For example, autonomous vehicles employ advanced algorithms and live data to navigate, significantly reducing human error—a leading cause of road accidents. There has been revealed research on new methods like the hybrid-open-closed-loop (HOCL) control strategy, which incorporates human-like reflexive responses into lane-keeping systems. The approach enhances stability and performance with tight tracking to actual driving behavior, eventually achieving safer means of transport in forthcoming autonomous vehicles (Gordon et al., 2016). Through minimizing accidents and probable hazards, autonomous systems enhance safety, with their use in such critical sectors as public safety and healthcare widened. Additionally, the economic benefits of using autonomous systems are tremendous. Since they operate around the clock and do not need rest or breaks, they can radically improve productivity for manufacturing, logistics, and other sectors that involve a heavy proportion of labor. By removing drudgery work, organizations can transfer human capital to more complex, innovative projects involving cognitive skills and innovative thinking (Hatherall et al., 2025). This shift not only optimizes the deployment of existing work resources but also opens up job prospects in new industries specializing in monitoring and managing autonomous technologies. Consequently, cost savings through enhanced efficiency and reduced operating costs help strengthen the value proposition of investing in autonomous system technology, thereby boosting their appeal to various industries' stakeholders. Moreover, ethical problems associated with using autonomous systems cannot be disregarded. It is escalating discussions regarding the liability of such systems, particularly as they assume roles conventionally demanding human intervention. Another method for maintaining responsible deployment of autonomous systems is through the concept of answerability, where focus is laid on establishing accountability mechanisms that guide their use (Hatherall et al., 2025). Through the incorporation of transparency and trustworthiness principles, organizations can make sure that there is public adoption of autonomous technology. This interaction not only contributes to their prestige but also accelerates broader application. As the discussion on ethics and autonomy matures, developing protocols for answerability will be at the heart of ensuring that the benefits of such systems are achieved within conformity with social concerns and expectations (Figure 1.10).

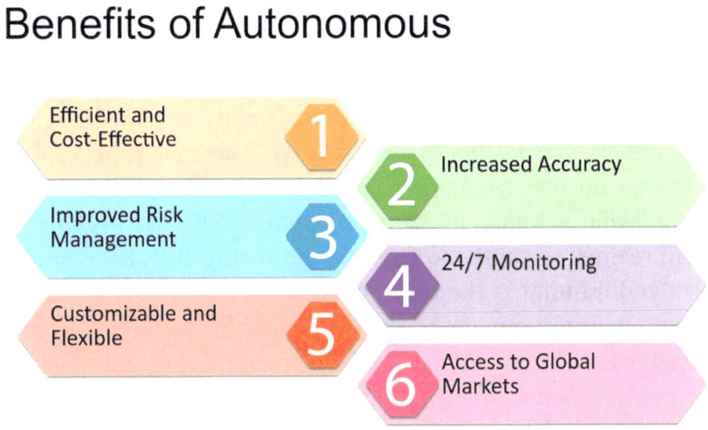

Figure 1.10. *Benefits of autonomous.*

Source: https://fastercapital.co/i/Asset-allocation--Optimizing-Asset-Allocation-with-Autonomous-Investment--Benefits-of-Autonomous.webp.

1.4.1. Increased Efficiency and Productivity

The emergence of autonomous systems represents a profound change in many industries, and it is leading to greater efficiency and productivity in many fields. Through the use of routine tasks automation and the streamlining of operations, these systems promote less labor expense and delay time in operations. For example, in manufacturing settings, autonomous robots can operate continuously without breaks, thereby delivering the highest output with less human error. As cited in local literature, areas such as Pacific Russia demonstrate how structured shifts in economic activities, particularly manufacturing and extractive industries, translate into better measures of productivity and the direct linkage of technological adoption with economic performance (Baklanov et al., 2016). When autonomous systems mature, their capacity to rationalize workflows and supply chains intensifies, making them strategic building blocks for enhancing efficiency operations. Moreover, autonomous systems' integration into organizational structures has brought about extensive changes in workplace dynamics and employee experience. Innovations in high-performance work practices (HPWPs), in general, have been promising towards maximizing productivity as well as job satisfaction. Empirical evidence shows that these innovations not only aid enhanced production methods but also enhance a more stimulating work culture for employees to prevent burnout and to boost retention (Cox et al., 2012). These gains can particularly be helpful in industries hit by high turnover rates, where enabling a motivated and efficient workforce is critical. As more and more of these technologies are embraced and implemented by organizations, the ultimate result is a synergy effect, where human and machine capabilities both become enhanced, leading to an overall increase in productivity. Finally, the destiny of autonomous systems in achieving efficiency continues to rest on the synergy of technology and human action. Through implementing such systems in complementary patterns, organizations can leverage both realms' strengths to reap the optimal results. Strategic employee development and training in the utilization of such advanced systems also can further extend operating outcomes, ensuring that the workforce is prepared to work with and maximize the abilities of automation. Lastly, as witnessed in the scenario of evolving economic paradigms in regions like Pacific Russia, the modification of industrial processes to more automated practices are testimony to a broader trend towards sustainability and long-term growth (Baklanov et al., 2016). Ultimately, as self-governing systems infiltrate all aspects of life, their role in facilitating more efficiency and productivity will be the hallmark of the contemporary industrial world.

1.4.2. Enhanced Safety and Risk Reduction

Integration of autonomous systems into modern-day transportation infrastructures is very promising since it has the ability to bring more safety and reduced risks of car travel. Traditional road safety systems, as effective as they have proven to be, have not been greatly innovated in recent years, particularly in restraint system design. To rectify this shortcoming, new approaches are being explored that involve sophisticated materials and new design ideas which can enhance vehicle-barrier interactions. By making roadside systems more efficient by improved design, e.g., for low-mass vehicles and traffic suitability, the severity of crashes can be minimized and the injury rate lowered. The potential for such innovation points to the original role of autonomous technologies in revolutionizing paradigms of safety on our roads because such technology can guarantee proactive risk control and avoidance of accidents and pave the way for safer transport conditions (Cantisani et al., 2017). Moreover, using sensors within safety barriers offers a compelling avenue for enhanced vehicle-to-infrastructure and vehicle-to-vehicle interaction and communication, ultimately enhancing total safety levels. These embedded sensors not only provide feedback on traffic flow but also enable vehicle-to-infrastructure (V2I) communication so that real-time information can be shared that can significantly influence autonomous system behavior.

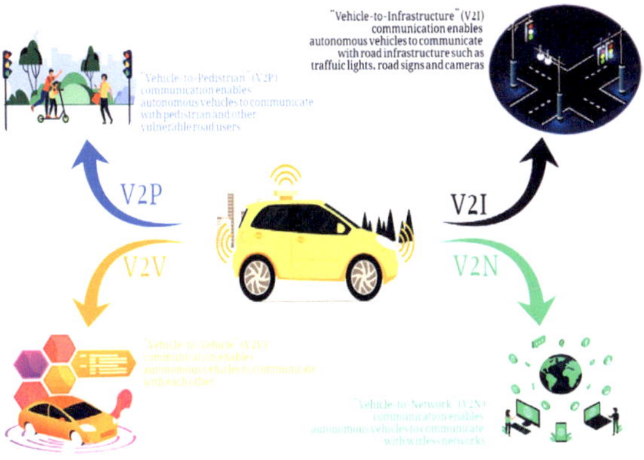

Figure 1.11. *Connected and automated vehicles.*

Source: https://www.mdpi.com/technologies/technologies-11-00117/article_deploy/html/images/technologies-11-00117-g001.png.

This shift to smarter infrastructure is a systemic shift wherein autonomous vehicles are able to tap into vital information that enhances their navigation and decision-making abilities. With such technologies coming to more areas, they strive to dynamically calibrate safety features to the unique conditions of the road, which leads to a drastic reduction in risk of accidents. The interface between technology and infrastructure highlights the underlying advancements that must be realized in creating a secure transportation system, with the ultimate aim towards larger objectives of minimizing risk in autonomous systems (Cantisani et al., 2017). Additionally, proof gathers for the hypothesis that organizational innovations have the capability to greatly increase operational safety and risk reduction in environments utilizing autonomous systems. High performance work practices (HPWPs), as characterized in emerging studies, have shown that work organization optimization can make better employee experiences possible while simultaneously fostering a culture of safety. Implementing HPWPs in autonomous system platforms triggers better interoperation between human operators and automated components, enhancing the workplace safety. Such coordination of technology and workforce management provides organizations with the capability not only to optimize productivity but to also enforce end-to-end safety practices which are responsive to the particular issues emerging out of autonomy. Hence, integrating work organization development is one of the most significant actions in offsetting the inherent risks of autonomous technology so that safety will remain at the top of this rapidly evolving technology (Figure 1.11).

1.4.3. Cost Savings and Economic Impact

As businesses increasingly adopt autonomous systems, the top priority of cost reduction and economic impact is made more and more apparent. Autonomous technology enhances operational efficiency by minimizing human mistakes and making optimal use of resources. For instance, automated supply chain management systems can significantly reduce logistical costs while improving delivery time. Such systems sort vast amounts of data to precisely predict demand, allowing companies to scale inventory accordingly. Therefore, these developments not only lower the cost of operations but also ensure higher profit margins, thereby improving economic performance in general. Further, autonomous systems are cost-effective by simplifying processes dependent on labor. In farming and other industries, automation minimizes dependence on human labor, which can be very expensive and prone to variation in availability. For example, autonomous farm equipment can sow, irrigate, and harvest accurately, minimizing wastage and resource consumption. This offers efficiency that helps lower operational costs as well as boost productivity, ultimately impacting individual farmers as well as the agricultural sector at large. In this sense, the overall economic impact is significant because it has the ability to lower food prices and improve food security in various regions. While, at the same time, adoption of autonomous systems will stimulate innovation and open new economic opportunities. As businesses take up these technologies, they normally find new sources of revenue that were not available previously. Innovation in developing and implementing autonomous systems creates need for skilled services and personnel, stimulating employment in the technological sector. Besides, initiatives such as infrastructure development, for example, the aqueducts that culminate in water management in projects as stipulated by Cowi et al. (2013), serve as an example of affordable solutions that produce long-term results. These infrastructure investments form the basis for strong systems that not only enhance economic efficiency but also produce a sustainable impact on surrounding communities, as highlighted by the evidence presented by Gerlagh et al. (2025).

1.5. CHALLENGES AND LIMITATIONS

With the rapidly evolving technology in autonomous systems, a number of challenges and restrictions impede their optimal potential. One key limitation is the technological constraint in information processing and decision making. For instance, processing medically relevant information to the maximum is extremely important for autonomous systems operating in critical conditions, i.e., space exploration. The ExMC project pinpoints areas of shortfall in current systems, for example, the lack of provision for computed medical decision support for long-duration missions. To fill these shortfalls, high-performance medical data management systems are needed that can facilitate autonomous medical care, illustrating a fundamental constraint towards achieving fully autonomous operations on the basis of real-time decision-making in conditions of uncertainty (Burian et al., 2025). A second major challenge lies in the gaps within organizations and systems that impede the evolution of autonomous systems across sectors. More specifically, research on municipal energy systems demonstrates that historical disconnection between heat and electricity supply utilities has constrained municipal engagement with integrated energy policy. Such organizational gaps echo those in autonomous systems because the lack of overall structures may impede autonomous technologies from operating optimally. Collaboration between municipality and utility companies to leverage collective resources and capabilities is a demonstration that partnership is critical to surmounting structural barriers. Hence, in the absence of strategic partnerships, development of isolated systems can be significantly undermined by decentralized infrastructures and clashing agendas (Ross et al., 1992). Secondly, interoperability and integration issues among various systems are monumental challenges in the backdrop of autonomous technology. Because autonomous systems more and more rely upon information gathered from a variety of sources, their ineffectiveness is a result of a lack of integrating and analyzing data effectively. For instance, the challenges pointed out in the ExMC project bring out the need for interoperability within clinical decision support systems, which must facilitate good communication and exchange of information to maximize operational capacity. Moreover, the development of machine learning algorithms contributes to the complexity because variations in data quality and structure can affect algorithm performance and confidence. Hence, interoperability and data integration are central to autonomous systems becoming mature, as they ultimately decide whether they can function independently and self-tune in dynamic environments (Burian et al., 2025).

1.5.1. Technical Challenges in Development

In self-adaptive systems, the design is affected by technical issues that call for innovative solutions in software engineering and quality assurance. Traditional methods may no longer be adequate in the presence of adaptability in self-adaptive systems. With adaptations in systems, ensuring quality assurance also has to be adaptive. Unless this disparity is addressed, autonomous system actions would be invisible to conventional testing techniques. To forestall this danger, the concept of scenario coevolution has been established, which suggests creating an environment for testing that coevolves alongside the system itself. This testing technique enables testing to adapt and become relevant when system capabilities change, and this is important in the drive to ensure reliability and performance in autonomous systems (Bernon et al., 2019).

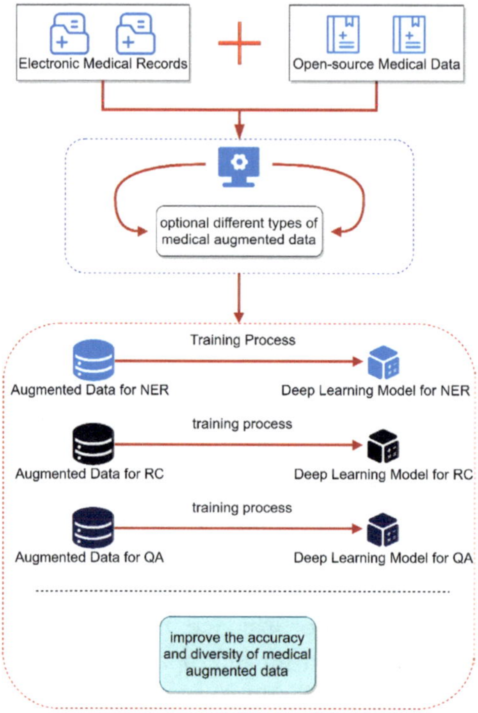

Figure 1.12. *An intelligent medical data augmentation.*

Source: https://www.mdpi.com/applsci/applsci-12-10655/article_deploy/html/images/applsci-12-10655-g005.png.

In addition, autonomous systems for application in high-priority areas such as medicine also reveal more sophistication in the sense of development. The Medical Data Architecture (MDA) project is a clear example of the necessity for sophisticated systems that can autonomously process medically relevant data in missions of exploration. Problems encountered in this project reflect the technical inadequacies of existing systems, particularly in their ability to provide sound medical decision support. Gap filling in clinical guidance and information processing calls for a robust decision-making infrastructure that draws upon sophisticated algorithms and machine learning platforms. Hence, the clinical decision support roadmap must describe current capabilities as well as proposed solutions to overcome interoperability issues and support real-time data integration, which are crucial to ensuring optimal health management during long-duration missions (Burian et al., 2025). As technology of autonomous systems advances even more, it is imperative to understand the multi-faceted nature of technical issues inherent in their development. With adaptive quality assurance measures to integrating complex data structures, the developers must navigate an utterly intricate matrix of technological issues. Interaction between system evolution and operating environment calls for continuous investigation and adaptation of software and hardware components. Such responsiveness is important to the success of the intended functional results of autonomous systems, particularly in life-critical activities like healthcare for exploration missions. Eventually, these technical challenges are overcome successfully in manners that not only enhance the functioning capabilities of autonomous systems but also provide them with acceptance and application in many industries (Bernon et al., 2019; Burian et al., 2025) (Figure 1.12).

1.5.2. Ethical Considerations and Societal Impact

As autonomous technologies permeate a variety of industries, ethical considerations about their design and deployment have become increasingly pertinent. Central to ethical discussions about these technologies is the provision of transparency, fairness, and accountability behind the algorithms driving them. Detractors argue that without strong regulatory frameworks, dangers of algorithmic bias and loss of personal data loom in the future. As a result, conceptual models like society-in-the-loop (SITL) have since emerged, advocating for a social contract that negotiates the values of various stakeholders affected by these systems. The paradigm sets forth that moral governance must catch up with the sophistication of modern AI to ensure human oversight is at the nucleus of machine learning processes and stakeholders have a voice in shaping the ethical framework of autonomous activities (Rahwan et al., 2017). The social consequences of autonomous systems go beyond the immediate effects of the technology, as it also redesigns labor markets and human-to-human interactions. As AI technologies are brought into sectors of production to healthcare, there are concerns about job loss and economic disparity. Autonomous systems can bring efficiencies through streamlining, but it can also widen current social disparities if not approached inclusively. Moreover, the application of consciousness to machines launches new avenues of perspective on social impact. Reframing autonomous systems as forms of complexity types—autonomous, cognitive, and social—is what allows us to ponder their actual potential to shape social organization and human relations (Arsiwalla et al., 2017). A systems approach will reduce negative impacts while capturing the benefits of these transformative technologies. Moreover, the ethical aspect of autonomous systems also involves responsibility in decision-making. With the power of machines making autonomously decided actions possible, questions about responsibility arise with how the systems can cause harm or act in unforeseen ways. The system of governance for autonomous systems then needs to provide whether developers, organizations, or machines are accountable. Such sophistication entails thoughtful considerations of regulatory actions that not only foresee potential risks but also place emphasis on ethical guidance and training in AI studies. From this point of view, it would be evident that an interactive strategy involving multi-disciplinary involvement is unavoidable for dealing with the ethical concerns as well as the fundamental social transformations caused by the advent of autonomous systems (Arsiwalla et al., 2017; Rahwan et al., 2017).

1.5.3. Regulatory and Legal Hurdles

With the creation and expansion of autonomous systems across various industries, legal and regulatory problems are significant concerns. The integration of autonomous technologies, such as autonomous vehicles, mandates the re-examination of existing legal frameworks that were not created to address such emerging technologies. Regulatory bodies have the daunting task of defining problems such as liability, standards of safety, and ethics. These issues are even more challenging on the transnational level, where varying standards and regulations may impede the global adoption of autonomous systems. Legal stakeholders must walk through this complicated landscape, balancing innovation, safety, and accountability, with regulation becoming inevitable for the successful adoption of autonomous technology. Among the most critical regulatory challenges is assigning liability in accident situations involving autonomous systems. As seen in the autonomous car debate, the traditional notion of liability can be made obsolete with driver error a less prominent consideration in accidents involving these technologies. Policymakers are challenged to reexamine existing systems of product liability so that they function in an increasingly driven-by-automation environment. Remarkably, the European Union Product Liability Directive articulates restraints that can hinder the implementation of effective liability schemes because the prevailing provisions are not optimally aligned with the operating conditions of autonomous vehicles. Without an adaptive and resilient liability framework, the efficacy of safety regulation and technological standards may be compromised and thus pose potential risks to consumers and society overall embracing these new technologies (Bruyne et al., 2018). Further, there is also the overlay of private contractual governance structures influencing the deployment of autonomous systems. The mounting reliance on transnational private regulation regimes puts into sharp contrast salient questions of legal legitimacy and the role of public regulation in an increasingly complex economic environment. Firms and innovators may be caused uncertainty by ambiguity regarding the legitimacy and enforceability of private regulations. Within this evolving regulatory landscape, stakeholders need to vigilantly examine the points at which public interests and private governance converge to ensure that the application of autonomous systems serves societal values and norms. Such a strategy would enable a more integrated regulatory culture that not only responds to immediate questions of law but also fosters an environment for innovation and the safe integration of autonomous technology into everyday practice (Zumbansen et al., 2012).

1.6. FUTURE TRENDS

As autonomous system technology becomes more advanced, future trends also indicate a significant shift towards developing robots with greater autonomy and flexibility levels. This is founded on technological advancements in artificial intelligence, machine learning, and sensor technologies that enable systems to operate within more advanced environments. Robot autonomy expectations are not only technological but also correspond to broader societal and cultural expectations. As the Epinet workshop highlighted, participants from various disciplines observed that although autonomy remains inconclusive in its applications outside the domain, it continues to be a dominant driving force for research support and priorities. This dialectical tension between social demand and technological advancement rightfully situates the debate on the future of autonomous systems, challenging researchers and implementers to continue striving for technological viability alongside ensuring ethical and pragmatic imperatives in their solutions (Dijk et al., 2014). The advent of Delivery as a Service (DaaS) illustrates the shift towards consumer-oriented business models that make use of autonomous systems. This model redesigns logistics by combining cutting-edge technological innovations into supply chain management and urban planning to meet the rapidly on-demand lifestyle of consumers. The progress of DaaS relies on the intersection of efficient transport modalities and IoT solutions, providing a blueprint for future development in logistics. As explained in the discussions on this model, autonomous delivery systems will be a primary force behind future urban logistics as well as customer engagement strategies. Through optimization of delivery operations through validity and speed of service, DaaS illustrates how autonomous systems can adapt to the nuance of the current economic and societal needs as well as drive progress in sustainability and improved urban living (Huschebeck et al., 2020). In the future, not only does the development of autonomous systems promise efficiency but also raises basic questions of regulatory infrastructures and ethical consequences of technology deployment. The policymakers, ethicists, and engineers will need to collaborate in clarifying the nature of autonomy and society's implications. Automation contains inherent risks, and there has to be debate around the responsibility of developers and users. As autonomous technology becomes further embedded in daily life—through the means of driverless cars or robotic assistants—there will be a critical need globally for standards to support responsible and safe adoption. Besides, as the combined findings gathered in workshops like the Epinet workshop also show, the establishment of integrated strategies for responding to the challenges and demands of future autonomous systems will be essential to unlocking their full potential while bringing them in securely into existing society (Dijk et al., 2014) (Figure 1.13).

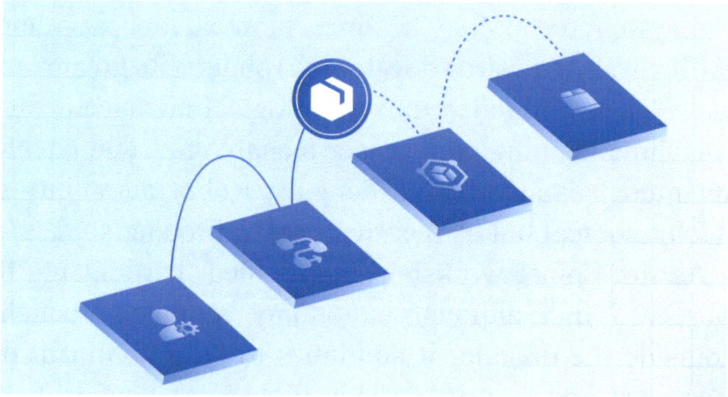

Figure 1.13. *Delivery as a service.*

Source: https://www.upperinc.com/wp-content/uploads/2024/10/your-guide-to-service-delivery-model-in-logistics.webp.

1.6.1. Advancements in AI and Machine Learning

The new developments in machine learning (ML) and artificial intelligence (AI) have facilitated a revolution in virtually every sector, particularly in the development of autonomous systems. The technologies have found their way deep into agriculture, resulting in concepts like precision agriculture and smart farming. Autonomous machinery used in farm operations aims at improving efficiency while reducing the costs of labor and use of resources. These technologies assist machines in performing complex tasks using improved algorithms that assist machines to move independently in various environments. For instance, farm field machines are being equipped with advanced path planning, mapping, and localization capabilities more and more that assist them in moving successfully in semi-structured and unstructured environments. These technologies are the clearest demonstrations of AI and ML technologies in reshaping operation paradigms among automation-dependent industries and reaping huge economic and operational benefits (Aghi et al., 2020). AI and ML development have also caused a significant shift towards more robust and flexible systems. This shift is quite evident in developing control algorithms that allow autonomous machines to respond in real-time towards ever-changing situations. By employing technologies such as double-layer control systems and RGB-D cameras, researchers have been able to create low-cost methods for making these systems more resilient. For example, the use of a backup algorithm not only improves the navigation ability of agricultural robots but also constructs a learning loop. As autonomous systems gather data in real time, they are able to use machine learning algorithms to iterate and improve at their task. Hence, as the amount of automatically labeled data keeps on growing, the systems get to utilize the transfer learning to optimize their performance. This versatility allows the autonomous systems to operate efficiently in various settings while having the capability to factor in real-world operations' nuances (Aghi et al., 2020). Further, the expansion of AI and machine learning brings to the fore the necessity of infusing ethical issues into autonomous system development. As these technologies become increasingly powerful, the resultant risks of their use multiply. The Value-Sensitive Design (VSD) idea emerges as a helpful conceptual framework in such situations, as it emphasizes the incorporation of stakeholder values from the beginning of AI production. With the application of VSD, developers are able to make proactive choices regarding ethical issues that the design and use of intelligent agents produce and thus reduce some of the built-in risks. This design strategy not only enhances the social acceptability of AI application but also unites technology development with societal values. Ultimately, as autonomous systems evolve further, integration of ethical considerations will be essential to guarantee their long-term, responsible evolution across industries.

1.6.2. Integration with Internet of Things (IoT)

Marriage between autonomous systems and Internet of Things (IoT) is the next big revolution towards developing smart technology. It enhances the capability of both autonomous systems and IoT to a point where data can be easily exchanged between devices that are connected. Since autonomous systems rely heavily on real-time data, the IoT provides the kind of infrastructure required to collect, analyze, and pass on the data.

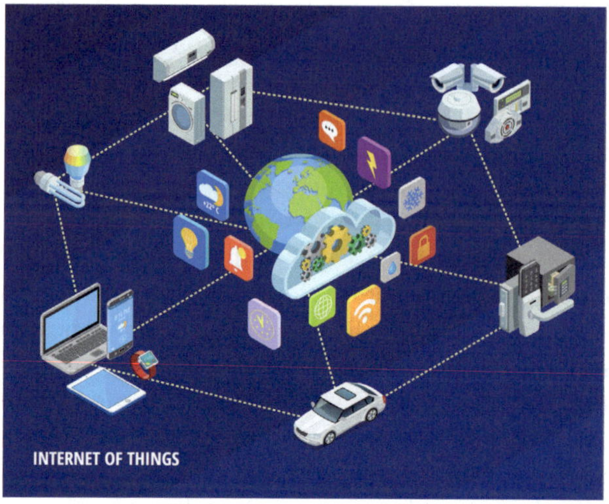

Figure 1.14. *IoT platform integration.*

Source: https://cdn.thingsup.io/wp-content/uploads/2023/09/11186-scaled.jpg.

As a result, these systems can learn and respond to their environment with greater precision and sagacity. The convergence of IoT and autonomous technology is paving the path towards advanced applications such as smart cities, where traffic management and resource allocation can be optimized in accordance with the data gathered from a collection of networked devices (Burattin et al., 2017). Yet IoT coupled with autonomous systems has not been without issues. Data privacy and security remain thorny issues with more devices getting interconnected. Having a huge surface area of potential cyberattacks requires establishing robust cybersecurity protocols to safeguard sensitive data. Moreover, the management of enormous amounts of data generated by IoT devices can overwhelm present infrastructure, necessitating improvement of data processing technology. In response to these challenges, the incorporation of Business Process Management principles can refine operations and enhance the security context of IoT applications. This strategic integration is essential in response to the challenges of the IoT ecosystem and enable secure and efficient deployment of autonomous systems (Abdolee et al., 2019). Additionally, the application of blockchain technology can help solve the challenges experienced with IoT integration with autonomous systems. The blockchain technology's decentralized nature can be employed to bring innovation in terms of security features through the offering of an open and tamper-free method of managing data exchange. Nevertheless, special attention is required in terms of the computational power and bandwidth consumption of blockchain technology within the IoT. Failure to do so would risk creating a new set of problems that would counteract the successful interoperation of such technologies. Understanding the symbiotic relationship between autonomous systems and IoT, and the probable role of blockchain, will be instrumental in deciding the future form of smart, connected technologies and how they are implemented in applications (Abdolee et al., 2019) (Figure 1.14).

1.6.3. Potential for Widespread Adoption

The potential for mass adoption of autonomous systems rests on an array of socio-technical factors that influence public perception and acceptance. As such technologies evolve, there is a need for elaborate frameworks that facilitate goal-oriented modeling of autonomous systems. The i* framework is a critical tool in this context, as it allows stakeholders to deal with the complexities of socio-technical interactions among autonomous agents. With this framework, developers can effectively model actor relationships, dependencies, and goals, thus enabling transparency and trust in autonomous systems. Such transparency is important for stakeholders like policymakers to comprehend the effects of mass deploying such systems. The increased acceptance and mainstreaming of autonomous technologies will rely on developing frameworks that put the collaborative aspect of these systems first while resolving issues of safety and reliability (Cares et al., 2016). Economic factors are also central to the future of widespread autonomous system adoption. The impact of self-driving vehicles on energy use and emissions of greenhouse gases is an example of the complex interdependence between technological progress and environmental sustainability. Research indicates that electrification, autonomy, and shared mobility simultaneously impact revolutionary transport dynamics. For instance, it is estimated that, by 2050, light-duty vehicle transportation can contribute up to 26% of United States electricity demand, with potential greenhouse gas emission reductions of up to 80% (Fox-Penner et al., 2018). Such potential for change resonates with the demand for policies that do not only allow rapid electrification but also address infrastructural challenges that increased vehicle autonomy may bring. Ultimately, integration of the economic benefits of these technologies with their environmental impact will be crucial to embedding widespread support and incorporation into established systems. Moreover, the social impacts of autonomous system adoption beyond just the pure economics cannot be overlooked. As these technologies become ever more deeply embedded in the fabric of everyday life, they represent opportunities and challenges that must be carefully weighed. For example, while autonomous vehicles hold the promise of lowered traffic congestion and improved safety, they can also lead to city sprawl if unchecked. The policymakers must design policies such that land gets utilized well and reverses anticipated increases in driving mileage so that adoption of autonomous systems does not go against broader city-planning goals. Moreover, promotion of energy efficiency in the context of autonomous systems can increase sustainability outcomes, enhancing public enthusiasm for their use. Therefore, such an all-encompassing approach on technical, economic, and social considerations is imperative to tap the total potential of autonomous systems in society (Fox-Penner et al., 2018).

SUMMARY

Autonomous systems encapsulate the key concepts described in the text and emphasizes their significance and implications for new applications. Autonomous systems, essentially, revolutionize traditional understanding of agency and control, allowing machines to perform tasks with minimal or no human intervention. This paradigmatic potential can be viewed in applications from autonomous vehicles to robotics, where advanced algorithms enable machines to traverse complicated spaces with efficiency and effectiveness. The small robot autonomous navigation of Caltech case study documented how sophisticated navigation algorithms, inspired by Professor Dickmanns' approach, could be employed to advance the transfer of robotics applications into practice, paving the way for automation (Feng et al., 1997). Ultimately, the potential for unmanned machines opens profound questions regarding responsibility, safety, and human-machine collaboration. Besides, the research into theoretical frameworks of autonomous systems indicates their extensive applications across scientific fields. While scholars delve into the intricacies of systems dynamics, they find that autonomous systems can contribute to our understanding of other phenomena, like cosmological models (Odintsov et al., 2018). For instance, the study of vacuum *f*(*R*) gravity and multifluid cosmology provides insights of relevance not only to theoretical physics, but also to the broader implications of autonomous decision-making within systems subject to multiple interacting elements. These cross-disciplinary implications further underscore the need for a unified approach to the study of autonomous systems, such that technological progress is linked with scientific exploration and ethical consideration. These integrative research works can create innovations that will enhance the performance and reliability of autonomous systems while reducing risks. In conclusion, research on autonomous systems is a world of potentialities, challenges, and ethical dilemmas. Being at the frontier of this technological revolution, it is crucial that one understands the implications of autonomous systems on society's institutions and on the lives of individuals. Ongoing development of sophisticated algorithms, here the ability of robots to maneuver, assures that before long autonomous technologies will be an integral part of daily life. But it is essential that researchers, developers, and policymakers act in concert to cope with the social, ethical, and safety issues arising out of introducing such technologies. By encouraging an active dialogue among various stakeholders and stressing a responsible path forward regarding innovation, we can guide autonomous system development towards a future that will benefit all of humanity.

The evolution of autonomous systems is fundamentally linked to technological advancement and more fundamentally to machine learning and artificial intelligence. With the advancement of the technologies, they further transform industries by enabling systems to operate with less human intervention. Assurance and resilience are critical since they directly influence the reliability and safety of autonomous cars. The technical topic notes (TTNs) prepared under the towards identifying and closing gaps in assurance of autonomous road vehicles (TIGARS) project provide insightful knowledge regarding these challenges. Part 1 of the TTNs outlines significant issues in assurance, where the call for resilience and safety requirements in ensuring the effective implementation of autonomous systems is emphasized. Through addressing these disciplines, researchers and engineers can set the foundation for safer, more efficient autonomous vehicles that meet stringent regulatory and societal constraints (Bloomfield et al., 2020). In addition, the value of dynamic testing and simulation cannot be overemphasized when it comes to autonomous systems. The techniques provide a simulated environment to evaluate and enhance the performance of autonomous vehicles before their operation in actual conditions. The latter part of the TTNs emphasizes this mandate by referring to simulated worlds, dynamic test practices, and defense in depth. Through rigorous testing of autonomous systems through simulation and embracing diverse strategies, stakeholders can ensure that vehicles will operate safely and efficiently even in the occurrence of unforeseen circumstances. This proactive stance towards safety regulation is necessary with growing public distrust over the reliability of autonomous technology (Bloomfield et al., 2020). Additionally, the application of clustering algorithms, for instance, those drawn from rough set theory, depicts advanced data handling in autonomous systems. The algorithms minimize reliance on individual human intervention and use local as well as global features of the data to arrive at more accurate clustering solutions. With autonomous systems increasingly required to handle mixed attribute data sets, the robustness of such approaches becomes ever more important. Through the provision of examples across disparate datasets, the proposed clustering algorithm shows efficiency capable of enhancing decision processes within autonomous vehicles. Collectively, such developments in data analysis not only enhance system functionality but also contribute overall safety and reliability of autonomous technologies as they enter everyday life.

REFERENCES

1. Abdolee, R., Maroufi, M., & Tazekand, B. M. (2019). On the convergence of blockchain and Internet of Things (IoT) technologies. http://arxiv.org/abs/1904.01936 (accessed on 10th September 2025).
2. Aghamohammadi, A., Chen, Y. F., How, J. P., Omidshafiei, S., et al. (2018). MAR-CPS: Measurable augmented reality for prototyping cyber-physical systems. https://core.ac.uk/download/156872644.pdf (accessed on 10th September 2025).
3. Aghi, D., Chiaberge, M., & Mazzia, V. (2020). Local motion planner for autonomous navigation in vineyards with a RGB-D camera-based algorithm and deep learning synergy. https://core.ac.uk/download/pdf/327177639.pdf (accessed on 10th September 2025).
4. Alba, A., Amato, A., Aplin, A., Axelrod, B., et al. (2018). Embodied evolution in collective robotics: A review. https://core.ac.uk/download/614587988.pdf (accessed on 10th September 2025).
5. Alves, J. D., Dias, J., Ferreira, J. F., Lobo, J., et al. (2015). Brief survey on computational solutions for Bayesian inference. https://core.ac.uk/download/153632959.pdf (accessed on 10th September 2025).
6. Ang, J. H., Flores Saldivar, A. A., Goh, C., Li, Y., et al. (2017). Energy-efficient through-life smart design, manufacturing and operation of ships in an industry 4.0 environment. https://core.ac.uk/download/82893542.pdf (accessed on 10th September 2025).
7. Arsiwalla, X. D., Herreros, I., Moulin-Frier, C., Sanchez-Fibla, M., et al. (2017). The morphospace of consciousness. http://arxiv.org/abs/1705.11190 (accessed on 10th September 2025).
8. Belpaeme, T., Cangelosi, A., Dautenhahn, K., Fadiga, L., et al. (2010). Integration of action and language knowledge: A roadmap for developmental robotics. https://core.ac.uk/download/1640754.pdf (accessed on 10th September 2025).
9. Bernon, C., Fraser, G., Andersson, J., Kephart, J. O., Hölzl, M., Črepinšek, M., et al. (2019). Adapting quality assurance to adaptive systems: The scenario coevolution paradigm. http://arxiv.org/abs/1902.04694 (accessed on 10th September 2025).
10. Biehle, T., Fischer, L., & Kellermann, R. (2020). Drones for parcel and passenger transportation: A literature review. https://core.ac.uk/download/323212148.pdf (accessed on 10th September 2025).
11. Bloomfield, R. E., Fletcher, G., Imai, K., Khlaaf, H., et al. (2020). Towards identifying and closing gaps in assurance of autonomous

road vehicles–a collection of technical notes Part 1. https://core.ac.uk/download/305120054.pdf (accessed on 10th September 2025).

12. Burattin, A., Di Ciccio, C., Gal, A., Janiesch, C., et al. (2017). The Internet-of-Things meets business process management: Mutual benefits and challenges. https://core.ac.uk/download/299377534.pdf (accessed on 10th September 2025).
13. Burian, B., Krihak, M., Schmitt, B., Shaw, G., et al. (2025). Medical data architecture prototype development–Summary of recent work and proposed ideas for upcoming work. https://core.ac.uk/download/pdf/286816747.pdf (accessed on 10th September 2025).
14. Burke, T. E., & Pugh, G. E. (2025). Automated control of hierarchical systems using value-driven methods. https://core.ac.uk/download/pdf/42822971.pdf (accessed on 10th September 2025).
15. Burov, O., Krylova-Grek, Y., Lavrov, E., Lytvynova, S., et al. (2020). Кибербезопасность в образовательных сетях. https://core.ac.uk/download/322373029.pdf (accessed on 10th September 2025).
16. Cantisani, G., Di Mascio, P., & Polidori, C. (2017). New research opportunities for roadside safety barriers improvement. https://core.ac.uk/download/153367220.pdf (accessed on 10th September 2025).
17. Cares, C., Colomer Collell, D., Franch Gutiérrez, J., López Cuesta, L., et al. (2016). The i* framework for goal-oriented modeling. https://core.ac.uk/download/46175391.pdf (accessed on 10th September 2025).
18. Colledanchise, M., & Ögren, P. (2020). Behavior trees in robotics and AI: An introduction. http://arxiv.org/abs/1709.00084 (accessed on 10th September 2025).
19. COWI, Directorate-General for Regional and Urban Policy (European Commission). (2013). *Quick Appraisal of "Sistema De Suministro Desde El Embalse Del Cenajo a la Mancomunidad De Canales Del Taibilla"*. https://core.ac.uk/download/232116437.pdf (accessed on 10th September 2025).
20. Cox, A., Rickard, C., & Tamkin, P. (2012). Work organization and innovation. https://core.ac.uk/download/16749552.pdf (accessed on 10th September 2025).
21. Das, N., & Yip, M. (2017). Robot autonomy for surgery. http://arxiv.org/abs/1707.03080 (accessed on 10th September 2025).
22. Dautenhahn, K., Dragone, M., & Saunders, J. (2015). On the integration of adaptive and interactive robotic smart spaces. https://core.ac.uk/download/29853781.pdf (accessed on 10th September 2025).

23. de Andres, E., Bernard, M., van Hoesel, L., La Civita, M., et al. (2007). AWARE: Platform for autonomous self-deploying and operation of wireless sensor-actuator networks cooperating with unmanned aerial vehicles. https://core.ac.uk/download/pdf/11468456.pdf (accessed on 10th September 2025).

24. De Bellis, A. F., & Umbrello, S. (2018). A value-sensitive design approach to intelligent agents. https://core.ac.uk/download/146502298.pdf (accessed on 10th September 2025).

25. De Bruyne, J., & Werbrouck, J. (2018). Merging self-driving cars with the law. https://core.ac.uk/download/158346992.pdf (accessed on 10th September 2025).

26. De Mol, J., Van Leeuwen, T., Vandenberghe, W., Vlassenroot, S., et al. (2008). Intelligente transport systemen its en verkeersveiligheid. https://core.ac.uk/download/55814175.pdf (accessed on 10th September 2025).

27. Dodig-Crnkovic, G., Holstein, T., & Pelliccione, P. (2018). Ethical and social aspects of self-driving cars. http://arxiv.org/abs/1802.04103 (accessed on 10th September 2025).

28. Fagerberg, I., Lindahl, T., & Ramström, A. S. (1999). Autonomous systems as legal agents: Directly by the recognition of personhood or indirectly by the alchemy of algorithmic entities. https://core.ac.uk/download/322820136.pdf (accessed on 10th September 2025).

29. Feng, X., & Liu, W. (1997). Robot autonomous navigation. https://core.ac.uk/download/4892777.pdf (accessed on 10th September 2025).

30. Finkelstein, A. (1998). Interoperable systems: An introduction. https://core.ac.uk/download/1669798.pdf (accessed on 10th September 2025).

31. Fox-Penner, P., Gorman, W., & Hatch, J. (2018). Long-term U.S. transportation electricity use considering the effect of autonomous vehicles: Estimates & policy observations. https://core.ac.uk/download/286679086.pdf (accessed on 10th September 2025).

32. Gerlagh, R., & van der Zwaan, B. C. C. (2025). A sensitivity analysis of timing and costs of greenhouse gas emission reductions. https://core.ac.uk/download/pdf/6362937.pdf (accessed on 10th September 2025).

33. Gordon, T., Martinez-Garcia, M., & Zhang, Y. (2016). Modelling lane keeping by a hybrid open-closed-loop pulse control scheme. https://core.ac.uk/download/76998863.pdf (accessed on 10th September 2025).

34. Gu, D., Wong, C., Yan, X., Yang, G., et al. (2017). Robots in industry: A shift towards autonomous and intelligent systems in the digital age. https://core.ac.uk/download/84146707.pdf (accessed on 10th September 2025).

35. Guizani, N., Sharma, V., & You, I. (2019). Security of 5G-V2X: Technologies, standardization and research directions. http://arxiv.org/abs/1905.09555 (accessed on 10th September 2025).

36. Hatherall, L., Kekulluoglu, D., Kokciyan, N., Rovatsos, M., et al. (2025). Making systems answer: A practitioner's handbook for trustworthy autonomous systems. https://core.ac.uk/download/666807483.pdf (accessed on 10th September 2025).

37. Husbands, P., Moioli, R. C., & Vargas, P. A. (2012). Synchronization effects on the behavioral performance and information dynamics of a simulated minimally cognitive robotic agent. https://core.ac.uk/download/9427880.pdf (accessed on 10th September 2025).

38. Huschebeck, M., Leonardi, J., Leonardi, J., et al. (2020). Approaching delivery as a service. https://core.ac.uk/download/305119210.pdf (accessed on 10th September 2025).

39. Jain, A. K., Jain, A. K., Skowron, A., Manly, B. J. F., Everitt, B. S., Bean, C. L., Bean, C. L., et al. (2008). Autonomous clustering using rough set theory. https://core.ac.uk/download/49393.pdf (accessed on 10th September 2025).

40. Ji, T., Nguyen, T. M., Wang, C., Xie, L., et al. (2018). Correlation flow: Robust optical flow using kernel cross-correlators. http://arxiv.org/abs/1802.07078 (accessed on 10th September 2025).

41. Kopardekar, P. H., & Sridhar, B. (2025). Towards autonomous aviation operations: What can we learn from other areas of automation? https://core.ac.uk/download/pdf/286393131.pdf (accessed on 10th September 2025).

42. Lorenz, E. (2025). Societal effects and the transfer of business practices to Britain and France. https://core.ac.uk/download/pdf/6583416.pdf (accessed on 10th September 2025).

43. Moniz, A. B. (2025). Synthesis about a collaborative project on "Technology assessment of autonomous systems". https://core.ac.uk/download/pdf/6815454.pdf (accessed on 10th September 2025).

44. Odintsov, S. D., & Oikonomou, V. K. (2018). Dynamical systems perspective of cosmological finite-time singularities in *f(R)* gravity and interacting multifluid cosmology. https://core.ac.uk/download/197479610.pdf (accessed on 10th September 2025).

45. Rahwan, I. (2017). Society-in-the-loop: Programming the algorithmic social contract. http://arxiv.org/abs/1707.07232 (accessed on 10th September 2025).

46. Ross, D. J. (1992). Making GATT dolphin-safe: Trade and the environment. https://core.ac.uk/download/62547848.pdf (accessed on 10th September 2025).

47. Rummel, K. (2019). The fundamentals of radar with applications to autonomous vehicles. https://core.ac.uk/download/213463684.pdf (accessed on 10th September 2025).

48. van Dijk, N., Gunnarsdottir, K., Rommetveit, K., Smits, R., et al. (2014). Making robotic autonomy through science and law? https://core.ac.uk/download/42414806.pdf (accessed on 10th September 2025).

49. Wang, F., Wang, S., & Zhu, H. (2010). On the convergence of autonomous agent communities. https://core.ac.uk/download/19784766.pdf (accessed on 10th September 2025).

50. Ya Baklanov, P., & Moshkov, A. V. (2016). Structural transformations of the economy in the Pacific Region of Russia and efficiency trends. https://elar.urfu.ru/bitstream/10995/47029/1/r-economy_2016_v2_1_05.pdf (accessed on 10th September 2025).

51. Zumbansen, P. (2012). Transnational private regulatory governance: Ambiguities of public authority and private power. https://core.ac.uk/download/62560183.pdf (accessed on 10th September 2025).

CHAPTER 2

EMBEDDED INTELLIGENCE

Contents

Introduction 43

2.1. Concept of Embedded Intelligence 47

2.2. Embedded Processors 53

2.3. Microcontrollers 59

2.4. Role of AI in Embedded Systems 63

2.5. Case Studies of Embedded Intelligence 68

2.6. Future Trends in Embedded Intelligence 73

Summary 78

References 80

INTRODUCTION

The concept of embedded intelligence is a marriage of hardware and software for the purpose of providing ordinary devices enhanced cognitive powers. Embedded technology is rooted in the blending of embedded processors and microcontrollers as the foundation of contemporary embedded systems. These systems allow devices to make decisions in real-time by processing data locally and without relying on continuous communication with remote servers. With the demand for improved, intelligent solutions mounting, particularly in segments like IoT (Internet of Things), the development of embedded intelligence is crucial. Here, advancements in machine learning algorithms further optimize these processors such that they can perform sophisticated functions while not easing the power consumption requirements, computational resources, and real-time processing needs outlined in standards like TinyML (Sakr et al., 2023). Besides, the role of artificial intelligence (AI) in embedded systems cannot be overemphasized.

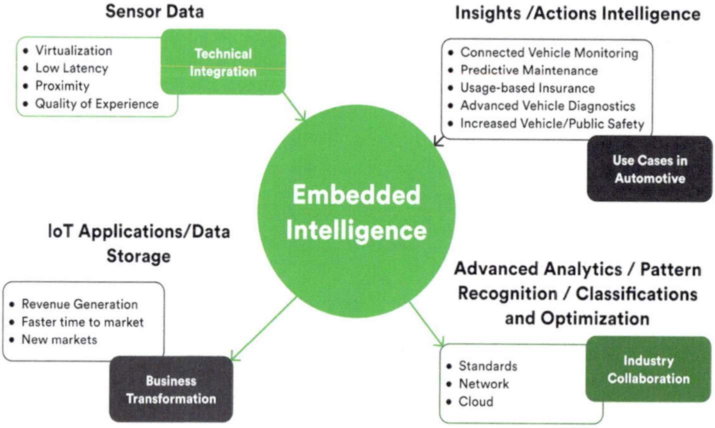

Figure 2.1. *Embedded intelligence.*

Source: https://www.rinf.tech/wp-content/uploads/2023/03/MicrosoftTeams-image-336.png.

AI enhances the performance and efficiency of devices through the ability to learn from data and acclimatize to new conditions. Such adaptability is particularly significant in uses ranging from smart domestic appliances to factory automation where the operational efficiency directly equates to performance outcomes. Embedded AI allows systems to execute complex algorithms without huge latency, and this is essential in making sure that systems can keep up with real-time responses under fluctuating conditions. Thus, the combination of AI and embedded systems does not only redefine the way single devices operate but also makes it possible to build intelligence ecosystems that are capable of addressing contemporary challenges like energy efficiency and resource allocation (Dokladal et al., 2023). Integration of energy-efficient solutions within embedded intelligence is a significant research and development area. As intelligent devices get increasingly deployed, power management has turned into a severe challenge, particularly for devices deployed in resource-scarce environments. Taking this into account, power-optimized hardware structures and software frameworks have been researched to ensure maximum power utilization without considerable performance loss. Thus, such solutions that will be decreasing the energy footprint, such as those that are being developed under the TinyML effort, are imperative. The efforts are working on enhancing the capability of microcontrollers to process high-end AI workloads with low power consumption and latency so that embedded systems can succeed in a large number of applications. With an increased synergy between embedded technology and AI, new avenues of sustainable and smart device deployment emerge, allowing for future growth (Figure 2.1).

Definition of Embedded Intelligence

Embedded intelligence is a name that has been used to refer to the integration of intelligence into embedded systems such that devices can make decisions based on the processing of information without human interference. This integration comes in handy in the use of today's modern technology where time and efficiency are essentials. Embedded intelligence includes the process of information processing, learning from it, and reacting thereto to varying conditions. This is as opposed to previous systems that were rule-based and outcome-specified. The growth of embedded intelligence has accelerated innovation across industries such as automotive, healthcare, and home automation technology where devices need to react in real-time to their environments. Embedded intelligence is therefore recognized not only for its technological benefits but for its capacity to transform user experiences and industry operating efficiencies. The implementation of embedded intelligence is totally based on embedded processors and microcontrollers, which are the components of the intelligent systems. These small computing units execute advanced algorithms and manage acquiring data from sensors, essentially transforming raw data to productive information. Embedded processors, through processing this information in real time, are able to enhance performance and ensure that devices operate under strict resource limitations, such as power and memory. Along with this, ongoing R&D work is focused on enhancing the abilities of these microcontrollers to process higher-order algorithms so that the boundaries of what can be done by embedded systems are continually being pushed. The TinyML achievement is an inspiring illustration of the way these technologies can enable machine learning models to be used efficiently on low-power devices, stressing the reciprocal wins synergy between processor technology and embedded intelligence as demonstrated in some studies (SAKR et al., 2023). The use of artificial intelligence (AI) in embedded systems is essential because it transforms passive devices into active, responsive products that can work autonomously. AI algorithms embedded in these devices allow important features such as pattern recognition, predictive maintenance, and anomaly detection, thus improving performance and user experience as a whole. More importantly, the ongoing miniaturization of devices as well as energy efficiency concerns have led to intelligent embedded devices that have to function under constraint power availability conditions. This is particularly the case where energy consumption is a perpetually critical aspect, so that pursuit of efficiency is an ongoing challenge (Dokladal et al., 2023). With increasingly complex embedded systems, the addition of AI enables the creation of smarter, more vibrant solutions that closely align with user requirements and environmental demands. Conversely, embedded intelligence is at the forefront of technology development, bridging the gap between hardware limitation and high-level computation.

Importance of Embedded Systems in Modern Technology

The position of embedded systems in today's technology cannot be overstated since they form a central component in a variety of devices that dominate both home life and industrial procedures. They integrate hardware and software to perform specialized tasks effectively and efficiently in larger systems. For instance, microcontrollers and embedded processors are frequently used at the heart of consumer appliances, automotive controls, and medical devices, enabling functionalities ranging from simple control tasks to sophisticated data processing. As technology continues to advance, embedded systems demand more efficiency and intelligence in power consumption and performance. As noted in recent research, computer processes' energy concerns are of crucial significance, where the balance is given to performance against the energy limitations of embedded computing environments (Dokladal et al., 2023). This demand generates innovation, which creates systems that not only are performative but also sustainable and efficient. Besides, the advent of artificial intelligence (AI) significantly enhances the functionality of embedded systems, particularly in the context of the Internet of Things (IoT). With the integration of AI algorithms in embedded systems, machines can learn from their environment and make intelligent decisions autonomously. However, machine learning on constrained devices is a challenge because it involves computational overhead, has limited memory, and lacks adequate power (Sakr et al., 2023). The arrival of TinyML solutions addresses these issues, offering the potential for advanced functionality in edge computing environments. AI in embedded systems is transforming applications such as intelligent agriculture, healthcare monitoring, and automation in industry. These advancements depict the critical convergence point of AI and embedded systems in providing smart functioning that enhances efficiency, optimizes resources, and generates real-time analytics. Finally, the insistent revolution in embedded systems heralds a broad trend towards increased connectivity and automation in various disciplines. As more devices are being drawn into smart networks, there is an increasing need for embedded systems to handle and manage data on the edge. This transformation not only provides capabilities to devices but also facilitates the development of smart applications that can be self-contained without excessive dependence on the cloud. New frameworks, as noted by studies, target the reduction of memory requirements and enhanced performance in microcontrollers, allowing for sophisticated functions to run without problems within confined environments (Sakr et al., 2023). This innovation reflects the necessity for embedded systems in today's technology, where they serve as a basis for innovations demanding low power consumption, high performance, and efficiency. As technology continues to advance, embedded systems will increasingly play a vital role in assisting the development of a smart, networked future (Figure 2.2).

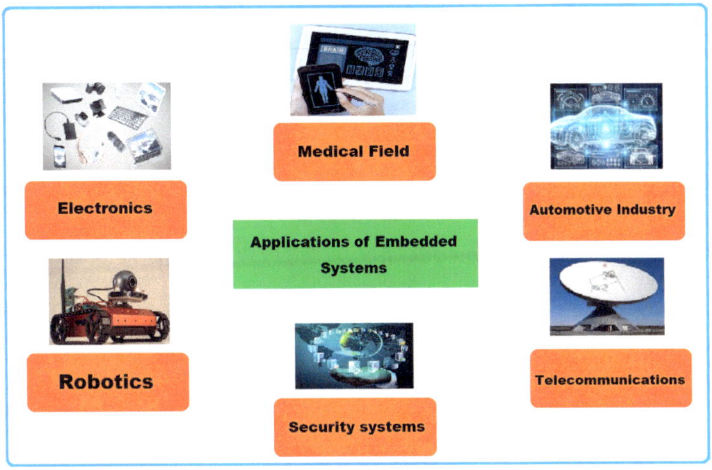

Figure 2.2. *Real-life applications of embedded systems.*

Source: https://images.theengineeringprojects.com/image/main/2021/06/applications-of-embedded-system.png.

2.1. CONCEPT OF EMBEDDED INTELLIGENCE

The concept of embedded intelligence is a milestone in the computing world, where intelligence is embedded directly within devices and systems that did not have such capabilities before. The embedding enables devices, such as sensors and microcontrollers, to perform tasks with complex processing without relying on big external computational abilities. The ability to execute machine learning algorithms on resource-limited devices is significant, especially considering computational overhead and scarce resources common in embedded systems. Such a framework not only facilitates real-time processing of data but also enhances efficiency and autonomous device operation to meet contemporary demands for smart technologies operating efficiently in the network edge (Sakr et al., 2023). With developing embedded intelligence, it promises to release new capabilities across various applications, from home automation to industrial monitoring systems.

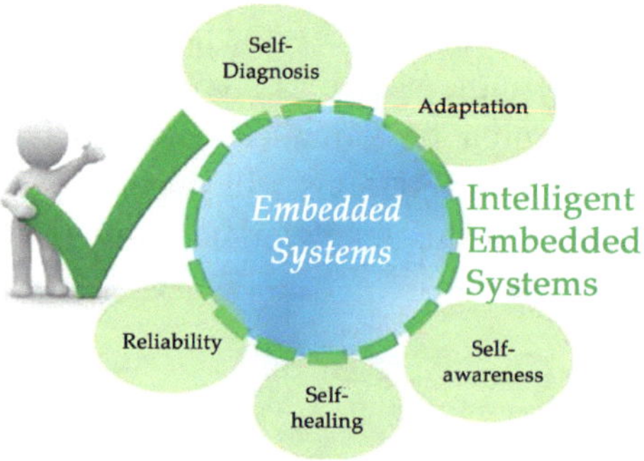

Figure 2.3. *Intelligent embedded software.*

Source: https://cdnintech.com/media/chapter/58179/1512345123/media/F3.png.

Embedded processors and microcontrollers are the backbone of embedded intelligence, with the computational power to implement sophisticated algorithms in environments where power is a constraint. The use of TinyML, a form of machine learning in low-resource environments, is the optimal manifestation of this confluence of hardware and software. TinyML applications can utilize the computational capabilities of microcontrollers to perform tasks such as data inference and even self-supervised learning, enabling devices to improve their own abilities independently (Sakr et al., 2023). Furthermore, the development of such environments as the Edge Learning Machine setting highlights how important it is to maximize performance in such constrained contexts. By balancing memory requirements against processing capabilities, embedded intelligent systems can yield performance levels comparable to more powerful configurations within the confines of the resource constraints characteristic of embedded environments. With increasingly pervasive uses of artificial intelligence (AI) in embedded systems, energy consumption and effectiveness implications become increasingly important. The combined challenge of limited power supply and heat management in embedded systems demands new approaches to energy efficiency in AI. Traditional computing platforms are confronted with the challenge of heat dissipation and energy consumption, but the limits in embedded systems aggravate these challenges. Therefore, it is essential to design energy-efficient algorithms that can execute on low-power devices without loss of accuracy. Advances in tool and framework technology aimed at efficiency of energy create the possibility of environmentally sustainable release of intelligent systems, turning embedded intelligence into not just a technological step forward but also an essential component in our effort to solve bigger environmental challenges and advance the state of smart devices (Figure 2.3).

2.1.1. Historical Development of Embedded Intelligence

The evolution of embedded intelligence is traced to early history of microcontrollers and embedded systems, paving the way for more complex applications. In the beginning, embedded systems were dominated by simple control functions based on limited-capability microcontrollers. As demands for smarter applications increased, hardware and software progress led to the incorporation of artificial intelligence (AI) methods into embedded systems. As microcontrollers evolved, for example, Arm® Cortex® series, more processing and memory power became available to developers, enabling the execution of more sophisticated algorithms and functions. The transition was a significant advancement as systems evolved from basic automation to intelligent decision-making devices with the ability to adjust to dynamic conditions, enhancing efficiency and functionality in applications. The embedding of AI in embedded systems has revolutionized the manner in which industries operate, particularly concerning predictive maintenance (PdM). Advances in embedded intelligence systems in recent times have employed machine learning algorithms for patterns comparison and the provision of predictive failure forecasting. The application of AI in microcontrollers, for instance, can now sustain real-time sensor data processing, which can identify anomalies and initiate preventive measures prior to experiencing a failure. This not only decreases downtime but also reduces operational costs significantly. The ability of such intelligent systems is also evident in innovations such as mini-edge Industrial Internet of Things (IIoT) devices using platforms such as NanoEdgeTM AI Studio and Edge Impulse. These developments are crucial in the creation of a more powerful and efficient industrial network through the provision of tailored PdM solutions through the capabilities of intelligent data processing (Coppola et al., 2022). Additionally, the shift towards Distributed Control Architectures (DCA) is a significant step towards the consideration and implementation of embedded intelligence. DCA enhances the ability to integrate various components of unmanned vehicles into an intelligent network through standardized communication protocols.

This design pattern not only enhances interoperability between heterogeneous sensors and actuators but also simplifies integration for programmers. The use of objects that present functionality through several interfaces exposes it to the programmer, making it possible for the DCA technique to facilitate simpler upgrades and promoting component reuse in different applications. Such advances mark a watershed moment in the development of embedded systems, paving the way for the development of intelligent systems with self-adaptation and reaction capabilities to operating environments. This is aligned with the new occurrence of AI in embedded systems and how it can transform traditional control mechanisms and enhance autonomy flexibility in autonomous technology (Rawlings, 2009) (Figure 2.4).

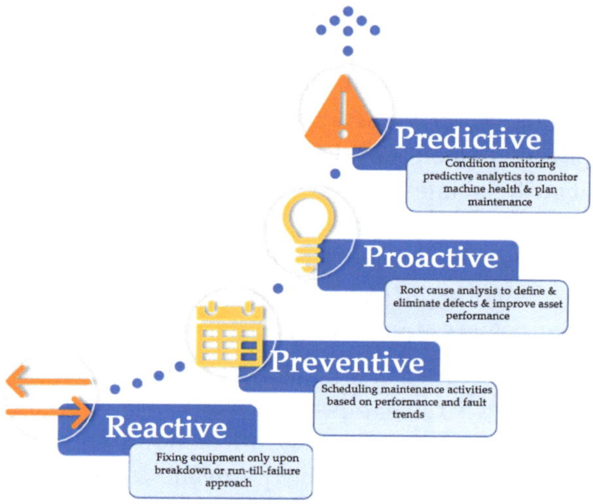

Figure 2.4. *Predictive maintenance and how it is transforming the manufacturing sector.*

Source: https://www.infinite-uptime.com/wp-content/uploads/2021/10/What-is-Predictive-Maintenance-banner-infinite-uptime-1-1.webp.

2.1.2. Key Characteristics of Embedded Intelligence

Embedded intelligence is intrinsically characterized as its ability to compute data locally on the device itself rather than to present data to cloud computing resources for processing. Decentralization enables real-time decision-making, which is crucial for applications that require instant responsiveness, such as autonomous transportation or medical monitoring systems. With sensors and computing capabilities embedded in one platform, embedded systems are able to process data quickly and execute complex algorithms without latency issues fueled by cloud computing. These capabilities are especially crucial in energy efficiency, where the devices must operate under highly constrained power. As discussed, in the former, power usage is critical since the available energy that can be dedicated to the devices is constrained (Dokladal et al., 2023). This focus on efficient energy use is representative of an underlying principle of embedded intelligence that seeks good performance subject to physical and operational limitations. A second striking characteristic of embedded intelligence is flexibility with the addition of ML algorithms designed especially for constrained-resources environments. This flexibility allows embedded systems to learn about their use environment and improve their performance with time. TinyML proves the potential for machine learning on edge devices, breaking significant barriers such as computational expense and resource limitations. Findings indicate that running machine learning algorithms on embedded platforms requires proprietary solutions and development tools to adapt to power and memory restrictions (Sakr et al., 2023). These innovations make it possible for embedded systems to alter their behavior dynamically based on feedback from data, which further enhances their functional capabilities, further making them more beneficial in a series of applications ranging from smart domestic equipment to advanced industrial automation. Furthermore, the combination of embedded processors and microcontrollers perfectly demonstrates the essence of embedded intelligence by virtue of their compact architecture and power-conscious performance. These components are specifically optimized to execute operations that require minimal power consumption with high functionality and performance. In artificial intelligence (AI), embedded processors allow the execution of complex algorithms on the device to minimize latency and improve reliability. The ability to run AI models locally provides equipment with the ability not only to respond to real-time information but also to engage in sophisticated analytics without drawing upon centralized infrastructure. The integration of hardware and software within embedded intelligence plays a pivotal role in enabling end-to-end efficiency and effectiveness of modern apps to be improved, furthering the idea that intelligence embedded in products can bring revolutionary transformation across domains.

2.1.3. Applications of Embedded Intelligence in Various Fields

Embedded intelligence has been used extensively in various industries, particularly augmenting functionality in automotive systems. In automotive, vehicles with embedded systems incorporate artificial intelligence (AI) to increase performance and safety. For instance, driver-assistance systems (ADAS) incorporate machine learning algorithms to translate sensor data and images from cameras to develop functions such as lane-keeping assistant and collision prevention. With optimal distribution of torque via innovative control structures, like one that has been set up recently, vehicles can exhibit improved handling and performance. Embedded platforms that integrate neural networks with fuzzy logic provide real-time decision-making needed in dynamic driving conditions. Apart from enhancing the performance of a vehicle, these applications address critical safety challenges, thus underlining the place of embedded intelligence in contemporary automobile designs (Jahnke et al., 2019). Embedded intelligence in industrial automation is a determining driver of maximum productivity as well as operational effectiveness.

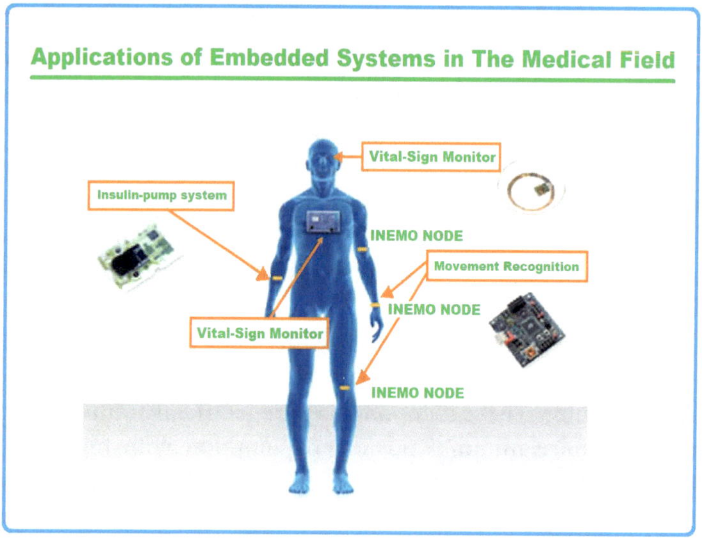

Figure 2.5. *Applications of embedded intelligence in various fields.*

Source: https://images.theengineeringprojects.com/image/webp/2021/06/Applications-of-Embedded-Systems-in-The-Medical-Field.png.webp?ssl=1.

Embedded systems are used by intelligent manufacturing processes for the monitoring of equipment and automation of processes. The systems apply AI-based predictive maintenance software that considers previous instances and current machine status to foretell breakdowns prior to their occurrence, thereby minimizing downtime. Predictive capability is particularly essential where time delays equate to enormous losses. Besides, the integration of embedded processors with IoT technology supports real-time communication and data analytics, facilitating a nimble manufacturing setup. The seamless merging of the technologies is the form in which embedded intelligence transforms traditional industrial processes into highly optimized smart systems (Jahnke et al., 2019). Healthcare is another sector where embedded intelligence is making a huge impact, especially in medical devices and patient monitoring. Wearable health technology, such as smartwatches and fitness trackers, utilizes embedded systems to track vital signs continuously and provide suggestions regarding the health of an individual. The devices use AI algorithms to predict likely health risks, thus enabling proactive healthcare management. Also, embedded intelligence assists in developing sophisticated medical devices that can carry out real-time diagnostics and customize treatment for patients. For example, smart insulin pens and glucose monitors employ in-built intelligence to tailor insulin release based on real-time measurement of blood glucose. These innovations underscore the revolutionary capability of in-built intelligence in enhancing patient care, health outcomes, and the future of medical technology (Jahnke et al., 2019) (Figure 2.5).

2.2. EMBEDDED PROCESSORS

Embedded processors are small microchips that are experts in performing dedicated tasks within bigger systems and significantly enhance efficiency along with reliability. Embedded processors are used in embedded systems that are prevalent in contemporary technology, ranging from consumer electronics to industrial as well as automotive applications. Embedded processors, in contrast to general-purpose processors, are architected around dedicated tasks so that they are able to perform operations on very little power as well as physical space. This focus on efficiency makes them highly suitable for application requiring real-time processing, where response time is critical. Additionally, as the face of technology evolves, newer architectures such as those introducing artificial intelligence capabilities are emerging, which allow embedded systems to learn and adapt from the environment, thereby enhancing their functioning and user experience. The processor architecture of an embedded processor typically incorporates features that respond to specific computation requirements, such as features for multithreading and parallel processing. They assist them in running different tasks at the same time, hence making them appropriate for complex embedded applications that have high-performance demands without losing any reliability. In addition, innovations in software and design techniques such as the described distributed agent architecture (DAA) demonstrate how traditional models of real-time computing can be used to enable artificial intelligence-based applications within embedded systems. Not only does this double capability ensure hard real-time processing requirements, but it also introduces flexibility in the form of cognitive agents that may operate under soft real-time. The versatility of embedded processors thus positions them on the leading edge of technology, combining computational capacity with innovative artificial intelligence algorithms to create smart systems that respond dynamically to real-world issues (Lee et al., 2025). With the progress made in embedded processors, their role in systems across various industries signifies their effectiveness in the era of augmented automation and networked devices. Particularly, initiatives in critical infrastructure and communication networks demonstrate the need to realize interoperability and consistent data processing. To provide an example, the architecture proposed for the Deep Space Network (DSN) demonstrates the necessity of a unified system with high automation and security levels during the processing of massive quantities of telemetry information. These advancements not only enhance operations efficiency but also allow more effective cooperation among different stakeholders, from spacecraft teams to non-NASA partners, by open systems standards. The advancement of embedded processors is thus beyond a technological evolution but a catalyst for radical innovations in how systems are designed, executed, and optimized for nascent challenges in data processing and communication. These developments make it imperative to introduce intelligence into the very fabric of processor design in commercial and experimental settings (Atkinson et al., 1992) (Figure 2.6).

Figure 2.6. *Embedded processors.*

Source: https://www.embedic.com/uploads/editor/watermark/20230506/1683353592156What%20is%20Embedded%20Processor.jpg.

2.2.1. Types of Embedded Processors

Embedded processors are the crucial devices within the embedded system ecosystem that facilitate the operation of diverse applications ranging from consumer products to industrial equipment. The three primary types of embedded processors include microcontrollers, digital signal processors (DSPs), and application-specific integrated circuits (ASICs). Microcontrollers are lightweight and designed to perform specific tasks in a system and thus well suited for simple applications with limited resource requirements. While DSPs are meant to execute fast mathematical calculations, which enable them to process communications, video, and audio signals at high efficiency. ASICs are application-specific chips optimized for a particular application, and thus they ensure maximum performance and efficiency but limited flexibility in comparison to general-purpose DSPs and microcontrollers. This classification proves the versatility of embedded processors, showing their application-specific designs to meet various requirements of applications within the framework of embedded intelligence. The introduction of artificial intelligence (AI) has brought with it a new dimension to embedded processors, specifically for internet-of-things (IoT) devices. With more and more AI functionality being embedded in the system, efficient and low-power processing is the priority. This has given birth to specialized architectures, such as those for Tiny Machine Learning (TinyML), where the priority is to execute AI code on ultralow-power (ULP) IoT devices. For example, the Dynamic AI-IoT architecture improves the processing efficiency by enabling effortless integration with the cloud without requiring huge firmware updates, thereby simplifying AI integration. All such improvements herald a paradigm shift toward more cognitive embedded systems that are capable of learning and adapting in real-time, and they are representative of the attributes of embedded intelligence. This evolution calls for a processor design rethink, with a focus on energy efficiency and flexibility to meet the requirements imposed by state-of-the-art AI applications on embedded platforms. Energy usage remains a critical consideration when embedded processors are being addressed, particularly against the background of embedded computing and supercomputing. The efficiency of embedded processors has a direct impact on their ability to perform required tasks without using too much energy, as most applications must operate within tight power budgets. Minimizing battery life in mobile phones and the demand for green operations in IoT deployments emphasize the importance of addressing energy efficiency. As recognized by current studies, the introduction of smart devices has increased energy concerns, which must be addressed from hardware and software perspectives (Dokladal et al., 2023). The embedded processors must be developed in a way that, along with performing computational operations efficiently, they also manage power usage for optimal performance. Consequently, the on-going improvement of embedded processor design, particularly with the inclusion of energy-aware elements, is a significant consideration in enhancing the functionality and longevity of modern embedded systems (Figure 2.7).

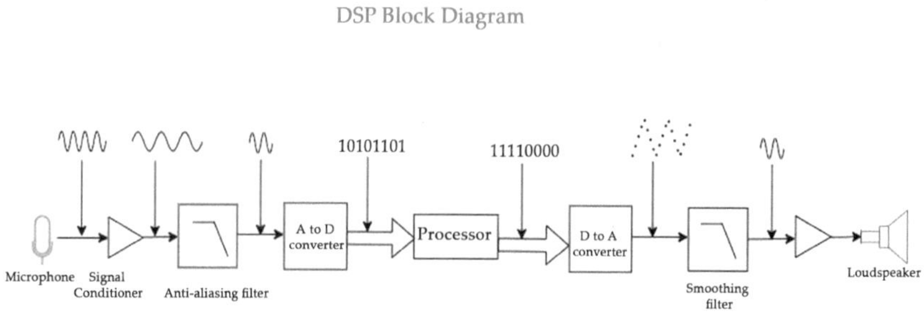

Figure 2.7. *Digital signal processing (DSP).*

Source: https://media.geeksforgeeks.org/wp-content/uploads/20240403130126/ Untitled-Diagram---2024-04-03T130110286.webp.

2.2.2. Architecture of Embedded Processors

The architecture of embedded processors takes a pivotal position in the creation of embedded intelligence, as it encompasses the concepts and operational models that allow these systems to perform unique tasks with efficiency and reliability. These processors, typically characterized by their specialized application and integration in the overall system, are fundamental to uses ranging from consumer products to car control systems. The architecture is typically designed to have low power consumption, compact size, and real-time execution support, which distinguish embedded processors from general purpose computing platforms. Their ability to perform tasks under harsh constraints such as time and resources is what is most important, especially in IoT devices in which several sensors and actuators operate concurrently. This design particularity enhances the overall performance of embedded systems, enabling them to respond immediately and effectively to real-world conditions. One of the key aspects of embedded processor architecture is the addition of artificial intelligence, which enhances the functionality and decision-making capabilities of the systems. As there is an increasing demand for intelligent behavior in devices, architecture also consists of AI techniques in an integrated manner. For instance, reactive agents can be constructed with the traditional computing approaches but can also utilize AI algorithms, such as Bayesian belief networks, to make more effective decisions, as outlined by Lee et al. (2025). The double capability allows embedded processors to execute pre-programmed tasks but also react to new data in real time. The ability to integrate hard real-time requirements with cognitive action enables more autonomous systems that can learn and adapt throughout their lifetime, important for applications like smart living houses and autonomous vehicles. Furthermore, the choice of system architecture—centralized or distributed— Plays a crucial role in embedded intelligence efficiency and scalability. A distributed simulation host computer architecture, one SwRI has studied for NASAs Space Station Freedom effort, describes this paradigm. This type of architecture is discovered to provide the scalability and flexibility essential to meet growing computational requirements with real-time operation efficiency, as noted by Bishop et al. (2025). Embedded processors, through distributed systems, can share and collaborate with resources freely and thus facilitate higher performance in dynamic and complex environments. These advances make it possible for embedded systems not only to run single functions but also to engage in complex interactions, making way for robust smart systems that are able to operate seamlessly within real-world applications and advance embedded intelligence (Figure 2.8).

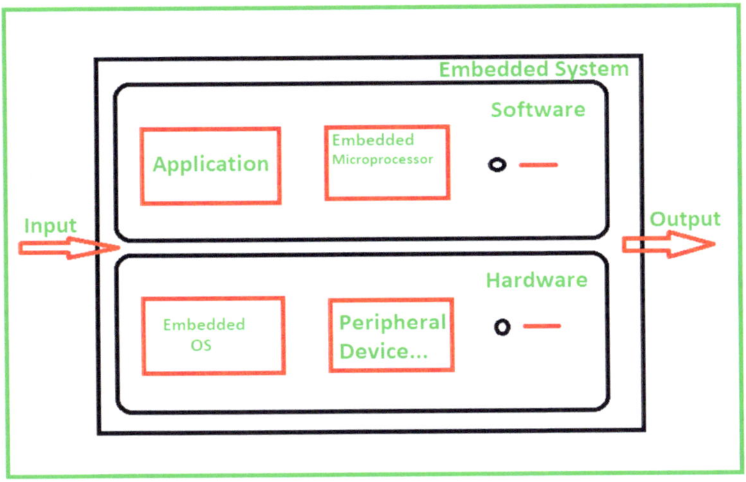

Figure 2.8. *Architecture of an embedded system.*

Source: https://media.geeksforgeeks.org/wp-content/uploads/20200607194233/embedded-2.png.

2.2.3. Performance Metrics for Embedded Processors

The performance factors of embedded processors play a major role in determining their efficiency in executing in multiple applications, particularly in the area of embedded intelligence. Speed of processing, power consumption, and computational performance are fundamental to maximally performing within the constrained resources of the environment. With the growing application of artificial intelligence (AI) in embedded systems, processor requirements increase. For instance, visual navigation in autonomous nano-drones powered by deep neural networks (DNNs) focuses on the necessity of high performance with low power usage. Real-time processing with complicated algorithms with real-time constraints requires processors that can handle low power as well as speed and efficiency. The challenge is to design embedded processors that are able to perform sophisticated tasks and be efficient and responsive, particularly in the case of increasingly miniaturized products such as drones and Internet of Things applications, providing an optimal blend of hardware performance and algorithmic optimization.

The integration of AI into embedded systems calls for a rethinking of conventional performance metrics, particularly for low-end Internet-of-Things devices such as microcontrollers. The majority of conventional machine learning methods consume large amounts of memory and computational resources, thereby posing significant problems for implementation within ultra-constrained contexts. Consequently, new approaches are necessary to ensure the applicability of AI to use within microcontroller constraints.

As highlighted in research, building an adapted model development process unique to resource-limited applications ensures that the attainment of desired performance without exceeding the devices compute and latency budget becomes feasible. This approach illustrates that, while energy efficiency and miniaturization are a priority with embedded systems, they are nonetheless capable of utilizing machine learning to enhance functionality and responsiveness, effectively transforming their functional abilities within the realm of IoT. Lastly, the success or failure of embedded processors in intelligent systems ultimately relies on their ability to balance performance measures against practical limitations.

The accomplishments achieved in platforms such as GAP8, which execute DNNs on-board with strict power constraints, represent a step in this direction. The concurrent focus on achieving high frame rates—albeit at the expense of low power utilization—emphasizes the critical intersection of algorithmic complexity and hardware design. By solving both the computational demands and the energy efficiency, embedded processors can make it possible for a wide range of applications within embedded intelligence, ranging from collaborative drones to intelligent sensor networks. With continued research breakthroughs in the formulation of innovative solutions to the challenges of integrating AI and machine learning into infrastructure, the ultimate development of these performance standards will serve as a building block for future innovations in the space, opening the door to broader implementation of sophisticated embedded systems into everyday technology.

2.3. MICROCONTROLLERS

Integration of microcontrollers into the world of embedded systems has been instrumental in propelling the concept of embedded intelligence forward. Microcontrollers are compact computing devices that act as the brain of most electronic systems, enabling them to perform certain tasks efficiently. With the integration of sensors, actuators, and communication interfaces, microcontrollers enable seamless communication between the hardware and software components of a system. This ability to process input data and execute control algorithms is the basis of underlying applications in a variety of industries, ranging from automotive technology and home automation controllers to industrial control systems. By providing an underlying foundation for embedded systems, microcontrollers make possible the utilization of advanced functionality and real-time processing and decision-making capabilities that mirror ideals of embedded intelligence. The evolving microcontroller ecosystem has sped up with the development of Tiny Machine Learning (TinyML), which has revolutionized artificial intelligence solutions in low-resource settings. TinyML facilitates machine learning algorithms to be executed on microcontrollers, enabling efficiency and lower power consumption at the expense of performance compromise. This aspect is particularly important in the context of the Internet of Things (IoT), where it takes millions of devices to possess robust yet effective solutions to execute and analyze locally. By integrating AI capabilities into embedded systems, TinyML makes it possible for microcontrollers to execute complex operations such as motion detection, voice recognition, and even predictive maintenance. As a result, with ever more advanced embedded systems, microcontrollers with TinyML are poised to drive the future of intelligent applications across different industries (Capogrosso et al., 2023). In addition, the integration of artificial intelligence in microcontrollers coincides well with wireless communication technologies as we move towards the rollout of 6G networks. Such next-generation wireless networks will heavily rely on microcontrollers to handle vast amounts of data generated by devices that are online real-time. Machine learning-based algorithms will reign supreme in the network performance optimization, promoting optimal usage of resources, and strong communications. Microcontrollers thus take the pivotal role in making possible the zero-touch optimization of wireless networks, indicating their importance within the overall ecosystem of embedded intelligence. As demand for efficient connectivity and autonomous decision-making grows, integrating advanced microcontrollers into embedded systems is essential to addressing the looming challenges of the next generation wireless communication (Abbas et al., 2020) (Figure 2.9).

Figure 2.9. *Microcontroller and its types.*

Source: https://media.geeksforgeeks.org/wp-content/uploads/20250501152116058488/microcontroller.webp.

2.3.1. Definition and Functionality of Microcontrollers

Microcontrollers form the heart of embedded systems and are the basis for a wide range of applications from home appliances to advanced robotic systems. A microcontroller is a small integrated circuit designed to manage some operations in an embedded system. It typically contains a processor core, memory, and programmable input/output peripherals. Microcontrollers can execute a list of commands and execute resources in a systematic manner. This combination of pieces into one chip not only saves space but also saves power—prime considerations in designing efficient energy systems. As embedded systems increasingly provide intelligent functions, the role of the microcontroller changes from simple control functions to complex decision-making typical of autonomous applications, thereby creating a building block in the development of smart technologies. Functionality is at the core of understanding the functionality of microcontrollers in embedded systems. Microcontrollers are programmed to perform an assigned set of activities from simple data processing to complex actions such as controlling interaction with sensors and actuators. Capability to interface with a large variety of peripherals makes microcontrollers suitable for different environments, with roles in industries such as automotive, health, and consumer appliances. Their adaptability is also boosted by advancements in artificial intelligence (AI), as microcontrollers become more commonly used to execute algorithms that enhance decision-making mechanisms. For instance, application of systems such as real-time fuzzy expert systems in microcontroller frameworks demarcates how AI can be utilized to manage imprecise information, hence enhancing the problem-solving ability of embedded systems in practical applications (Sarpe et al., 2025). Recent developments have also taken into account microcontrollers as key platforms for embedding AI-based capabilities, especially in autonomous systems. New research points out the capability of microcontrollers to enable sophisticated tasks, such as visual navigation for drones, without violating power limits. The advent of ultra-low-power computing platforms, such as represented by chips like GAP8, allows for microcontrollers to execute high-level deep neural networks (DNNs) onboard with the expenditure of limited energy budgets (Benini et al., 2018). It not only benefits the deployment of smart behavior in small devices, like nano-drones but also is a revolutionary shift in the work potential of embedded systems. With the development of microcontrollers, their integration with AI technologies heralds a future of embedded intelligence founded upon increased autonomy and greater functional capability (Figure 2.10).

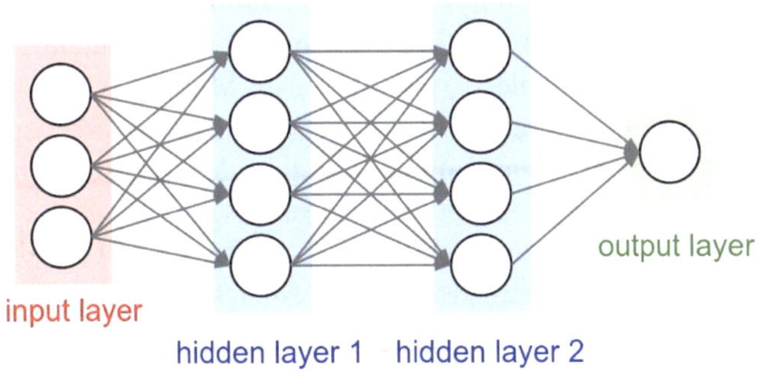

Figure 2.10. *Deep neural network.*

Source: https://s7280.pcdn.co/wp-content/uploads/2020/07/Two-or-more-hidden-layers-comprise-a-Deep-Neural-Network.png.

2.3.2. Comparison Between Microcontrollers and Microprocessors

Under the umbrella of embedded systems, microcontrollers and microprocessors have a profound distinction. Microcontrollers are low-profile computing platforms with the essentials of a CPU, memory, and input/output peripherals, all in one chip. It is this integration that reduces power usage and physical space, thereby the application of microcontrollers to specific purposes like home appliances and sensor networks. Conversely, microprocessors that are typically used in personal computers are designed for performance and versatility, enabling them to carry out sophisticated tasks and run multiple processes simultaneously. These distinctions have an impact on their differing application in embedded systems, where the precise processing capacity and efficiency needed dictate the use between these two architectures. The application of artificial intelligence (AI) to embedded systems also refers to the distinctions between microcontrollers and microprocessors. Microcontrollers are trending in developing smart devices requiring real-time processing of data with low latency, making them suitable for machine learning on the edge. For instance, adding an Artificial Neural Network (ANN) to a microcontroller can facilitate energy-efficient smart home applications such as indoor temperature prediction, controlling HVAC conditions, and enhancing user comfort with reduced power consumption (Botella-Rocamora et al., 2015). In return, the advanced computational powers of microprocessors support more compute-heavy AI applications requiring excessive processing capabilities, like high-resolution image recognition or huge data analysis processes. These variations reveal the specific roles each processor plays within the continuously changing world of embedded intelligence. With the further technical development of embedded technology ongoing, particularly with the TinyML effort, microcontrollers are evolving to support more advanced machine learning processes. TinyML supports the deployment of machine learning models on constrained devices, with an emphasis on performance and conservation of resources (Sakr et al., 2023). Breakthroughs in frameworks, such as the Edge Learning Machine environment, enable microcontrollers to perform near-real-time data processing and self-supervised learning, thereby bridging the gap between microprocessors and microcontrollers in terms of AI capabilities. These advance underlines the extremely crucial role played by microcontrollers in the modern embedded system philosophy, ever refashioning the future interaction between AI and embedded intelligence. Thus, it is imperative to understand the comparative merits and applications of microcontrollers and microprocessors for optimal design and use of smart technologies.

2.3.3. Common Applications of Microcontrollers

Microcontrollers are at the center of a wide variety of applications, particularly in control and automation systems. They are the basis for household appliances like refrigerators and washing machines, allowing for complex operations like energy-saving functionality and programmable settings. Manufacturers can implement features like remote control, automatic cycle adjustment, and sensor functionality through microcontrollers, thereby enhancing convenience for users and operational efficiency. For instance, smart home spaces highly rely on microcontrollers that regulate various devices to communicate freely with one another. The ability to program and reprogram them not only results in enhanced functionality but also offers future upgrades without structural changes to hardware, showcasing the malleability of microcontroller technology in current use. In environmental monitoring and automation, microcontrollers find application in the domain of wireless sensor networks (WSN). Such systems prefer to utilize low-cost microcontroller technology for monitoring temperature and humidity parameters in real time. A classic example is the application of Artificial Neural Networks (ANN) on microcontrollers in the context of predictive analytics in HVAC systems. With the integration of such capabilities, microcontrollers can not only manage building conditions but also learn from real-time data streams. The self-learning mechanism that does not depend on detailed history databases optimizes energy efficiency. As highlighted, the ability of the system to react to timely environmental changes brings a prudent depth of smartness to otherwise uncomplicated devices, showcasing the remarkable agility of microcontrollers in smart home applications (Botella-Rocamora et al., 2015). Also, the integration of AI in microcontroller uses extends to more complex decision-making in systems such as real-time fuzzy expert systems. These systems apply artificial reasoning to deal with and process uncertain or imprecise information, which is at the core of application where human-level decision-making is essential. Such application indicates the possibility of embedded systems in applying fuzzy logic in making judgments on situations and providing best results from varying inputs. By putting such systems into microcontrollers, device developers enable devices to carry out tasks that demand a refined sense of conditions ranging from industrial automation to individual medical monitoring systems. This development in knowledge management systems is a pointer to the greater need for embedded devices to provide intelligent outputs, thereby making them more valuable and efficient in different scenarios (Sarpe et al., 2025).

2.4. ROLE OF AI IN EMBEDDED SYSTEMS

The integration of artificial intelligence (AI) into embedded systems is a quantum leap that enhances their functional effectiveness and operational efficiency. Embedded systems, which are known for their singular functions within large systems, have increasingly incorporated AI algorithms to support complicated decision-making processes. This is especially true in real-time data-heavy applications, such as autonomous vehicles and home automation technology. AI enables such systems to learn from the environment, make informed predictions, and adapt in a dynamic environment, ultimately leading to improved user experiences and system reliability. Utilization of AI in embedded systems carries the advantages of continuous learning and adaptability, which are important for applications where accuracy and punctuality of response are necessary in dynamic situations.

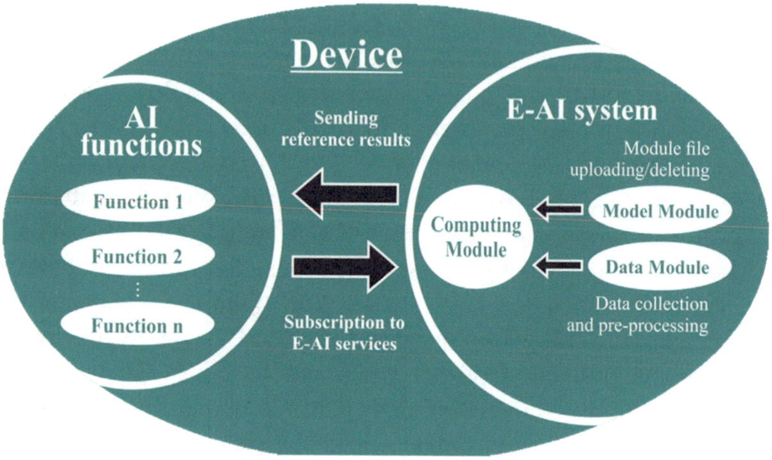

Figure 2.11. *Embedded artificial intelligence for business purposes.*

Source: https://dac.digital/wp-content/uploads/2022/05/e-ai--1200x687-optimized.png.

As AI technologies advance, the role of these systems becomes increasingly important in addressing power and efficiency challenges. Energy-efficient solutions take priority, particularly for embedded computing applications whose power sources are typically constrained. AI can optimize energy usage through smart load management and predictive maintenance, reducing operating costs and device lifespan. This is specifically applicable in settings like power distribution and management systems within space modules, where maximum energy utilization is of utmost priority. With the application of AI-controlled strategies, the systems can adaptively adjust to shifting power loads, allowing effective energy distribution and the assurance of critical functions with the available limited energies, as exemplified by recent advancements in spacecraft power systems (Hall et al., 2025). Besides, the combination of embedded systems and AI is full of potential for enhanced capability in a broad set of applications. Embedded AI has the potential to enable more sophisticated functions, such as proactive anomaly detection and fault recovery. Such functionality plays a critical role in considering the operational concerns related to embedded devices in mission-critical applications, such as industrial automation and healthcare monitoring systems. In such scenarios, AI-enabled embedded systems can detect abnormalities in operations immediately, take corrective measures, and minimize downtime, thereby making the entire system more robust. Additionally, the use of AI serves to solve the increasing energy problem both associated with embedded and supercomputing technologies as reducing energy consumption becomes the biggest concern in technical design across these technologies (Dokladal et al., 2023). With such advances, AI is set to reimagine the efficiency and efficacy of embedded systems such that they become more integral to modern technological networks (Figure 2.11).

2.4.1. Integration of AI Algorithms in Embedded Systems

The integration of AI algorithms in embedded systems marks a watershed in technology, particularly as these systems begin to execute smart actions typically native to higher-order computing. In enabling embedded processors and microcontrollers to reason locally from the processed data and execute real-time actions, companies are able to enhance functionality in many applications. For example, the trend towards edge AI has worked to curb latency, bandwidth optimization, and even improve service reliability by means of proximal-to-source data processing. Besides making operations easier, this also makes it possible for sophisticated applications to be built across sectors like healthcare and industrial automation, thereby cementing the idea that embedded intelligence is the key to transforming the manner in which devices communicate with their environment and users as well (N/A, 2025). The architecture of AI-driven embedded systems must be improved to manage the computational load of machine learning algorithms. Embedding AI into resource-limited systems means coming up with innovative ways to ensure performance and energy consumption while still being efficient. Researchers have explored the use of lightweight models and techniques such as quantization to enable AI tasks without saturating the limited amount of resources typically present in embedded systems. A good case is the use of MobileNet-SSD in real-time video monitoring, where it is shown how light models can deliver great accuracy in challenging situations. This innovation emphasizes the utmost importance of the optimization of the deployment of AI algorithms within embedded systems so that systems may effectively operate even under extremely stringent resource constraints (Parro et al., 2025). Furthermore, the integration of AI algorithms into embedded systems needs to address lifecycle management and continuous evolution of the underlying technology. As edge AI system development continues to evolve, solutions no longer need to merely counter the deployment of algorithms but must also ensure their scalability, maintenance, and security. The implementation of MLOps practices is essential to allow machine learning models to function well in collecting new data inputs while continuously adapting in the long term. As applications such as autonomous and intelligent healthcare devices become more prevalent, the demand for high-capacity, real-time processing solutions will likely rise. By prioritizing these aspects, the embedded systems community can leverage the transformative power of AI, opening new avenues for innovation and reshaping industrial practices (N/A, 2025).

2.4.2. Benefits of AI in Enhancing Embedded System Performance

Artificial Intelligence (AI) is pivotal in enhancing the performance of embedded systems, presenting sophisticated abilities that extend beyond standard programming methodologies. Integration with AI provides intelligent decision-making processes with embedded systems, enabling them to learn and adapt according to changing environments and user desires. This adaptability is particularly required for environments like smart home devices, autonomous vehicles, and industrial control, where real-time processing and efficient use of resources are paramount. With the ability to process enormous amounts of data within a brief time, AI algorithms optimize system functionality, reduce power consumption, and improve overall performance. With embedded systems becoming increasingly complex, the necessity for intelligent processing solutions that enable multitasking while maintaining minimal power needs is more significant than ever. Using AI to optimize embedded systems not only increases functionality but also makes devices more responsive and reliable. Through machine learning, systems can learn from previous interactions and therefore optimize their operations to be more efficient. For instance, AI can predict upcoming failures in hardware, allowing for preventive maintenance as well as reducing downtime. Moreover, just like advancements using image processing software like digital steganography, sophisticated CPU architectures can leverage AI-based algorithms to maximize data management without compromising high-performance output. Here, AI integration with high-performance processing elements and parallel computation techniques is a convincing way towards high-performance embedded systems and reiterates the fundamental role of AI in designing novel solutions (Bramas et al., 2020). Additionally, the developments in AI technologies in embedded systems reflect a broader movement towards open-system standards and interoperability. As outlined in revising architecture guidelines of systems having complicated nature, like the revising guidelines to the Deep Space Networks information system architecture, AI has a significant role in establishing consistent working frameworks that integrate various parts of a system. The pressure towards automation and improved human-interface technology allows embedded systems to perform industry function and user convenience also. Interoperability provides for the ability of various systems to communicate optimally, while advanced AI features facilitate effortless monitoring and control operations, consistent with the huge technological leaps envisioned for the future. Such synergy underscores the revolutionary ability of AI to drive the capabilities of embedded systems towards greater strength, efficiency, and user-friendliness (Atkinson et al., 1992) (Figure 2.12).

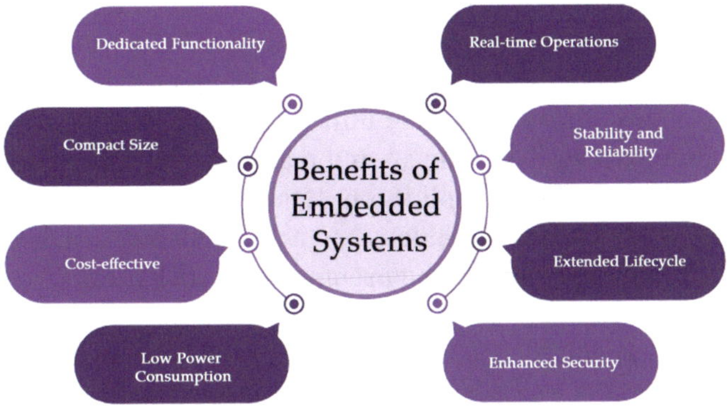

Figure 2.12. *Benefits of embedded systems.*

Source: https://www.theknowledgeacademy.com/_files/images/Benefits_of_Embedded_Systems.png.

2.4.3. Challenges in Implementing AI in Embedded Systems

The integration of artificial intelligence (AI) into embedded systems is accompanied by many challenges, the foremost of which are rooted in the hardware limitations of such systems. Embedded systems often possess limited processing capacity, memory, and power budgets, which are necessary for running advanced AI algorithms. For example, while deep neural networks (DNNs) offer robust capabilities for activities such as visual navigation, the computational expense can prove unaffordable for minute devices like nanodrones. As literature indicates, the introduction of novel low-power computation platforms, e.g., GAP8, shows the potential for parallel execution of DNNs in resource-restricted settings. However, striking a balance between performance and power efficiency in such applications is a daunting task, particularly where real-time processing is the very core tenet of business success and customer satisfaction (Benini et al., 2018). Compounding the hardware issues are the complexities involved in carrying out effective machine learning algorithms on embedded systems. The operating constraints—i.e., power budgets and real-time requirements—demand highly optimized algorithms that make use of the limited computational resource available. The TinyML vision is a reaction to these operating constraints to emphasize light-weight machine learning capabilities optimized for microcontrollers and low-energy devices. Research indicates the excessive overhead of traditional ML algorithms, which necessitates new frameworks like the Edge Learning Machine (ELM) for embedding shallow and self-supervised ML on IoT devices. These frameworks are meant to enhance memory efficiency and processing speed for embedded systems to make accurate decisions without overstepping their natural boundaries (Sakr et al., 2023). Furthermore, the utilization of AI in embedded systems is also an issue regarding development and maintenance, particularly relating to the lifetime of the deployed application. Because of the rapid development in the approach of AI, it needs continuous updates and tuning, which turns out to be logistically difficult in embedded environments where hardware updates are not common. Furthermore, adequate data management and security turn out to be the real issues, particularly in applications that work with sensitive information. With more and more embedded systems connected to each other in larger networks, the problem of incorporating AI without sacrificing data privacy and integrity is another daunting challenge to be overcome. These interconnected problems call for collaborative inputs by groups from different disciplines, such as software professionals, hardware engineers, and security specialists, to create sustainable solutions that advance the objectives of embedded intelligence without compromising on performance and security levels.

2.5. CASE STUDIES OF EMBEDDED INTELLIGENCE

The integration of intelligence in modern devices has revolutionized many industries, especially where power usage is the priority factor. The paradigm case studies of this phenomenon attest that what embedded systems do but need to be efficient and optimized in doing with a minimum amount of power consumption, a consideration that is increasingly important with increasing environmental concerns. Particularly, intelligent devices leverage sophisticated algorithms that are computationally heavy in their requirements, enabling them to perform real-time processing and decision-making. This need for efficiency is especially important in embedded computing, where low-energy supply translates to the need for battery optimization and avoiding waste. As mentioned by Dokladal et al. (2023), the interplay of hardware capabilities and software efficiencies is crucial to guarantee operational integrity in energy-critical applications. Automotive engineering witnesses the use of machine learning and embedded intelligence, whose case studies reveal how the vehicle's dynamics are enhanced. Notably, the development of torque vectoring controllers is one of the best examples of using AI in embedded systems effectively, illustrating the potential of such applications to maximize torque distribution in individual electric motors. The novel control topology relies on neural networks employed as virtual sensors that enable the system to anticipate and react to current driving conditions effectively. Not only does this approach provide an enhanced driving experience, but it also addresses pertinent safety and efficiency concerns as outlined by Jahnke et al. (2019). Responding rapidly to rapidly evolving circumstances, they embody the strength of embedded intelligence to redefine traditional automotive use cases while meeting industry requirements and limitations. Additional consumer electronics case studies provide better insight into the pervasive impact of embedded intelligence in various domains. A good instance is smart household appliances, which employ artificial intelligence algorithms to simplify operations and enhance user experience via automation. Such devices, occasionally founded upon microcontrollers and embedded processors, operate suitably within constrained environments while leveraging network connectivity to offer real-time updates and functionalities. The implications of energy consumption are uncovered as the need for developers to exchange performance for sustainability emerges, respecting the ecological footprint of smart technology. The findings of the case study emphasize fusing advanced computational techniques within embedded systems, ultimately presenting an enhanced intelligent and responsive ecosystem that is rewarding to end users and engages with overall energy efficiency goals, as noted in findings referenced by Dokladal et al. (2023).

2.5.1. Smart Home Devices

Smart home product integration is a huge step in the field of embedded intelligence, the very definition of which rewrites the interface of human beings with their homes. Smart thermostats, security cameras, and voice control devices employ various embedded processors and microcontrollers to perform data processing effectively and enable the user to interact with them. By applying the capabilities of artificial intelligence (AI), the smart home technology enhances the experience of the user through automation and personalization. The technology is also Internet of Things (IoT) enabled, making it simple to interface with other devices and cloud services. This interconnectivity facilitates live data analysis, resulting in enhanced energy efficiency, increased security, and optimized day-to-day operations, thus showcasing the capability of embedded systems to transform home environments while fostering a smarter habitat (Fallucchi et al., 2017). AI plays a critical role in the functioning of smart home devices, spearheading the intelligent operations that users have grown accustomed to. AI software uses massive quantities of data generated from user interactions and sensor readings in the world to inform predictive analysis and perform tedious tasks. A classic example is thermostats that are intelligent, which learn users' behavior over time and adjust temperature and air conditioning to provide optimal comfort while saving energy. Such a use of AI is important to enhance the performance of embedded systems since it entails the development of light-weight machine-learning models that need to be applied on low-power microcontrollers.

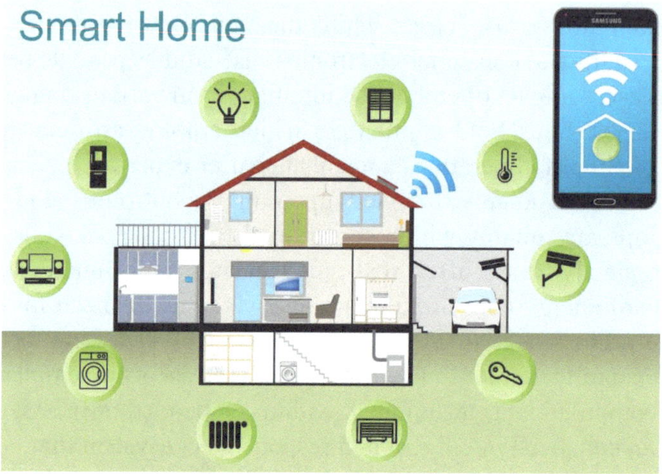

Figure 2.13. *Smart home devices.*

Source: https://www.providenthousing.com/wp-content/uploads/2019/05/Smart-Homes-ProvidentBlog8.jpg.

This is the problem that has resulted in innovative solutions like TinyML that allow machine learning algorithms to be run on limited devices and allow real-time decisions with no reduction in performance (Sakr et al., 2023). Accordingly, these developments in AI not only improve functionality but also improve the general user experience of smart home devices. The future of smart home devices will indeed be based on further developments in embedded intelligence and AI capabilities. With such technologies further developing, they will increasingly focus on matters of security, privacy, and interoperability. For example, use of more secure encryption methods will become essential to ensure protection of sensitive information exchanged among devices. In addition, ongoing innovation in edge computing will make it possible to process data near its source, minimize latency, and enhance responsiveness. Integration of smart home systems with next-generation technologies such as 5G networks will further advance their capabilities to support more devices to be connected and collaborate. Lastly, with microcontrollers and embedded processors improving, the future is towards an increasingly integrated and automated life, where homes can learn and respond to the needs of their residents (Fallucchi et al., 2017) (Figure 2.13).

2.5.2. Automotive Embedded Systems

Automotive embedded systems development is a massive shift in the automobile industry, wisely integrating technology into nearly every vehicular operation. Embedded systems not only enhance traditional mechanical qualities but introduce a factor of sophistication in the form of real-time data processing and regulation. With the advent of electric cars and higher demands for safety systems, the systems employ a series of sensors and signal processors that monitor vehicle movement and driver behavior. This sophisticated architecture invites innovations such as adaptive cruise control, lane-keep assist, and park assist. By leveraging the processing capacity of embedded platforms, performance and efficiency can be optimized, responding dynamically to changing road conditions and driver inputs, providing a more responsive and secure drive experience aligned with the ambitions of today's technology (Jahnke et al., 2019). Artificial intelligence (AI) is a transformative driver behind the evolution of automotive embedded systems, facilitating complex real-time decision-making. The application of machine learning algorithms, for instance, allows such systems to predict maintenance needs, perform traffic pattern analysis, and even optimize energy distribution in electric vehicles. The new use of deep learning techniques and neural networks enables the development of systems that improve their performance with increasingly cumulative data. This is realized by heterogeneous embedded platforms that combine processors with high computational power, thereby realizing maximum performance and minimum energy consumption. The applications of AI manifest in the area of ensuring safety, fuel efficiency, and boosting overall customer satisfaction through intelligent and adaptive systems (Jahnke et al., 2019). As embedded systems in automobiles become more sophisticated, they also pose special challenges that must be thoroughly examined, primarily in relation to safety regulations and industry standards. Stringent use of safe-critical components is also a must, as these systems must guarantee operational reliability under various driving scenarios. This demands a multi-dimensional solution that combines rigorous testing and certification with innovative design methods. Moreover, the employment of future technologies, such as IoT connectivity, adds a degree of complexity, in the sense that the systems are not merely supposed to be independent but also exchange information with external worlds. The continuous evolution of embedded intelligence in car systems positions them at the very forefront of technological change, at a point where technology meets regulatory context in redesigning future means of transportation (Jahnke et al., 2019).

2.5.3. Industrial Automation Solutions

In contemporary industrial automation solutions, the incorporation of embedded intelligence plays a crucial role in enhancing operational efficiency and system dependability. Embedded microcontrollers and processors form the foundation for such intelligent systems with support for real-time data processing and control in various industries, including manufacturing and logistics. With industries heading towards smart factories and the Internet of Things (IoT), the role of advanced embedded solutions is gaining increasingly more importance. The systems not only monitor equipment performance but also collect and analyze massive amounts of data to make intelligent decisions in an attempt to optimize production activities. Moreover, the use of AI algorithms in embedded platforms facilitates predictive maintenance, minimizing downtime and maximizing total productivity. This sophisticated interaction between industrial processes and embedded intelligence ultimately results in a more industrially responsive and adaptable manufacturing environment. The application of AI in embedded systems introduces a new dimension to industrial automation solutions that provides enhanced capabilities in data-driven decision-making processes. Based on machine learning methodology, these systems can adapt to diverse operation parameters as well as adapt to patterns in historical data. This transformation leads to more advanced control mechanisms that can dynamically optimize the use of resources and energy consumption, eventually lowering operating costs. The union of AI with embedded intelligence not only automates processes but also participates in safety-critical operations, where rapid and accurate reaction to dynamic conditions is of paramount concern. For instance, vehicle dynamics application areas utilize neural networks to design high-end control systems that manage torque distribution in real-time, ensuring stability and performance for different driving scenarios (Jahnke et al., 2019). This is merely one example of how the synergy between embedded systems and AI is transforming sectors. Additionally, industrial automation systems with embedded intelligence enable critical issues of safety and regulatory compliance. As more and more critical operations make use of these systems, their safety compliance with standards and procedures is essential to prevent risks of system failure or malfunction. By applying architectures with redundant processes and fail-safes, manufacturers are able to provide customer needs and safety concerns and maximized performance. Advancements in fuzzy logic and optimization techniques further enrich embedded intelligence through the ability of such systems to distribute priorities based on real-time data feeds and operational needs (Jahnke et al., 2019). Industries therefore must not only ride on the potential of embedded intelligence but also navigate the complexity of safety norms while being innovative and compliant in the new industrial automation world (Figure 2.14).

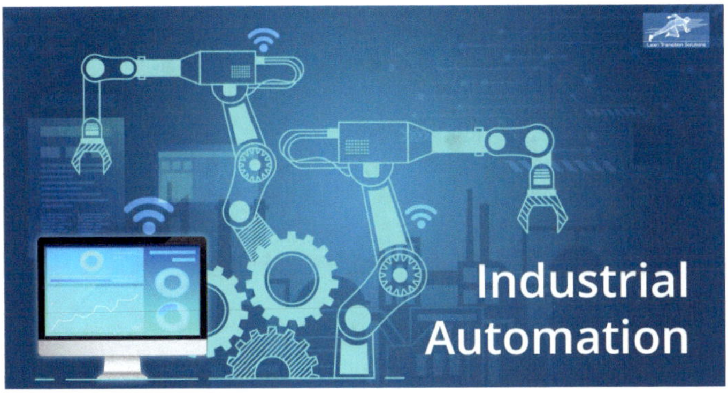

Figure 2.14. *Industrial automation.*

Source: https://leantransitionsolutions.com/assets/img/blog/Industrial_Automation.jpg.

2.6. FUTURE TRENDS IN EMBEDDED INTELLIGENCE

The future embedded intelligence is, in effect, characterized by the increasing presence of artificial intelligence (AI) in embedded systems. As industries strive to be increasingly efficient and automated, the demand for intelligent embedded applications is expanding exponentially, thereby stimulating innovation around embedded processors and microcontrollers. This revolution enables devices to not only perform what they have been programmed to do but learn from the environment as well and adapt. The scope of the contribution of AI to powering this revolution cannot be overstated; it makes it possible to use complex algorithms in real-time, facilitating innovations such as predictive maintenance in manufacturing and enhanced personalization in consumer electronics. The adoption of AI within embedded systems is not a passing fancy, but a wave of change which can transform the capabilities of all devices into making them more responsive and easier to use. As the technology landscape continues to change dynamically, some trends are emerging that make the usability of embedded intelligence even better. For instance, Field Programmable Gate Arrays (FPGAs) are gaining usage in order to speed up the adoption of AI-based controllers in embedded solutions.

Figure 2.15. *The future of AI in embedded systems.*

Source: https://www.evolute.in/wp-content/uploads/2023/03/Web_Evolute-Group_Blog_The-Future-of-AI_Image_27-Feb-2023.jpg.

Their programmability makes it possible to tailor hardware to the specific requirements of complex tasks, such as those involving sensor fusion or real-time data processing. Furthermore, research has demonstrated the flexibility of FPGAs to computationally demanding applications and pinpointed their future scope in fields like robotics and industrial automation (Bahri et al., 2011). This trend is not just reflective of the need for customization at the design level but also encourages the development of more powerful and efficient embedded systems capable of executing more sophisticated algorithms. Another significant development happens in the field of TinyML, where machine learning models are implemented on devices with a constraint in resources. As the Internet of Things (IoT) is developing at a fantastic rate, the need for efficient algorithms that can perform operations within strict resource constraints is vital. TinyML addresses these issues by bringing training and inference on the device, reducing the reliance on cloud computing and enhancing real-time response (Sakr et al., 2023). Development of such platforms as the Edge Learning Machine environment makes this trend possible with software that simplifies machine learning functionality for embedded applications. Embedded intelligence can then implement more sophisticated capabilities within the constraints of memory, power, and latency, hence making broader usage of AI capabilities possible across various industries (Figure 2.15).

2.6.1. Advancements in Processor Technology

The evolution of processor technology has significantly influenced the development of embedded systems by offering a platform for the execution of artificial intelligence (AI) in the devices. Modern embedded microcontrollers and processors offer more computational power and efficiency in energy consumption, enabling them to execute complex algorithms required for intelligent functions. The transition from simple task-specific chips to advanced multicore processors is one indication of this expansion. This conversion allows for parallel processing abilities, which are also necessary for industrial automation through to smart home systems applications. As the AmbienNet environment illustrates, these advancements support intelligent systems that aid individuals with disabilities by providing tailored functionalities aimed at making independence and quality of life achievable. The role of the enhanced processors in such applications emphasizes the role of modern technology in establishing varied environments (Abascal et al., 2009). Moreover, the embedding of AI is facilitated by the new capabilities of the processors, which enable real-time processing of data and experience-based learning. The embedding is witnessed across various applications, including wearables and autonomous vehicles, where speed of decision-making matters. Integrated processors with neural network-based designs allow for on-device processing, significantly reducing latency and enhancing performance compared to cloud computing. These technologies allow devices to perform operations such as image recognition and speech processing autonomously without relying on over-the-air servers. The search for efficiency and responsiveness in such systems emphasizes the position of processor technology in promoting innovation as well as the range of embedded systems in most fields (Sułek et al., 2024). In addition, advances in processor technology are not only expanding the boundary of computational power but also enabling modular designs that maximize the scalability and flexibility of systems. The advent of user-programmable hardware platforms, such as those utilized in drone technology, demonstrates the flexibility of modern embedded systems to be configured to specific operational needs without a range of redesigns. Such flexibility opens up sophisticated systems to hobbyists and professionals alike to engage with, lowering the barriers of entry into domains such as unmanned aerial vehicle (UAV) design. As the discipline of embedded intelligence advances, the ability to design robust systems based on modular architecture and AI-based characteristics becomes increasingly vital. Such advancements potentially further develop the integration of intelligent systems into everyday life, impacting various sectors and applications even more intensively (Sułek et al., 2024).

2.6.2. The Impact of IoT on Embedded Intelligence

The implementation of the Internet of Things (IoT) into embedded intelligence represents a revolutionary shift in the nature of how devices operate and communicate within various environments. This integration enables devices with embedded processors and microcontrollers to communicate and exchange information in an efficient manner, creating a network of systems that communicate and work harmoniously together. When devices are endowed with the immense resources of the cloud, their potential to collect, process, and analyze vast amounts of data considerably increases. Consequently, this revolution allows these devices to perform complex tasks such as real-time monitoring and predictive maintenance with unmatched accuracy and efficiency. The synergies created by combining embedded intelligence with IoT not only enhance device functionality but also usher in ground-breaking applications across numerous fields, including healthcare, industrial automation, and smart cities. To counter the limitations that come with executing machine learning algorithms on resource-constrained IoT devices, the emergence of TinyML becomes a much-needed solution. This field focuses on optimizing machine learning potential on embedded platforms so that they are permitted to operate within power and memory limits. As research has put it, the establishment of platforms such as the Edge Learning Machine (ELM) enables the deployment of various machine learning services tailored to microcontrollers. These advancements enable local data processing, which is highly positively contributory in reducing latency and conserving bandwidth. Further, placing self-supervised learning on microcontrollers broadens the embedded intelligence scope to support devices to enhance their learning capability independently. In fact, mitigating memory constraints while maintaining performance is the core focus of TinyML research, as reflected in innovations that have created optimized architectures for specific uses like e-skin systems (Sakr et al., 2023). Introduction of a Dynamic AI-IoT architecture also reflects the impact of IoT in embedded intelligence, particularly in ultralow-power (ULP) devices. The architecture is designed to enable dynamic integration of AI functionality without the firmware update process taking so much time and thus ensuring seamless operation across various platforms. Through the implementation of Narrowband IoT (NB-IoT), it facilitates efficient cloud interaction and real-time update of models, enhancing the responsiveness of embedded systems to the changing technological landscape. Empirical results validate the efficiency of the system through timely model updates and efficient management of tasks on ULP devices in a 5G testbed environment. This transformation is a pivot towards the exploitation of the full potential of IoT in combination with advanced embedded intelligence to accommodate a new generation of intelligent applications that can adapt dynamically to user need in real-time (Alcaraz-Calero et al., 2023).

2.6.3. Predictions for AI Integration in Future Embedded Systems

The future of embedded systems is in for a monumental shift with the implementation of artificial intelligence (AI), but in more advanced vision navigation features. From recent advancements, all the research that has gone into developing self-sustaining tiny robots, like nano-drones, reflects how well AI can be applied to limited resources environments. For instance, visual navigation using deep neural networks (DNNs) is a cutting-edge move where complex algorithms can be executed directly onboard to provide real-time adaptability (Benini et al., 2018). Such advancements signify an era in the future where embedded systems not only process data but also intelligently interpret and act upon it, thereby increasing their operating autonomy and functional versatility. Moreover, this technology would create new applications in various industries, thereby fostering a more intense human-machine interaction in areas such as intelligent transportation systems and smart homes. Furthermore, the application of Field Programmable Gate Arrays (FPGAs) in embedding AI cannot be downplayed. The hardware is reliable and therefore presents a fantastic opportunity for researchers and developers to implement complex algorithms in better ways. As an example, FPGAs can be programmed to address specific applications, i.e., a sensorless motor controller with an Extended Kalman Filter, which shows the symmetric integration of AI in industrial applications (Bahri et al., 2011). The flexibility and efficiency of this make FPGAs extremely desirable to the design of next-generation embedded systems, where rapid prototyping and flexibility can be extremely important. With increasing advancements in AI, the integration of these technologies can potentially enhance computational power while maintaining energy efficiency, as preferred by modern industrial and consumer markets. Additionally, increasing requirements for AI-based embedded solutions will pave the way for new design techniques that emphasize energy efficiency without losing out on performance. Future embedded systems will possess highly optimized architectures for enabling dynamic processing capacity adaptation in alignment with workload demands. Integrating AI algorithms with the capability to perform predictive analysis will enable systems to learn and adapt in their environments, providing a new kind of intelligence (Benini et al., 2018). This advancement will propel the creation of systems that respond not just to real-time data but also forecast changes in user activity or system circumstances, thereby enhancing user experience as well as business efficiency.

SUMMARY

The embedded intelligence environment is defined by the synergistic fusion of embedded processors, microcontrollers, and the burgeoning relevance of artificial intelligence (AI). The unification results in systems with the capability to perform sophisticated functions within limiting resources, significantly enhancing operating efficiency in a wide range of applications. Hardware-software interaction in embedded systems not only facilitates the processing of real-time information but also allows adaptive solutions for dynamic environments. With industries and businesses yet to evolve, embracing such technologies is then the solution to maintaining competitive grounds. Embedded intelligence's hurdles will need to be overcome through innovative approaches, empirical research, and cohesive interdisciplinary teamwork. Furthermore, advances in artificial intelligence, especially in the field of embedded systems, have led to the emergence of TinyML that seeks to tailor AI algorithms for low-power, energy-aware environments. The integration of AI in hardware not only drives capability but also releases new prospects for use across industries like healthcare and intelligent farming. The development of smart sensors remains imperative, as attested to by ongoing challenges around market adoption and deployment in real-world applications. The exploration of green solutions which counter technological impediments based on environmental as well as ethical perspectives is of paramount significance in rendering such innovations deployable on a massive scale with ease (Luo et al., 2023). Finally, the future of embedded intelligence relies on a support framework of collaborative research as well as development strategies that bridge technical and non-technical perspectives. As set out in existing literature, there is a pressing need to synchronize advances in hardware technologies with societal needs and legislative frameworks (Abadade et al., 2023). By maintaining solutions holistic in nature as well as focusing on performance improvements, stakeholders can achieve sustainable innovation in embedded systems. Cross-pollination of ideas among disciplines will most certainly yield breakthroughs that fulfill the ultimate potential of embedded intelligence to deliver greater quality of life and reduce critical worldwide challenges.

Within the quest to uncover embedded intelligence, certain fundamental points emerged that highlight its importance in today's technology. The combination of embedded processors and microcontrollers has helped bring about devices that not only perform tasks but also sensibly adapt to their working conditions. This is made possible on the back of advanced algorithms that handle real-time data to make informed decisions. With the addition of artificial intelligence to such systems, the potential for additional functionality becomes apparent, in that devices can be tailored according to varied conditions. The research illustrates that effective embedded intelligence has the ability to bridge the gap between machine learning theoretical models and practical use in many industries, outlining how embedded systems go beyond automatons in encompassing learning as well as adaptive properties (Jahnke et al., 2019). The other significant field of embedded intelligence is machine learning and soft computing's role in enhancing vehicle dynamics applications. One such case study detailed the development of a Torque Vectoring Controller for electric vehicles, outlining how embedded platforms might utilize high-performance algorithms to optimize torque distribution between individual wheels. The new control system employs neural networks as virtual sensors to complement the conventional sensing technologies for unmeasurable forces. Moreover, the application of fuzzy logic allows for goal prioritization based on driving situations in the moment and thereby optimizes safety and vehicle performance. It is a testament to the revolutionizing capability of embedded intelligence to provide precision and responsiveness requirements needed by industries such as the automotive sector where performance and safety are factors (Jahnke et al., 2019). The integration of embedded intelligence with high-performance computational platforms, such as Field Programmable Gate Arrays (FPGAs) and Graphics Processing Units (GPUs), marks a further development in the ability of embedded systems. The ability to leverage heterogeneous computing environments means more complex algorithms can be run efficiently in real time, therefore enhancing the overall efficiency of embedded solutions. With the demands for more performance and efficiency increasing, the findings of numerous research endeavors highlight the need to continue innovating in embedded intelligence. As businesses strive for improved solutions, the convergence of embedded processors, microcontrollers, and artificial intelligence sets the stage for a future where systems become intelligent and react proactively to their world and user needs, shifting the technology applications paradigm (Jahnke et al., 2019).

REFERENCES

1. Abadade, Y., Temouden, A., Bamoumen, H., Benamar, N., Chtouki, Y., & Hafid, A. (2023). A comprehensive survey on TinyML. *IEEE Access, 11*, 96892–96922. https://doi.org/10.1109/access.2023.3294111 (accessed on 10th September 2025).
2. Abascal, J., Bornail, B., Casas, R., Cascado Caballero, A., et al. (2009). Towards an intelligent and supportive environment for people with physical or cognitive restrictions. https://core.ac.uk/download/158964615.pdf (accessed on 10th September 2025).
3. Abbas, R., Ahmad, I., Ali, S., Amuru, S., et al. (2020). 6G white paper on machine learning in wireless communication networks. http://arxiv.org/abs/2004.13875 (accessed on 10th September 2025).
4. Alcaraz-Calero, J. M., Alselek, M., & Wang, Q. (2023). Dynamic AI-IoT: Enabling updatable AI models in ultra-low-power 5G IoT devices. https://core.ac.uk/download/603274589.pdf (accessed on 10th September 2025).
5. Atkinson, D. J., Beswick, C. A., Cooper, L. P., Crowe, D., et al. (1992). Deep Space Network information system architecture study. https://core.ac.uk/download/pdf/10438278.pdf (accessed on 10th September 2025).
6. Bahri, I., Cirstea, M. N., Idkhajine, L., Monmasson, E., et al. (2011). FPGAs in industrial control applications. https://core.ac.uk/download/77283019.pdf (accessed on 10th September 2025).
7. Benini, L., Conti, F., Flamand, E., Loquercio, A., et al. (2018). A 64mW DNN-based visual navigation engine for autonomous nano-drones. http://arxiv.org/abs/1805.01831 (accessed on 10th September 2025).
8. Bishop, A. M., Montag, B. C., & Redfield, J. B. (2025). Advanced manned space flight simulation and training: An investigation of simulation host computer system concepts. https://core.ac.uk/download/pdf/42826139.pdf (accessed on 10th September 2025).
9. Botella-Rocamora, P., Pardo, J., & Zamora-Martinez, F. (2015). Online learning algorithm for time series forecasting suitable for low-cost wireless sensor networks nodes. http://arxiv.org/abs/1504.05517 (accessed on 10th September 2025).
10. Bramas, B., Clerc, M., Cox, I., Jongen, H. T., Katzenbeisser, S., Price, K. V., Raggo, M., et al. (2020). JPEG steganography with particle swarm optimization accelerated by AVX. https://core.ac.uk/download/323112813.pdf (accessed on 10th September 2025).
11. Capogrosso, L., Cheng, D. S., Cristani, M., Cunico, R., et al. (2023). A machine learning-oriented survey on tiny machine learning. http://arxiv.org/abs/2309.11932 (accessed on 10th September 2025).

12. Cob Parro, A. C. (2025). Contributions to low cost and real time smart video surveillance systems and their industrial deployment. https://core.ac.uk/download/659609956.pdf (accessed on 10th September 2025).
13. Coppola, M., & Vermesan, O. (2022). *Embedded Edge Intelligent Processing for End-to-End Predictive Maintenance in Industrial Applications*. https://core.ac.uk/download/588316341.pdf (accessed on 10th September 2025).
14. Dendaluce Jahnke, M. (2019). Real-time multi-domain optimization controller for multi-motor electric vehicles using automotive-suitable methods and heterogeneous embedded platforms. https://core.ac.uk/download/547397035.pdf (accessed on 10th September 2025).
15. Dokladal, P., Mesri, Y., Nana, L., Tadonki, C., et al. (2023). Energy concerns with HPC systems and applications. http://arxiv.org/abs/2309.08615 (accessed on 10th September 2025).
16. Fallucchi, F., Di Nunzio, L., Pieroni, A., Scarpato, N., et al. (2017). E-health-IoT universe: A review. https://core.ac.uk/download/296922213.pdf (accessed on 10th September 2025).
17. Hall, D. K., Lollar, L. F., & Walls, B. (2025). Augmentation of the space station module power management and distribution breadboard. https://core.ac.uk/download/pdf/42819039.pdf (accessed on 10th September 2025).
18. Lee, S. D. (2025). A distributed agent architecture for real-time knowledge-based systems: Real-time expert systems project, phase 1. https://core.ac.uk/download/pdf/42815389.pdf (accessed on 10th September 2025).
19. Luo, Y., Abidian, M. R., Ahn, J. H., Akinwande, D., Andrews, A. M., Antonietti, M., Bao, Z., et al. (2023). Technology roadmap for flexible sensors. *ACS Nano, 17*, 5211–5295. https://doi.org/10.1021/acsnano.2c12606 (accessed on 10th September 2025).
20. Rawlings, T. (2009). Distributed control architecture. https://core.ac.uk/download/288387683.pdf (accessed on 10th September 2025).
21. Saha, S. S., Sandha, S. S., & Srivastava, M. (2022). Machine learning for microcontroller-class hardware–A review. https://core.ac.uk/download/612398083.pdf (accessed on 10th September 2025).
22. Sakr, F. (2023). Tiny machine learning environment: Enabling intelligence on constrained devices. https://core.ac.uk/download/566650308.pdf (accessed on 10th September 2025).
23. Sarpe, D., & Mazilescu, V. (2025). The relationship between fuzzy reasoning and its temporal characteristics for knowledge management. https://core.ac.uk/download/pdf/6372384.pdf (accessed on 10th September 2025).

CHAPTER 3

SENSORS, ACTUATORS, AND PERCEPTION SYSTEMS

Contents

Introduction 83

3.1. Types of Sensors 87

3.2. Data Acquisition 93

3.3. Signal Processing 97

3.4. Sensor Fusion Techniques 102

3.5. Actuators and Their Role 107

3.6. Applications of Perception Systems 112

Summary 117

References 119

INTRODUCTION

The endless speed of technological development has made the automation sector into increasingly sophisticated systems that are able to execute complex activities on their own. Central to this revolution are sensors as the primary means by which machines perceive their world. Several types of sensors, including visual sensors, LIDAR, inertial measurement units (IMUs), and GPS, are used to collect information about surroundings and therefore construct the foundation of modern perception systems. These sensors make critical data capture possible for environmental cue interpretation, facilitating more intelligence-based decision-making. With these technologies, further development takes place, and sensor measurement reliability and accuracy become a highlight of maximum performance, particularly in mission-critical use cases such as autonomous driving systems, where accuracy is a case of life or death (Yurtsever et al., 2020). The integration of different sensory information enhances the functionality and performance of automated systems through improved signal processing and sensor fusion processes.

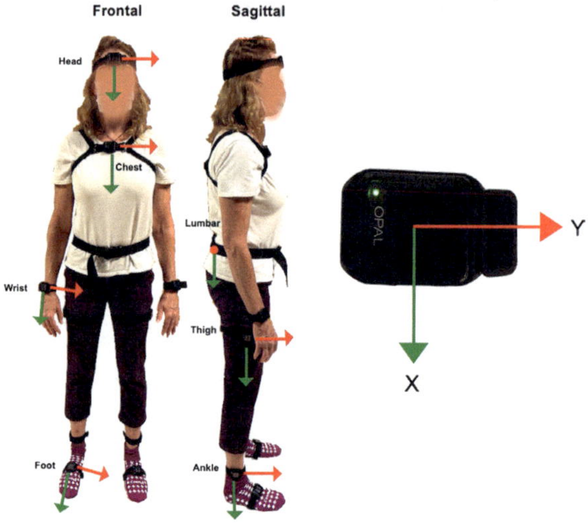

Figure 3.1. *IMU sensor orientations.*

Source: *https://www.researchgate.net/publication/358577746/figure/fig1/AS:1123212758913027@1644806198356/IMU-sensor-orientations-Eleven-inertial-measurement-units-IMUs-APDM-Wearable.png.*

By integrating data from multiple sensors, systems can avoid the inherent flaws that come with the utilization of a single type of sensor. Sensor fusion leverages the best of all sensors—increasing robustness, reducing uncertainties, and improving the overall accuracy of the information obtained. For instance, combining LIDAR data and vision sensor data can allow for a deeper level of understanding of an environment with less likelihood of misinterpretation by sensor noise or occlusions. According to contemporary research, overcoming the issues of sensor fusion is not only vital in terms of developing current systems but is also the gateway to future advancements in advanced automated systems (Guang-Yang et al., 2018). Besides, one can observe the impact of robust sensors and effective data processing algorithms in real-world applications, particularly in automated driving systems (ADSs). The synergistic nature of the sensors, actuators, and perception systems enables vehicles to navigate complex environments without impairing safety and efficiency. However, despite all these developments in these technologies, there are yet unresolved issues of fundamental scale, including reliability and integration concerns. A comprehensive understanding of sensor architectures, data interpretation methodologies, and human-machine interfaces is paramount for advancing ADSs. The continuous research into these domains presents opportunities to harness technological breakthroughs that could redefine our interaction with automated systems, ultimately translating to safer and more efficient transportation solutions (Yurtsever et al., 2020) (Figure 3.1).

Definition of Sensors and Actuators

The intricate topology of modern technical systems relies on the accurate readings and outputs, which are based on the definitions of sensors and actuators. Sensors are the primary gates to the ambient environmental information of a system, collecting data via a range of modalities such as vision, LIDAR, inertial measurement units (IMUs), and Global Positioning System (GPS) signals. Vision sensors, for instance, employ cameras to gather visible data, which is essential for applications such as object detection and recognition in autonomous cars. LIDAR systems, on the other hand, offer a more extensive spatial awareness through laser measurement, which offers precise topographical data. Such sensors' collected data are subsequently processed, and this requires complex signal processing techniques to remove noise and extract useful patterns for effectuating system behavior and actions. This multi-perspective approach stresses the importance of sensors in establishing the foundation for automatic systems and communicating with the real world to reach their performance and reliability. While sensors extract data and manipulate it, actuators turn it into real actions and thereby complete the cycle of feedback required for dynamic systems. These components convert control signals indicated by the data analysis into physical reactions—whether motor movement or adjustments to valves. Sensor coordination with actuators is central in fields ranging from robotics to automotive engineering, where responsiveness and precision give optimal performance. Contemporary systems now involve sophisticated sensor fusion techniques that integrate data from multiple sensors, enhancing richness and accuracy of the data. Successful fusion of sensors not only improves detection and identification of objects in an environment but also supports navigating complex scenarios in which redundancy and robustness are key. With actuators, sensors coordinate the smooth functioning of perception systems, enabling automation and safety breakthroughs in increasingly automated environments.

With the evolving paradigm of autonomous systems, the roles and duties of sensors and actuators become increasingly significant since they become entangled with methods such as systems engineering and model-based design. The integration challenges of high-end vehicles, particularly those operating at SAE Level 3+, require a rigorous methodology for reliability and safety. The ability to monitor interactions between heterogeneous elements becomes critical, establishing the required accountability between design and operating stages. Additionally, the fusion of sensor feedback and actuator actions often requires a cross-layer framework combining safety, timing, and functional architecture. By calling on a nuanced comprehension of such factors, engineers are better able to create more stable perception systems that not merely respond elegantly to stimuli but also uphold a vigilant balance between functional efficiency and user security. Descriptions of sensors and actuators thus extend beyond basic components; they are the foundations of intelligent systems that interact effectively with their operating environments, as highlighted through ongoing advances in microsystems technology (Fluitman et al., 1996).

Importance of Perception Systems in Technology

The complicated interaction between various sensing modalities is the basis of highly advanced perception systems enabling machines to comprehend the world around them with remarkable accuracy. Vision sensors such as cameras and LIDAR sensors take center stage, recording extensive, multivariate data in environments. The process of collecting data, however, is only the beginning of an even more complex system of reading stimuli. Supported by signal processing techniques, these systems can allow noise through, enhance noteworthy features, and ultimately generate noteworthy representations of the environment. The complexity of contemporary perception systems is brought about not just by the use of specific sensors but also by the combination of multidimensional data inputs, leading to more correct interpretations, and hence significantly boosting automation and decision-making capabilities in many technological applications (Fluitman et al., 1996). Sensor fusion techniques extend the capabilities of perception systems by fusing data from multiple sources, hence extending situational awareness. Fusing allows systems to compensate for the shortcomings of individual sensors, leading to a more robust and dependable perception of the outside world. As an example, the combination of GPS position data with inertial measurement units (IMUs) provides both high-accuracy location tracking and comprehension of objects' dynamic movement. Additionally, the combination of spatial measure and visual-sensing sensors can create accurate maps for autonomous cars, which would directly influence their navigation plans. The convergence of these varied inputs not only supports greater data to process but also facilitates complex analyses and operating functions essential in today's technology settings (Lachmayer et al., 2012). With evolving technology, perception systems remain of utmost importance across varied fields ranging from healthcare to transportation to robotics. The application of advanced perception systems promotes increased machine-to-human interaction, practically bridging the gap between artificial intelligence and real-world applications. Through the transmission of legal and functional outcomes, these systems not only provide functional benefits but also compliance and ethical effects, redefining structures in which these technologies are implemented. The ability of machines to replicate human potential through perception systems ensures that they can respond rationally to diverse stimuli and requirements, marking a direction towards technologically interactive interfaces with a humanness factor. Therefore, investment in the development and perfecting of such perception systems is paramount to the innovations of tomorrow, particularly ensuring reliability, safety, and efficiency in technologically integrated settings (Fluitman et al., 1996) (Figure 3.2).

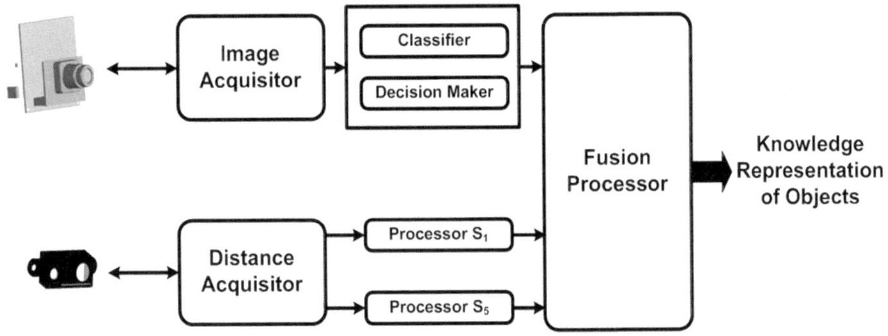

Figure 3.2. *Architecture of the perception system.*

Source: https://www.researchgate.net/profile/Luciano-Oliveira-7/publication/220773858/figure/fig3/AS:305417033928711@1449828503881/Architecture-of-the-perception-system.png.

3.1. TYPES OF SENSORS

In the background of upcoming technologies, sensors in various types are leading the way to enhance perception systems. Vision sensors like computer vision capture images of the surroundings and interpret visual information to allow objects to be detected and tracked. LIDAR (Light Detection and Ranging) is another fundamental sensor technology that employs laser light to create precise, three-dimensional representations of areas. This technology finds extensive application in self-driving cars, making them travel by giving them real-time spatial information. In addition to this, inertial measurement units (IMUs) work in cooperation with these features by sensing accelerations and angular rates, which are essential in tracking motion and stability. The integration of these sensors creates a general sense of the environment and hence enhances overall functionality and reliability in tasks from robotics to smart grids as proposed under the realms of IoT-supported systems (Copeland et al., 2019). Data acquisition and signal processing techniques can be very effective for enhancing the efficiency of sensor systems. After they collect data, sensors convert raw signals to useful information through algorithms to remove noise and enhance signal quality.

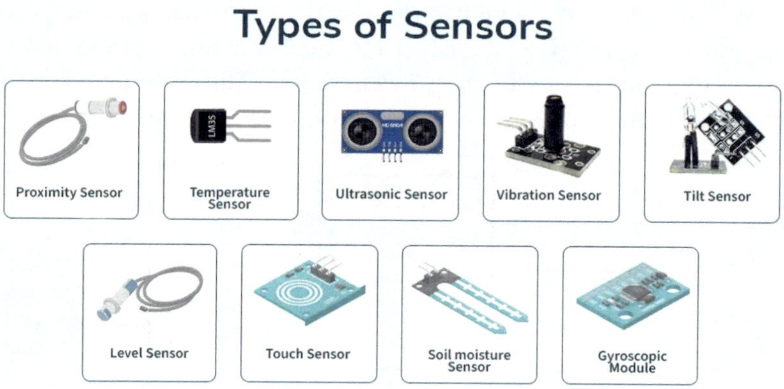

Figure 3.3. *Types of sensors.*

Source: https://media.geeksforgeeks.org/wp-content/uploads/20240605014033/Types-Of-Sensors.webp.

For instance, two-dimensional images from cameras can be processed for better vision and useful feature extraction, for instance, edges or colors. Similarly, with LIDAR, sophisticated algorithms are able to transform raw distance measurements into drivable maps. This process, in addition to producing clearer outputs, also maximizes the effectiveness of sensor data for different applications. Such sophisticated data processing techniques are crucial in the use of sensors in dynamic systems where adaptability and real-time analysis are crucial to successful performance, particularly in next-generation smart grid systems (Copeland et al., 2019). Sensor fusion techniques also reinforce the resilience of various types of sensors by combining data from multiple sources to produce a composite perception of an environment. This approach reinforces reliability and accuracy since the strengths of one sensor can counteract other weaknesses. For example, combining GPS and IMUs' data can significantly increase the positional location accuracy, especially where there is a chance of satellite signal interference. In addition, vision integration with LIDAR facilitates stricter object detection and classification, thereby ensuring safer navigation for autonomous cars. As technology keeps evolving, the place of sensor fusion in coming up with human-machine interaction that is seamless becomes clear, particularly as it is a key feature in the development of high-performance perception systems that have the ability to efficiently and intelligently move around in rough terrain (Figure 3.3).

3.1.1. Vision Sensors and Their Applications

The incorporation of vision sensors in modern perception systems is a milestone for applications such as automation, robotics, and augmented reality. With rich visual information provided, the sensors allow machines to perceive and grasp their surroundings just as human beings do, enhancing interactions with physical and virtual worlds. Visual sensors take pictures, which can be processed to recognize objects, detect motion, and find patterns and thus develop useful information. This capability is particularly crucial in designing Tangible User Interfaces (TUIs), where the blending of physical interaction and digital information relies significantly on accurate visual information. TUIs research centers on bridging the physical and virtual, symbolizing the growing need for advanced vision sensors that can deal with dynamic complex scenes to promote user experience and interaction (Zhang et al., 2017). The functionality of vision sensors is naturally enhanced by the intersection of data acquisition and sophisticated signal processing techniques. With the use of image interpretation and analysis algorithms, it becomes feasible to filter raw visual data in an attempt to obtain useful information. For instance, machine learning algorithms, like convolutional neural networks, have been effective in vision sensor platform object detection and classification issues. Furthermore, multi-modal sensor fusion techniques improve data interpretation by integrating vision sensor data with other sensors, i.e., GPS or LIDAR, in order to form an overall perception of the working environment. This cross-disciplinary setup not only enhances the precision of system output but also fosters innovation in many fields, from autonomous vehicles and robotics to rely upon the ability of sensing and mobility in complex environments more efficiently and precisely (Zhang et al., 2017). As vision sensors evolve, their applications become increasingly diversified and span across healthcare, agriculture, and surveillance. In the healthcare setting, vision sensors provide non-invasive imaging-based diagnosis to identify diseases at an early stage.

Similarly, in agriculture, vision sensors provide crop monitoring and optimal utilization of resources based on imaging techniques determining plant health. Besides, in security, vision sensors provide real-time monitoring and anomaly detection, significantly enhancing security. The expanding uses of these sensors necessitate continuous R&D efforts to address associated problems, including data privacy and algorithmic prejudices. Continuous research in vision sensor technology holds the promise of not just improving existing applications but also creating new ones that can be adapted to address new society requirements (Shaer et al., 2009) (Figure 3.4).

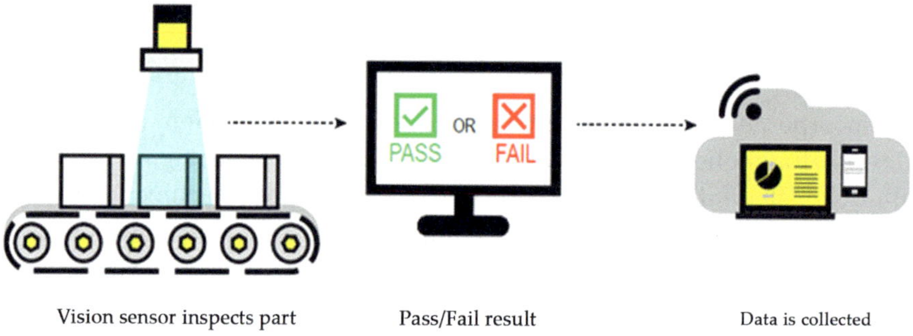

Figure 3.4. *Vision sensors work.*

Source: https://www.cognex.com/library/media/intro-topics/vision-sensor/vision-sensor-process.jpg?h=231&w=647&sc_lang=en&hash=675C9C61252DA03D696A51BA22AA2F78.

3.1.2. LiDAR Technology and Its Significance

Advanced technologies have been incorporated inexorably into the area of automated vehicle systems, particularly to optimize sensor, actuator, and perception system interaction. LIDAR technology takes center stage in this revolution given its capability to deliver precise three-dimensional map functionality essential for autonomous movement. It facilitates high-resolution mapping to detect obstacles and environmental features, a key requirement for creating safe autonomous cars. While industrial and academic research and development efforts intensify, LIDAR has attracted attention not only because of its immediate applications in car automation but also due to its potential to revolutionize the whole transport system as a cutting-edge technology, which showcases the extensive impact of sophisticated sensor technology on systemic effectiveness and safety (Gordon et al., 2015). The superior capability of LIDAR systems is their ability to generate timely and accurate environmental data, a vital parameter for effective decision-making in autonomous vehicles. By employing laser pulses for measuring distances, LIDAR systems create highly resolved maps of their surroundings with unbelievable spatial resolution. This is an essential foundation for sensor fusion approaches, under which LIDAR measurements can be combined with camera, GPS, and inertial measurement unit (IMU) feedback to create a generalized understanding of the environment. This data combination not only offers a more reliable perception system but also assists in addressing fundamental challenges associated with vehicle dynamics and motion control to propel further advancements in automated driving technologies (N/A, 2025). The influence of LIDAR technology transcends the aspect of vehicle automation only, with influences in various fields including city planning and environmental observation. Precision and accuracy offered by LIDAR enable it to be applied for surveying topography and infrastructure, leading to effective decision-making in such an industry. Besides, with the rising necessity for intelligent transport systems, the use of LIDAR in safeguarding safe and efficient travel is even more crucial. The meeting of LIDAR with other modalities of sensors promises to enhance operational capacity, resulting in new technologies of autonomous navigation. These advancements further underpin the revolutionary character of LIDAR technology in the broad context of perception systems, actuators, and sensors, fueling the demand for continued development and research (Gordon et al., 2015) (Figure 3.5).

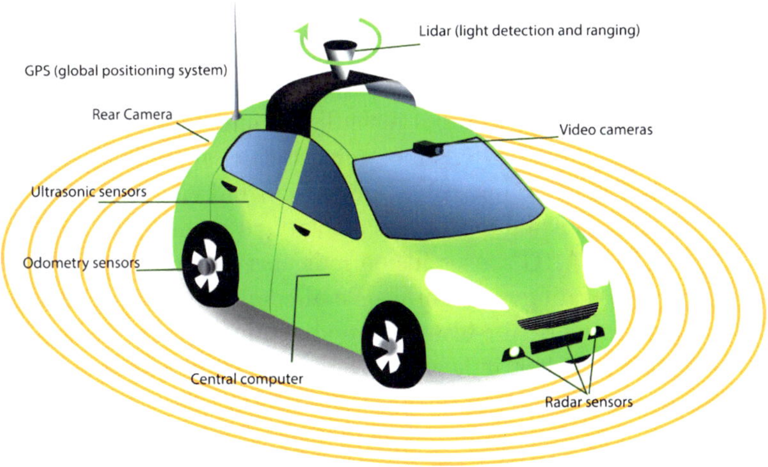

Figure 3.5. *LiDAR and how does it work.*

Source: https://circuitdigest.com/sites/default/files/field/image/What-is-LiDAR-and-How-does-it-Work.png.

3.1.3. Inertial Measurement Units (IMUs) and GPS Integration

The intricacy of navigation through both planetary and interplanetary mediums demands robust sensing capacities that can compensate for the deficiency of a one-system sensor. Inertial Measurement Units (IMUs), commonly consisting of accelerometers and gyroscopes, provide valuable information regarding acceleration and spin velocities. IMUs are plagued by cumulative errors since sensors drift after some time, and on their own become unreliable, particularly in the long run. In order to offset these constraints, the application of Global Positioning System (GPS) data is critical. GPS provides geolocation accuracy and gives periodic readings that can redefine the built-up errors present in IMUs. Such sensor fusion enhances position accuracy as well as maintains perception systems with high fidelity in dynamic environments, such as those found in varied applications from robotics to human navigation around cities. Following on from the integration of GPS and IMUs, advanced algorithmic methods like Kalman filtering have become the core sensor fusion method. By employing these statistical techniques, the optimal possible combination of IMU inertial data and GPS positional data can be generated in order to minimize the effect of noise and uncertainty from both sets. This combination leads to an improved estimation of the state of an object, which is specifically important in applications where precise trajectory control is required, such as autonomous cars or spacecraft missions. In addition, sensor fusion algorithms in integrated systems enhance the overall system robustness by minimizing the risks of sensor faults or external interference. To this effect, the variable vector countermeasure suit (V2Suit) highlights the potential for utilizing integrated systems in hostile environments, as given in previous research (Duda et al., 2025). This suggests that there is a necessity to move the focus from individual sensors towards an end-to-end solution using the complementary strength. Also, the design of new algorithms for MARG sensor fusion indicates ongoing innovation in this field, further demonstrating the significance of combining IMU and GPS. New algorithmic developments, for instance, modifications to the extremely well-established Madgwick filter, indicate not only computational performance but also improved accuracy and robustness with adverse sensor conditions. These enhancements are of utmost significance in applications where full orientation is critical, such as teleoperation of humanoid robots, as shown by experiments (Eberle et al., 2019). By integrating IMU data with other sensing modalities, such as mechanomyography in a human-machine interface, researchers are pushing the limits of sensor utilization. This paradigm of combining different sensor data emphasizes the significance of IMUs and GPS integration in perception systems and stresses the imperative of continuous technological advance to facilitate a wide variety of applications.

3.2. DATA ACQUISITION

Continuous research and development of sensor technology have broadened the range and impact of data acquisition processes in many fields. In the field of perception systems, cameras and LIDAR are two sensors that are part of an essential element that captures spatial and environmental information crucial for decision-making. Vision sensors, for instance, allow for image processing and object recognition, and LIDAR systems enable enhanced depth perception through laser scanning. Besides, IMUs and GPS provide critical information of motion and position that, when merged, create a full description of an object's trajectory and environment. Merging data from these heterogeneous sensors is central to applications ranging from autonomous vehicles to robotics such that systems are maintained adaptable and responsive to varying conditions. Data acquisition is not merely a matter of gathering information; it also involves meticulous processing and analysis to extract meaning from the large number of signals that are collected.

Figure 3.6. *Sensor technology.*

Source: https://www.patsnap.com/wp-content/uploads/2023/03/Sensor-technology-.png.

Signal processing techniques play a critical role in refining raw data, removing noise, and identifying significant patterns to enable better interpretation of sensor data. For example, filtering algorithms can be used to improve the accuracy of noisy sensor measurements, thus ensuring quality in real-time systems. Furthermore, advanced processing techniques facilitate the translation of information gathered into informative forms supportive of decision-making, in agriculture, environmental monitoring, or industrial automation. Thus, the relationship between signal collection and signal processing is a critical determinant in optimizing sensor performance and achieving desired functional outcomes in modern perception systems. One such robust data acquisition method is also depicted by means of sensor fusion techniques, which synergistically integrate data from different sources to infer deeper insights than received from an individual sensor. This technique enhances accuracy and believability of information obtained by reducing the limitations of every sensor. In some applications, such as in the design of high-performance aircraft systems, the Rotorcraft-Aircrew System Concepts Airborne Laboratory (RASCAL) demonstrates how data and sensor fusion could enable control and maneuverability in challenging environments (Aiken et al., 2025). Similarly, transparent pressure sensor arrays demonstrate the possibility of sensing and controlling haptic information, pushing the boundaries of data acquisition to include touch interaction (Brochu et al., 2013). This intersection of disparate sensory modalities emphasizes the promise of fusion algorithms to transform the function of sensors, actuators, and perception systems generally (Figure 3.6).

3.2.1. Methods of Data Collection from Sensors

On the existing technology front, many sensors have been developed in an attempt to support the process of data collection on different applications. The most critical among these are vision sensors, LIDAR, inertial measurement units (IMUs), and global positioning systems (GPS). Vision sensors are a crucial element for tasks involving imaging because they enable automated systems to interpret visual content and respond accordingly. Alternatively, LIDAR sends pulses of light and measures the time taken to return, thus creating accurate three-dimensional representations of the environment. IMUs, which combine accelerometers and gyroscopes, provide useful movement and orientation data, while GPS sensors provide location data needed for navigation. Effective sensor fusion does not just enhance the accuracy of information but also extends the applicability of perception systems across contexts. Therefore, the selection of the types of sensors is an elementary process in the development of an effective data collection infrastructure required for successful information-gathering operations. Data acquisition and signal processing are equally crucial in sensor data gathering. The transformation of raw sensor data into useful information requires sophisticated techniques that eliminate noise and enhance the quality of signals. For instance, data acquisition systems read out sensor outputs at discrete points in time, ensuring timely and correct capture of dynamic events. With the data collected, signal processing algorithms take it from there, solving many problems such as removing unwanted signals and eliminating distortions. Use of these techniques not only improves data quality but also has it prepared for further analysis and interpretation. Correct and unambiguous rendering of information is essential for subsequent decision-making actions in perception systems, opening the doors to innovation from autonomous vehicles to health monitoring systems. In addition, sensor fusion techniques play a key role in enhancing the quality of data obtained from an array of sensors. Sensor fusion merges multiple streams of data from different sensors to produce a more comprehensive view of the surroundings. This method enhances reliability and accuracy, particularly in complex cases where single sensor readings might not be sufficient. For example, integrating data from IMUs and GPS can produce positioning data that is accurate in terms of motion dynamics and geographical location. Such methods have been critically noted in the literature, which has resulted in the development of more sophisticated systems that achieve sensor efficiency (Fluitman et al., 1996). Additionally, tangible user interfaces (TUIs) are among the new areas where sensor data collection is combined with user interaction synergistically, showing the future promise of unconventional approaches in advancing technology (Shaer et al., 2009). From continued research and development of such approaches, future research will deliver significant progress in sensor applications, giving the Sensors, Actuators, and Perception Systems field a more prosperous outcome.

3.2.2. Challenges in Data Acquisition

In the highly complex sensor environment of perception systems and actuators, the data acquisition process is riddled with various problems that have broad-reaching consequences for system reliability and performance. With many different types of sensors such as vision, LIDAR, inertial measurement units (IMUs), and GPS in widespread use, accurate data gathering is of the utmost importance. Each sensor modality has its own environmental and noise characteristics, so the acquisition phase is complicated. For instance, vision sensors are capable of extracting rich imagery but are sensitive to lighting changes and occlusion. Meanwhile, LIDAR can provide precise measurements of distance but may be lackluster in unfavorable weather. The need for a robust solution to counter sensor limitations becomes even more pressing as these systems transition out of laboratory environments into dynamic, uncontrolled real-world scenarios, putting the inherent complexities of multi-sensor data fusion into real-world contexts. Moreover, data acquisition difficulties are not merely founded upon the physical limitations of the sensors themselves but also upon the intricate signal processing and sensor fusion algorithms required to create useful interpretations of the gathered data. Proper signal processing to remove noise and enhance data quality is necessary but typically calls for phenomenal computational power, particularly when dealing with bulk data generated by high-res sensors. Sensor fusion, the combination of data from multiple sources to create an integrated perception of the environment, is inherently complicated. It is made even more so when considering differential time delays as well as synchronization issues that arise in between heterogeneous sensors. Inefficient fusion of sensor data will lead to misleading interpretations and thereby faulty operating decisions, particularly in safety-relevant applications such as autonomous navigation or combat operations. As systems employ increasing autonomous capability, procurement of predictive information is an essential aspect to optimizing learning and decision-making processes. Intrinsic motivation due to the need to maximize predictive information enables advanced models of learning to evolve independently by exploring environmental regularities. This advanced interaction brings additional challenges related to balancing exploration and exploitation. Research has shown that while maximizing can provide great predictive data, optimal performance will require a balanced system which incorporates this inbuilt motivation with task extrinsic rewards. The balancing of such concerns is particularly relevant in uses such as military and strategic interventions, where the integration of lethal autonomy into tactical frameworks needs empirical foundations behind the evolving disciplines of hybrid threats and countermeasures, as discussed in modern debates around such systems (Ay et al., 2013; Nyagudi et al., 2013) (Figure 3.7).

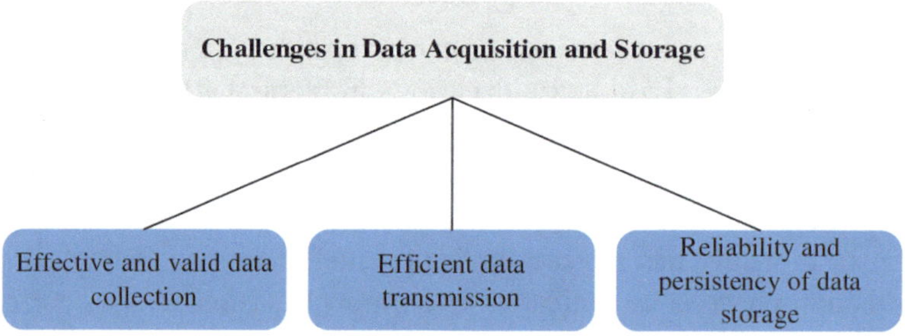

Figure 3.7. *Challenges in data acquisition and storage.*

Source: https://www.researchgate.net/publication/361290554/figure/fig2/AS:1143 1281233514716@1712056887136/Challenges-in-data-acquisition-and-storage. png.

3.2.3. Importance of Real-Time Data Processing

The marriage of various sensor technologies significantly enhances the capability of systems to process data in real-time. Vision, LIDAR, and IMUs (Inertial Measurement Units) are such sensors that collect a wealth of information from the environment, allowing richer understanding and interaction. While the sensors collect data, signal processing concepts are employed to filter out noise and retain meaningful information to facilitate better decision-making in dynamic environments. For instance, LIDAR can provide precise measurement of distance, which, when combined with visual information through sensor fusion techniques, can generate comprehensive environmental maps. The importance of real-time processing has both to do with the instant response it enables in such systems as well as to future national space programs, as are also underscored by OAST's critical technologies of high priority on automation and very data-dependent systems (Beck et al., 2025). Besides, real-time data processing is essential for the automation of operational activities, particularly in applications that call for high accuracy and dependability. Autonomously operated ground robots such as those examined under the GRASP project showcase the immense advantage that accompanies real-time data handling. Such robots perform tasks that were earlier assigned to man at the Kennedy Space Center, thus enhancing safety as well as reducing operating expenses. The real-time processing facilitates these robots to quickly react to unplanned alterations or issues as they arise, facilitating an uninterrupted flow. The alignment of timely data analysis with the operation efficiency indicates the necessity of real-time processing toward achieving reliable, cost-effective outcomes across a range of applications in space technology and robotic automation (Grissom et al., 2025). Real-time data processing has a significant impact on the development of perception systems in robots and automation. Using real-time response to sensory data, they can perform operations with improved speed and precision. Interaction between physical phenomenon detection sensors and command-executing actuators relies solely on real-time-processed data. In addition, with improvements in sensor technologies, the role of combining data from multiple sources using sophisticated algorithms is complementing the overall robustness of perception systems. This synergy of sensors, actuators, and real-time processing defines the foundation of creating intelligent machines. Lastly, the importance of processing real-time data cannot be overstated, since it is the basis for the capability and performance of high-level systems for increasingly critical uses in applications from industrial automation to space exploration (Beck et al., 2025; Grissom et al., 2025).

3.3. SIGNAL PROCESSING

The sensing interaction and ability to handle data contribute significantly towards the effectiveness of perception systems. Various sensors like vision systems, LIDAR, IMUs (Inertial Measurement Units), and GPS all possess various data characteristics which require certain signal processing techniques. Signal processing is essential to noise filtering, enhancing signal quality, and rendering the unprocessed data captured by these sensors understandable and meaningful. For instance, vision sensors must employ image enhancement as well as object recognition algorithms, while LIDAR systems must utilize sophisticated point cloud processing techniques. This multi-faceted approach to signal processing allows for the generation of cleaner, more actionable intelligence, which is critical to robots operating in complex environments. Lastly, the efficacy of such operations has a direct bearing on the performance of autonomous systems, impacting their navigation and interaction capabilities. An important aspect of signal processing is achieved through integration and fusion of data obtained from different sensor modalities. Sensor fusion techniques are employed to merge the various types of information acquired using sensors to create a better understanding of the world. By processing complementary information, for instance, combining visual information from a camera and range data from LIDAR, systems can overcome the deficiencies of an individual sensor.

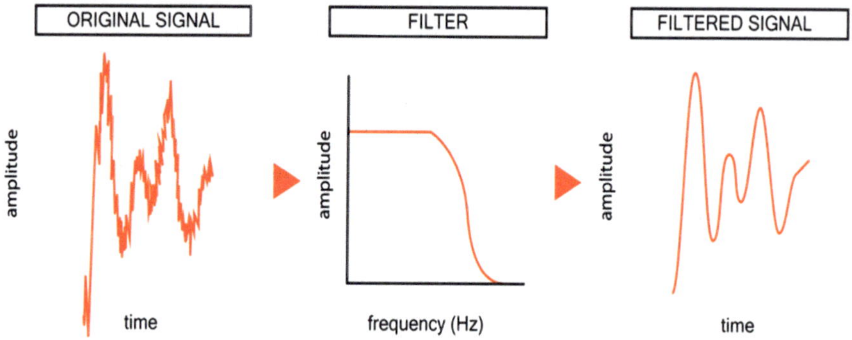

Figure 3.8. *Signal processing.*

Source: https://www.datocms-assets.com/53444/1661860595-filtered-signal-graph-2.png?auto=format&fit=max&w=1024.

For example, conditions of low visibility have the potential to weaken camera functionality, whereby LIDAR can continue to perform at its best, thereby compensating for visual sensor frailty. As outlined in novel cognitive robot control architecture CORBYS, effective information sharing and processing across various control modules illustrate the significance of advanced signal processing techniques in robotic framework ruggedness designs (Badii et al., 2013). The combined ability achieved through sensor fusion not just improves perception but also facilitates more informed decision-making in dynamic settings. The development of new sensors with specialized functionality further illustrates the significance of sophisticated signal processing techniques in modern robotics. New developments such as a photoelastic effect-based tactile sensor demonstrate how specific material properties can augment sensor functions. The ability of this sensor to detect slip on an object while providing normal force information at the same time allows for full feedback to be gained in interaction (Crowder et al., 2006). Such dynamic sensors refer to the importance of effective signal processing in translating complex mechanical signals into meaningful information. The calibration and modeling processes involved with these new sensors clearly demonstrate the importance of precise signal manipulation in order to produce reliable output. As robotics continues to develop, integration of sophisticated signal processing methods will be crucial in the establishment of systems with the capacity to handle the intricacies of real-world issues to enhance human-robot interaction and performance in a wide range of fields (Figure 3.8).

3.3.1. Techniques for Filtering and Enhancing Sensor Data

In the area of sensor data enrichment, integration of various sources of sensors is necessary in achieving accurate and reliable output. Vision sensors such as cameras provide rich visual information but are susceptible to noise and variations in light. LIDAR sensors, however, measure distances accurately by sending laser pulses and providing hardened data where vision sensors are insufficient. With sensor fusion technologies such as Kalman filtering and complementary filtering, the disparate sources of data can be successfully combined to create a single image of the environment. The strengths of each sensor are able to compensate for individual vulnerabilities to give a more cohesive interpretation of the data gathered. This synergy approach is essential in applications ranging from autonomous cars to robots, in which situational awareness is essential for reliable function and operation. Sensor readings are enriched with sophisticated data acquisition and signal processing techniques to counteract built-in errors and uncertainties. For instance, sensor readings are prone to deformation due to environmental noise or mechanical oscillations, and hence filtering algorithms are used to mask the noise. Techniques such as low-pass filtering allow noise reduction at high frequencies without compromising the signal's inherent characteristics. Besides, adaptive filter techniques are capable of adapting parameters dynamically to changing conditions in such a way that they operate at the best in a wide range of situations. The integration of these signal processing techniques not only enhances the fidelity of the sensor data but also makes subsequent analysis and decision-making easier. By systematic refinement of sensor inputs, systems can acquire higher accuracy and reliability in outputs, which are essential for effective actuator responses within perception systems. Real systems, however, have to cope with variability in sensor data due to operators, which complicate filtering and upgrading this information. In industrial environments, for example, operator actions can alter data records and labels in ways that can misrepresent the actual state of processes. To address this issue, techniques like iterative relabeling schemes may be applied to consistently refine dynamic features and optimize classification performance. These involve developing strong fault detection models capable of learning from operator-process interactions and, therefore, reduce the uncertainty involved in sensor measurements. By integrating these adaptive techniques into sensor data processing architectures, engineers can come up with systems that not only filter and refine information effectively but also are immune to variability caused by human input, thereby producing more trustworthy insights and operational efficiencies in complex applications such as the ones addressed by Askarian et al. (2016).

3.3.2. Role of Algorithms in Signal Processing

Algorithm-based signal processing is a central aspect of modern sensor technology significantly improving the performance of most systems such as autonomous vehicles, robots, and environmental monitoring. Multimodal data from sensors such as vision, LIDAR, and IMUs pose a need for sophisticated signal processing techniques to get useful information from raw data. Algorithms play a vital function in noise removal, signal improvement, and feature extraction of significant information from intricate data sets. For example, in autonomous vehicles, real-time signal processing algorithms process streams of data from multiple sensors to deliver precise situational awareness, allowing safe navigation. This intricate interaction of algorithmic calculation and data acquisition emphasizes the requirement for designing efficient algorithms that can operate within the constraints of real-time systems and yet ensure reliability and accuracy in various applications. In the field of sensor fusion, algorithms become the fulcrum by combining information from more than one sensor to create a more correct and holistic picture of the environment. Sensor fusion techniques enhance perception system reliability through reducing potential inconsistencies generated by the weaknesses of individual sensors. An example is combining GPS and IMU sensor data for more accurate positioning and motion tracking through the leveraging of IMUs capability to provide high-frequency rates and GPSs global coverage. By using advanced algorithms, it is feasible to balance the input of each sensor with high precision based on how well every sensor will perform in a specific environment to maximally optimize sensor output overall. All these methods not only improve the reliability of data utilized in making decisions in applications, such as robotics and driverless vehicles, but also raise the systems' resistance to sensor faults or malfunctions, improving operation safety. In addition, the hierarchical framework proposed in robot control structures better captures the fundamental contribution of algorithms towards sensor coordination as well as signal processing. These models have lower levels directly interacting with sensors and actuators to deal with motor commands and raw data to lay the foundation for upper-level decision processes. Usage of Boolean Artificial Neural Networks on the rule level allows real-time correct operations in accordance with the processed sensor information to achieve quick adaptation to environmental changes. Hierarchical structure not only allows efficiency in processing information but also demonstrates algorithmic capacity to make robotic systems responsive to dynamic, complex environments adaptively. Through better utilization of algorithms in signal processing and sensor fusion, systems will be capable of achieving higher accuracy, responsiveness, and capability in numerous applications, as evidenced in creating sophisticated projects like the RASCAL helicopter to enhance maneuverability and pilot assistance in performing missions in challenging environments (Aiken et al., 2025; Chang et al., 2025) (Figure 3.9).

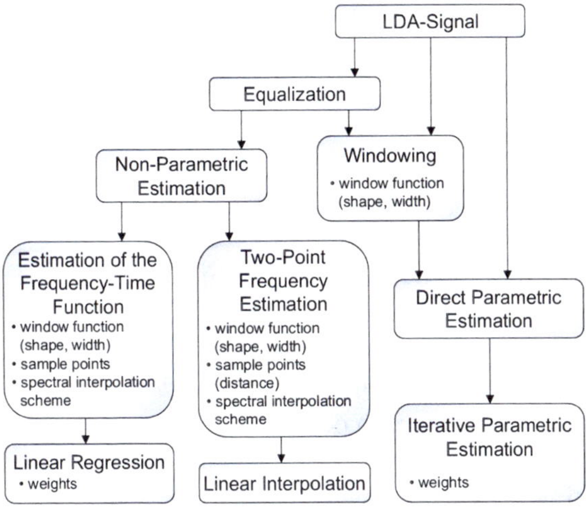

Figure 3.9. *Classification of the signal processing algorithms with specific options.*

Source: https://www.researchgate.net/profile/Cameron-Tropea/publication/224801980/figure/fig2/AS:668944184668171@1536500130508/Classification-of-the-signal-processing-algorithms-with-specific-options.png.

3.3.3. Impact of Signal Processing on System Performance

Signal processing technologies have played a crucial role in enhancing the performance of numerous sensor systems, which are a component of modern technology. By processing data from a sensor network, like vision, LIDAR, IMU, and GPS, systems get more responsive and accurate. The technologies not only sense raw data but also interpret it using advanced algorithms that filter out noise and enhance relevant signals. To illustrate, when used in autonomous vehicles, effective signal processing facilitates real-time environmental mapping and obstacle detection, which leads to improved navigation and safety. The ability to derive actionable knowledge from massive amounts of sensor data ultimately optimizes the operational efficiency and resilience of perception systems, rendering them critical in applications as diverse as aerospace and robotics. In addition, the incorporation of sensor fusion methods plays an important role in maximizing system performance through developing a consistent and complete understanding of the environment. Signal processing algorithms can compensate for limitations of individual sensors by combining information from multiple sensors, thus enhancing overall system accuracy and resilience. This is particularly required in complex situations where single sensor inputs can be insufficient to be useful, for instance, during aircraft landing operations or space exploration. Advanced signal processing techniques, as evidenced by the GRASP research findings, uncover that operations automation through better data interpretation has lower costs and better safety measures when robotic systems are implemented in essential environments (Grissom et al., 2025). Thus, sensor fusion by good signal processing is a paradigm of performance measures in most technological systems. Furthermore, signal processing efficiency is not all about merging data; its core to developing new technologies that directly respond to the needs of future explorations. As discovered in the In-STEP workshop, foundational areas of technology require cutting across numerous disciplines, with leading fields where signal processing may be able to contribute significantly, particularly in automation and robotics fields (Beck et al., 2025). As systems mature to perform more complex tasks autonomously, the ability to process and interpret sensor data in real time correctly will be essential. Effective signal processing capabilities will not only assist better decision-making but also facilitate useful operation of sophisticated systems in adverse environments such that missions can be conducted securely and efficiently. Therefore, the impact of signal processing on system performance is impactful, determining the course of technology development for sensor-based applications.

3.4. SENSOR FUSION TECHNIQUES

The ability of various types of sensors to communicate with each other is a key determinant of perception system effectiveness. Vision sensors such as cameras, for instance, provide rich contextual data by capturing high-resolution images, while LIDAR systems provide precise distance measurement through laser scanning. Coupled with inertial measurement units (IMUs) and global positioning systems (GPS), all these sensors enable creating robust situational awareness. Sensor fusion techniques seem to be pivotal techniques for integrating these disparate data streams in order to facilitate more robust and reliable environmental perception. By the integration of knowledge from various sources, sensor fusion techniques have the capability of overcoming the limitations of individual sensors—such as poor visibility in vision sensors during inclement weather—thus providing an expanded view of the environment. This unification ultimately facilitates advanced applications, like autonomous cars and robot systems, which require advanced navigation and decision-making in real-time environments. The data acquisition and signal processing stages are integral parts of effective sensor fusion. These stages are responsible for converting raw sensor inputs into meaningful information that can be utilized in real-time analysis. Several algorithms, such as Kalman filters, Bayesian networks, and deep learning methods, are employed to analyze sophisticated data streams from heterogeneous sensors. The correctness of the fusion process is influenced by the timeliness and accuracy of these algorithms, which must consider sensory noise and uncertainty. Optimizing both performance and responsiveness, sophisticated signal processing functions are able to identify useful features from heterogeneous sources, strengthening system reliability. Furthermore, the need for adaptive filtering methods is critical, particularly in dynamic environments where numerous factors can influence sensor performance and reliability. Continued advances in these disciplines are crucial since they directly impact the efficacy of sensor fusion in most applications that are critical to modern technology. In the midst of evolving technologies, the use of sensor fusion techniques becomes increasingly relevant across different industries, including the Internet of Things (IoT) and robotics. With the pervasive presence of connected devices and autonomous systems, there arises new set of challenges related to transparent integration and real-time processing of information. For instance, the Internet of Robotic Things (IoRT) and similar paradigms necessitate fusion-based usage of different sensor data to facilitate effective collaboration among single systems. The employment of sophisticated fusion technologies becomes essential to facilitate stability and reliability in operation in use cases like cooperative robots (COBOTS) that need to operate in challenging environments with safety protocols. Therefore, the development of these techniques not only enhances performance but also addresses fundamental problems such as reconfiguration of resources and self-healing capability in dynamic IoT scenarios. With the stimulus of a robust understanding of sensor fusion techniques, researchers and engineers can prepare themselves adequately for future tasks in working towards intelligent and autonomous systems (Figure 3.10).

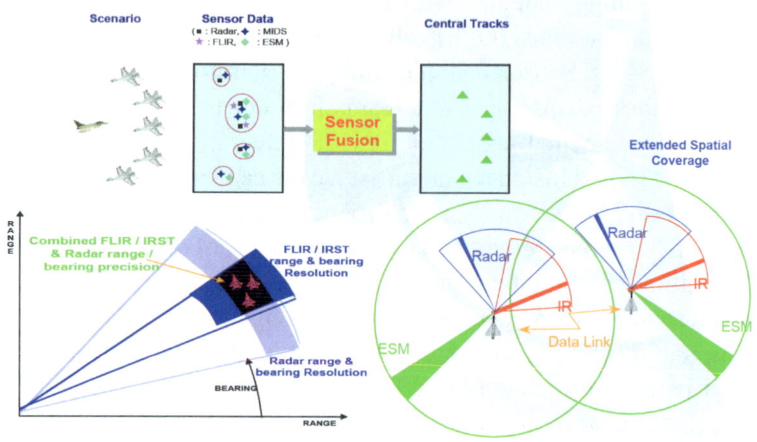

Figure 3.10. *Sensor fusion.*

Source: https://upload.wikimedia.org/wikipedia/commons/f/f9/Eurofighter_sensor_fusion.png.

3.4.1. Definition and Purpose of Sensor Fusion

With the employment of multiple sensor modalities in contemporary robotics and automation, perception and interaction ability of intricate environments is highly enhanced. Sensor fusion is a method that plays a pivotal role in maximizing the reliability and precision of robotic systems. By combining data from various sensors—such as cameras to perceive, LIDAR to determine distance, IMUs to detect orientation, and GPS to monitor location—sensor fusion enables richer perception of the world. All sensor types have something unique to contribute, and through signal processing and data collection, these incongruent inputs can be combined to yield a synthesized and actionable representation of the world. This capability is particularly valuable in application scenarios where precision is a necessity, such as autonomous vehicles and robot limbs, where real-time comprehension and reaction are required to maintain proper functionality. Sensor fusion's function transcends data acquisition; it is actually about enhancing system ability and decision-making. Through the implementation of algorithms that aggregate sensor data, systems can remove noise, reduce uncertainty, and complement gaps where a single sensor may not be adequate. This combined approach enables more robust navigational and operational capabilities. For instance, in robotic systems like anthropomorphic hands, the hierarchical organization makes lower-level computations handle raw sensor information while the higher levels handle more abstracted tasks like planning and execution with focus on real-time control. Hierarchical processing in such a way helps the reactive components of the system to react in time to stimuli, thereby facilitating smoother functioning and improved interaction with dynamic environments (Chang et al., 2025). With technology advancement pushing the development of autonomous systems, the use of sensor fusion becomes the prominent issue, particularly in specialist markets like space exploration. Projects like the Italian National Space Plans SPIDER program illustrate the use of sensor fusion in the development of autonomous integrated robots for complex repair missions in space environments. The cooperation between sensors in these systems enables effective inspections and operation control according to the demand for high perception detail under difficult conditions (Dipippo et al., 2025). Using sophisticated artificial intelligence techniques, these robotic systems have the capability of effectively processing information from various sources to enable complicated tasks as well as dealing with the various challenges faced in space. These applications emphasize sensor fusions significant contribution not only to enabling new capability but also to creating new opportunities for robotic systems to perform both terrestrial and extraterrestrial environments.

3.4.2. Common Algorithms Used in Sensor Fusion

With cutting-edge technologies pushing the scale of sensor system complexity, the need for effective sensor fusion algorithms has never been greater. Merging various kinds of information from various sensors—vision, LIDAR, and inertial measurement unit (IMU) data, to give a few examples—is making sophisticated perception and decision-making capabilities available to systems ranging from autonomous vehicles to robots. Sensor fusion techniques are confronted with the task of combining measurements that may differ in accuracy and credibility in respect of the unique character of each type of sensor. Sensor fusion algorithms, for instance, can use probabilistic approaches that analyze information based on the performance of sensors in the past, resulting in a more robust output that reinforces overall system reliability. These techniques not only enhance the fidelity of the perceived environment but also find application in real-time scenarios where timely information is critical to operation success. The Kalman filter is one prominent technique utilized in sensor fusion, which accurately estimates the state of a dynamic system based on noisy sensor measurements over time. It operates recursively, predicting a system's future state before it is updated according to new sensor data. It is this capacity that enables the Kalman filter to be very effective in handling uncertainty and dynamical changes, which is common in motion tracking and navigation systems. Apart from Kalman filters, other techniques such as particle filters have gained much popularity, particularly in nonlinear systems where traditional methods may not work.

The flexibility offered by particle filters allows them to deal better with non-Gaussian noise, and as such, areas of application for sensor fusion are extended successfully, particularly where maximum accuracy as well as flexibility is required in complex scenarios. Furthermore, the ion fusion process can heavily benefit from deep learning techniques, which are rich with capabilities for extraction of patterns and features from raw sensor data. By training on big sets of neural networks, these algorithms can learn to learn the capability of being able to predict accurately states and results from multi-modal sensor readings. This is especially important in use cases like autonomous driving, where camera visual interpretation must be combined with LIDAR distance reading and GPS data. Deep learning-based methods can complement traditional means in signal processing and interpretation efficiently, bypassing challenges with high dimensionality and variation of data rates. As is realized with advances in robot tactile sensing and computer vision, implementation of advanced algorithms continuously redirects the function of sensor fusion in perception systems toward autonomy and performance improvement in intelligent systems (Bacciu et al., 2017; Liñán et al., 2005) (Figure 3.11).

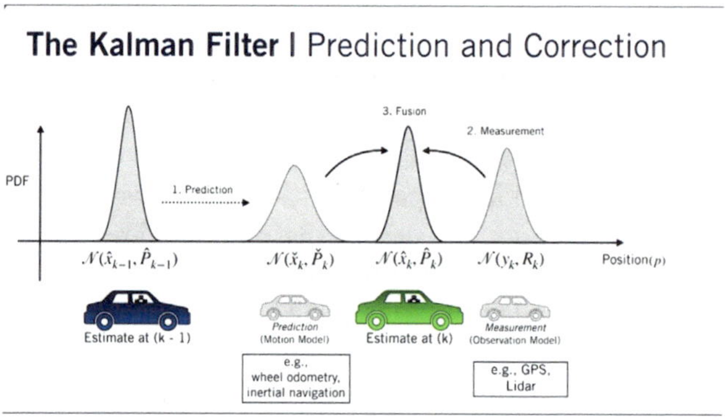

Figure 3.11. *Kalman filter.*

Source: https://i.ytimg.com/vi/LioOvUZ1MiM/maxresdefault.jpg.

3.4.3. Benefits of Integrating Multiple Sensor Data

The integration of different sensor data is a key part of enhancing the operational performance of systems across a variety of systems, particularly in robotic and smart grid applications. With the use of different sensor types such as vision systems, LIDAR, Inertial Measurement Units (IMUs), and GPS, systems can gain an enhanced level of knowledge about their environment. Various sensors possess different advantages; vision systems provide rich spatial information while LIDAR offers precise measurements of distances. In robotics, the CORBYS architecture is a sterling case of such integration and supports the integration of high-level cognitive modules that allow robots to operate in dynamic environments where interactions with human beings have a central role. By such integration, not only does the robot have more autonomy but also its adaptability over different situations is enhanced, thereby its functional utility in real-world applications like autonomous movement and performance of tasks (Badii et al., 2013). Moreover, application of sensor fusion techniques has a profound impact on improving data reliability and accuracy. Data from multiple sensors can be combined to allow systems to overcome the weaknesses of single sensors and provide an overall view of the operational world. For instance, fusing IMU information with GPS information provides a more stable and accurate positioning solution—something critical for robotics and smart grid applications. Smart grid environments require the use of combined sensor data through IoT devices to enhance real-time management and control, leading to improved energy efficiency and management. Dynamic adaptation of the systems to increase or decrease levels of energy demand and supply enhances reliability. This addition of data not only enhances operations but also allows for smart decision-making, which further results in lower operation costs and improved delivery of services (Copeland et al., 2019). Moreover, benefits of integrating various sensor data translate to real-time analysis and response use cases. In robotics, collaboration among various sensor inputs offers robust autonomous navigation methods, wherein a robot is able to effectively decipher complex environments, clear obstructions while conducting tasks. In smart grids use cases, sensor fusion offers enhanced predictive maintenance methods through continuous monitoring and processing of data. With sensors throughout the grid delivering real-time information, operators can quickly sense anomalies or inefficiencies, allowing for timely corrective actions. The overall effect of such benefits underscores the revolutionary potential of sensor integration data, where augmented situational awareness is converted into better performance outcomes in diverse industries, transportation, energy, and sophisticated manufacturing systems (Badii et al., 2013; Copeland et al., 2019).

3.5. ACTUATORS AND THEIR ROLE

In the world of modern-day automation and robotics, actuators have a major role to play in converting electrical or mechanical signals into motion. This is an indispensable process that plays a crucial role in numerous applications, from basic mechanical operations to highly complex interaction systems in smart spaces. Actuators serve as the final component in the chain of control, enabling devices to respond dynamically to sensor-sensed inputs. For instance, in agricultural settings, IoT-based platforms integrate various sensors that monitor soil moisture, temperature, and pH level content and actuators for the control of irrigation systems for optimal crop health and resource utilization. Such interplay is essential in facilitating remote monitoring and management in different agricultural scenarios, as discussed in the research of flexible IoT-based platforms with the incorporation of actuators for real-time data-driven decision-making (Mohamed et al., 2022).

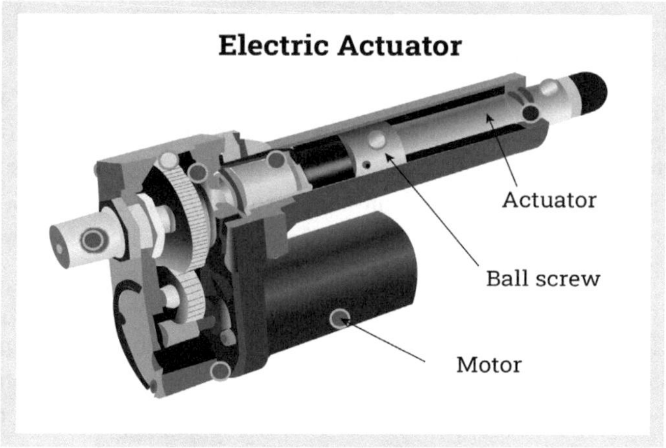

Figure 3.12. *Applications and benefits for electric actuators.*

Source: https://www.iqsdirectory.com/articles/linear-actuator/electric-actuators/electric-actuator-2.jpg.

Actuators play a crucial role in enhancing the performance of electric connected autonomous and shared (ECAS) automobiles in the automotive industry. ECAS systems are heavily reliant on actuator technology for interpreting information created by hundreds of sensors—e.g., LIDAR, cameras, and GPS—into actionable activity. As an example, when sensor data detect obstacles, actuators that are part of steering and braking systems make infinitesimal-scale maneuvers to prevent collisions. The communication of perception systems through sensors and reactive responses made possible through actuators illustrates a complex web where open communication between components is the safeguard of safety and efficiency. The digital backbone unifying these abilities is also crucial, as it serves as the backbone for data sharing and processing towards autonomous functionality and navigation (Vermesan et al., 2021). Furthermore, the changing paradigm of automation demands continuous advancements in actuator technology to keep pace with the complexity of newer perception systems. Technological advances in smart actuators, which come with more flexibility and responsiveness, are required for application across a range of applications from industrial robots to home automation systems. Not only must such systems process sensory data in an efficient manner but execute actions compatible with functional goals as well in an immaculate way. As IoT uses increase, demand for actuators that can communicate with different sensors in real-time will also pick up, fueling actuator precision and responsiveness R&D. Along with sensor fusion techniques, which integrate data from sensors' sources, is growing significance placed on developing actuators that are capable of providing efficient and consistent physical actions, which consequently will lead to even smarter and more flexible systems across industries (Mohamed et al., 2022) (Figure 3.12).

3.5.1. Types of Actuators Used in Perception Systems

In the realms of robotics and automatic systems, combining sensory input with actuator response is paramount to realizing wished-for operating outcomes. The quality of actuators used in perception systems depends largely on the effectiveness of robotic action in evolving environments. Electric motors, which are widespread actuators utilized in acts, are important in converting electrical energy into precise mechanical motion. This capability is manifested by the movement of robotic arms or wheels that can enable agility and responsiveness to sensory input derived from integrated sensors such as cameras or LIDAR. Mechanical agility is necessary for systems to interact with individuals or respond to varying environments, pointing to an inherent characteristic of control architectures such as the CORBYS system favoring cognitive processing and coordination of actuators to enhance robotic functionality (Badii et al., 2013). Hydraulic and pneumatic actuators are also part of complete elements in certain perception systems, particularly those requiring high force or specific actuation speed. Both actuators are ideally employed in applications like industrial automation, wherein material handling and precision tasks require quick and firm action. Intelligent sensors built into the actuators make them quicker to respond with real-time data. For example, a smart actuator, in combination with pressure or position sensors, can dynamically adapt its operation to maintain optimal performance levels. This convergence of sensor technology and high-performance actuation facilitates the emerging trend towards reactive and flexible machine tools, where actuators adapt cutting parameters depending on sensory feedback, foreshadowing innovation in smart manufacturing environments (Archimède et al., 2006). With the growing complexity of perception systems, research into high-end actuator types such as shape memory alloys and electroactive polymers has gained momentum. Both of these materials have properties that are conducive to offering lightweight, compact structure in robotic applications and yet achieving significant actuation performance. Utilization of such new actuators is also further supported by high-level signal processing and sensor fusion technology, which can offer the possibility of improved accuracy and responsiveness in real-time applications. The integration of disparate actuators with sophisticated sensory technologies enables the development of robots that not only sense their environment well but also act in rich manners to enhance human and system interaction. Such synergy heralds a promising future in research and operational deployment, calling for continuous innovation in actuator technology in perception systems.

3.5.2. Interaction Between Sensors and Actuators

Sensors and actuators' integration without a point of failure is pivotal in ensuring maximal autonomous systems effectiveness and functionality. The extent of sensors, including vision systems, LIDAR, inertial measurement units (IMUs), and global positioning systems (GPS), plays a key role in collecting environment information. All sensor types contribute unique information to have a complete grasp of the operating context. For example, vision sensors can be used to sense obstacles, and LIDAR can be used to take accurate measurements of distances, essential for the navigation of challenging environments. But data acquisition alone is not enough; it requires efficient signal processing and data fusion algorithms. Employing algorithms to combine data from different sensors guarantees resilience to uncertainties, hence enabling accurate decision-making. This intricate dance of sensor data and the associated actions performed by actuators is the foundation of interaction dynamics that define how systems react to real-time challenges and execute tasks programmatically. Actuators are dynamic units that translate processed information to physical movement or action, thereby directly influencing system behavior. Actuator performance is a function of the responsiveness sensitivity of the response to sensor data, speed, and accuracy. For example, an accurate motor-based robot arm can change its location based on feedback from proximity or vision sensors to execute tasks with high reliability. This reactivity is followed by the integration of advanced control algorithms that consider feedback provided by sensors to facilitate a closed-loop system. Particularly noteworthy in this regard is sensor-actuator interaction, offering a one-to-one relationship between sensed data and performed responses, a feedback system which optimizes sensing and action plans perpetually. This collaboration is particularly valuable in robotics and vehicle control systems, where it is now possible to make compensations in real time for changing environmental conditions. Technological progress has significantly improved the sensor-actuator correspondence, providing a new paradigm for intelligent system design. Powerful approaches, such as empowerment, have been formulated to quantify the degree to which an agent controls its world, identified as the channel capacity between actuators and sensors (Glackin et al., 2013). These measurements allow control strategies autonomous systems employ to be evaluated, and hence predictive modeling is central in designing actuators. The CORBYS project has demonstrated that it is possible to integrate a cognitive framework within the control architecture for enhanced human-robot interaction (Glackin et al., 2014). The creation of task-specific and generic cognitive algorithms allows these systems to be able to better predict human action, and this leads to smoother cooperation. As these interactions continue to advance, the potential for developing increasingly adaptive, intuitive, and more efficient perception systems grows, creating opportunities for breakthrough applications across various industries (Figure 3.13).

Figure 3.13. *The global positioning system.*

Source: https://cdn.oceanservice.noaa.gov/oceanserviceprod/facts/gps.jpg.

3.5.3. Applications of Actuators in Robotics and Automation

The history of automation and robotics is greatly intertwined with the utilization of actuators that have the capability of injecting advanced movement and functionality into machines. These devices transform energy into mechanical motion, enabling them to be controlled flexibly, which is paramount in most applications. In applications where automation is essential, such as in manufacturing lines or even in space mission exploration, actuators are the foundation. They do not just enhance operation efficiency but also lower the risk of human error. Ground Robotic-Hand Applications for the Space Program (GRASP) research affirms this by indicating specialized end-effectors that can amplify or mechanize physical work at the NASA Kennedy Space Center. Such automation leads to reduced expenses, improved safety, and increased dependability in the handling of logistics and payload operations, emphasizing the critical contribution actuators make to high-risk contexts (Grissom et al., 2025). Moreover, the application of actuators in robotics intersects with advances in various sensors. Such complementarities are necessary to achieve sophisticated automation outcomes. For example, the integration of vision sensors and actuators enables robots not just to sense but also to act upon their surroundings effectively. Signal processing and data acquisition are needed to process the sensor input, decoding it into useful commands for the actuators. Such interaction is crucial in making exact movements, particularly in robotics applications such as robotic surgery or automated manufacturing, where accuracy is the only option. Secondly, sensor fusion techniques enhance the reliability of robotic systems to make them operate fluently within changing environments while harnessing the capacity of the actuators. Thirdly, the actuation and advanced sensor technologies together optimize the exploitation of automation applications across numerous industries, enhancing overall system performance as well as operational efficiency (Beck et al., 2025). Continued advances in automation and robotics include further studies in actuator technologies. As increasingly more industries apply these systems to carry out everything from mundane to highly sophisticated operations, actuator design breakthroughs are required. R&D agendas, like those for the Office of Aeronautics and Space Technology's In-Space Technology Experiments Program, address the necessity of demonstrating actuator technologies in a variety of environments. Not only does it demonstrate that they can function, but that they are also flexible enough for a multitude of applications, from autonomous unmanned vehicles to sophisticated robotic arms. The ongoing advancements in actuator technology have the potential to enlarge their range of adaptability and usability, ensuring that they meet the prerequisite demands of various operation environments. The shift not only ensures optimal productivity but also opens up new avenues for integrating robotics into everyday use, thus radically altering operation models in healthcare, manufacturing, and space exploration (Beck et al., 2025; Grissom et al., 2025).

3.6. APPLICATIONS OF PERCEPTION SYSTEMS

The growth of modern technology has brought a paradigm with which perception systems play a key role in numerous applications across many fields. Perception systems integrate various forms of sensors—i.e., vision systems, LIDAR, IMUs, and GPS—to sensorially observe and make sense of the environment continuously. Through the use of advanced data acquisition and signal processing techniques, perception systems can transform raw sensor data into more applicable knowledge. Such transformation enhances operations like autonomous vehicle navigation, where stable environment perception is needed for safety and efficiency. Sensor fusion techniques also allow such systems to combine input from diverse sources, remove uncertainty, and increase the overall accuracy of perception, thus meeting advanced requirements of contemporary functionality.

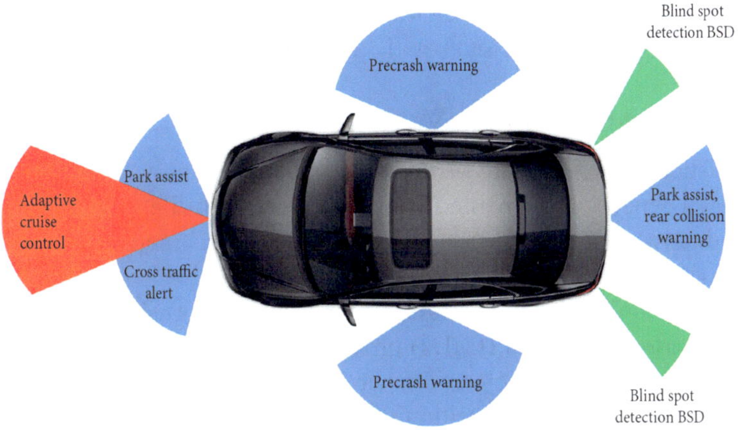

Figure 3.14. *Advanced driver assistance systems (ADAS).*

Source: https://www.researchgate.net/publication/328920450/figure/fig1/AS:692681546141697@1542159558581/Advanced-driver-assistance-systems-ADAS-for-active-passive-safety-comfort-functionality.ppm.

In robotics, perception systems are the basis for creating machines that are intelligent enough to interact with their world. Through the use of a combination of LIDAR and cameras as sensors, they can construct high-resolution, three-dimensional maps of the world in real-time. These capabilities are essential for applications as varied as industrial automation and service robots because they enable precise navigation and task execution. The use of smart algorithms makes robots able to understand sensory data and make smart decisions without direction. As seen through various studies, effective design of such perception systems has the potential to significantly enhance operation effectiveness, even in cases of uncertainty in environments. The capability of perception systems to react and adjust to situations shows their flexibility and usability in other technological areas, contributing even more to the importance of advanced sensing techniques. The evolution of perception systems is of significant use in the discipline of vehicle dynamics and control as well. Advanced methodologies, such as model-free control strategies, facilitate precise control of vehicle longitudinal and lateral motion irrespective of tedious mathematical modeling. Based on intelligent controllers, these systems enable optimal path following of vehicles from sensor readings in real time. For instance, the integration of actual data from car instrumentation and simulated inputs enables an adaptive process to generate Advanced Driving Assistance Systems (ADAS). The integration of the real and simulated data forms a flexible system that is capable of formulating robust control strategies for autonomous vehicles, eventually delivering safer and more efficient transport systems in urban cities, as stated in current studies (D'Andréa-Novel et al., 2015) (Figure 3.14).

3.6.1. Use in Autonomous Vehicles

The sophisticated architecture of self-driving cars is largely reliant on a variety of advanced sensors that enable them to interact with the world around them in a productive manner. Vision systems, such as cameras, provide useful information regarding road signs, lane markings, and other cars nearby, and LIDAR technologies give precise distance measurements using laser reflections. Inertial Measurement Units (IMUs) provide added stability to automobiles by tracking orientation and acceleration, necessary for the management of intricate driving scenarios. Global Positioning Systems (GPS) provide geographic context, aiding in path planning and real-time guidance. The integration of these diverse sensors allows for entire situational awareness, where the automobile will precisely analyze its surroundings and make appropriate driving choices. Lastly, integration of such technologies is the basis of autonomous vehicle functionality, ensuring safe and efficient transportations. Signal processing and data acquisition form the backbone of autonomous vehicle functionality. Once data has been acquired, signal processing employs several algorithms to cleanse and transform raw sensor data into information that can be exploited. For instance, image processing algorithms enhance the quality of visual inputs, whereas noise reduction algorithms improve the fidelity of LIDAR data. All such processing is important for reliable object detection, classification, and tracking, which influence the way an autonomous vehicle perceives its surroundings. Furthermore, utilization of machine learning models enables vehicles to learn their surroundings in an adaptive manner and enhance their performance with time. The effectiveness of these processing techniques forms the basis for advanced decision-making capabilities, wherein cars can perform maneuvers securely and with confidence. Optimization of data gathering and signal processing is therefore of profound significance in autonomous driving system development that can be trusted. Sensor fusion techniques are an important aspect of building perception systems in autonomous vehicles. By integrating data from multiple sensors, vehicle systems can have a richer and more accurate understanding of their environment than any individual sensor. Merging LIDAR data with vision data, for example, can improve object detection in low visibility conditions by avoiding some of the intrinsic limitations of individual systems. Also, model-free control methods, as discussed in recent research (D'Andréa-Novel et al., 2015), enhance vehicle response and trajectory following without using complicated mathematical models. Path and trajectory planning are also elementary, as they allow for safe navigation through various and changing environments (Maurer et al., 2017). The cutting-edge integration of sensor methods, thus, not only upgrades perception in abilities but also overall safety and reliability in autonomous vehicles.

3.6.2. Role in Smart Home Technology

Smart home technology development is highly reliant on a sophisticated web of sensors designed to properly sense the environment. A broad spectrum of types of sensors, from vision systems and LIDAR to inertial measurement units (IMUs), enable the acquisition of real-time data upon which reactive automation can act. Sensors are combined with data acquisition processes that transform raw data into actionable information. The expert integration of sensors allows sophisticated signal processing to detect user intent and environmental change. For instance, vision sensors can detect movement to enhance security systems, while LIDAR provides spatial awareness, assisting in navigation and obstacle avoidance. Leverage of strengths offered by various sensor technologies enables smart homes to achieve a level of interactivity and responsiveness which significantly boosts user experience and satisfaction. The concurrent and perpetual functioning of home automation ecosystems is also rendered smoother by the actuation function, which is the executing arm of control systems. Actuators translate the processed data into real-life actions like switching lights on and off, manipulating temperature, or running security systems. This harmonious interplay between sensors and actuators is instrumental to unleashing the full potential of home automation. For example, if a motion detector detects activity in a room, the relevant actuators can be designed to adjust lighting levels automatically to enhance brightness, creating an adaptive space. Furthermore, actuator systems can be tailored to learn a user's routines and preferences, thereby enabling more personalized interactions. This integration leads to an integrated environment where comfort, functionality, and security are harmonized, ultimately enhancing the users' overall quality of life. The Cyber-Physical Social Machines concept is a unifying idea for describing the issues with human interaction in smart homes, where the integration of sensors and actuators is critical. The new paradigm illustrates the way humans and technology unite to facilitate natural interactions and sophisticated automated systems. As technology advances, concerns regarding security and privacy in such systems cannot be overlooked. A critical aspect of smart homes is the need for effective security systems that protect confidential information alongside meeting the needs of convenience and efficiency in automation. A proper understanding of human-robot interaction (HRI) is crucial as it factors into the adoption of such smart systems by consumers. This technological capability and user experience interaction demonstrates the ongoing necessity for innovation in HRI incorporating flexibility and personalization, ultimately dictating the success of smart home technology within modern society.

3.6.3. Applications in Industrial Automation

In industrial automation, the integration of high-level sensor systems has transformed operational efficiency and precision. Various types of sensors, such as vision systems, LIDAR, and inertial measurement units (IMUs), are of utmost importance here. The sensors collect data about the environment and the health of machinery, enabling real-time monitoring and decision-making. For example, vision systems can detect flaws on a manufacturing line, while LIDAR provides high-resolution spatial awareness, enabling navigation and collision avoidance in self-driving cars. Such capabilities for data collection allow automated systems to adapt dynamically to changes, creating environments in which productivity and quality can be optimized through rapid adjustments to processes. Data interpretation from sensors taken is made possible by sophisticated signal processing techniques transposing raw data into useful information. Industrial automation systems utilize algorithms utilized in signal filtering and interpreting as well as eliminating noise to reveal meaningful information. Data fusion techniques also intervene by correlating data acquired from various sensors to yield a more profound perception of the environment. Through sensor fusion, a composite view may be constructed, enhancing accuracy in object position and condition detection, and hence optimizing processes. Through effective aggregation of various streams of data, automation systems can advance machine synchronization and reduce errors in tasks, as demanded in high-hazard manufacturing environments where reliability comes first. In addition, the Cyber-Physical Social Machines phenomenon is transforming human-machine interactions in industrial settings. These systems employ Internet of Things (IoT) actuators and sensors to have an active relationship that goes beyond simple automation. Integrating these systems involves overcoming the problem of communication security as well as privacy of data because the systems actively engage in real-world processes. As organizations adopt these kinds of systems, the robust capabilities of sensor networks enable more robust cooperation among workers and automated systems. This evolution not only improves productivity but also new innovation opportunities, further increasing the demand for overall safety controls and ethical considerations in future industrial automated environments (Creese et al., 2018; Crowder et al., 2006).

SUMMARY

The advancement of actuators, sensors, and perception systems is a groundbreaking step in the evolution of automation and robots, serving as the foundation for more efficient and reliable applications. Such technology enables complex machine-world interactions by facilitating data acquisition in various forms like vision, LIDAR, and IMU sensors. Such diversification not only enhances situational awareness but also allows systems to process and respond to changing conditions in real time. Through the use of sensor fusion technologies, heterogeneous data sources can be reconciled, allowing for enhanced decision-making capabilities. The technology continues to advance, offering increasingly precise solutions for solving real-world problems associated with automation, thereby transforming the terrain of autonomous systems and their applications. The variation in performance of various actuators also complements the effectiveness of perception systems in real-world usage. Different actuators, hydraulic, pneumatic, or electrical, provide certain advantages in terms of responsiveness and adaptability in operations ranging from straightforward motion to highly sophisticated interactions. As exhibited in soft robot advancements that capitalize on compliant mechanisms for grasping and manipulating objects, ongoing improvement in actuation technologies brings forth increased performance in dynamic situations. In particular, soft robot grippers with bistability enable sensor-less dynamic grasping, achieving remarkable speed and efficiency in delicate operations with minimal computational burden (Abidi et al., 2019). These innovations are a good representation of the fine synergy between functionality and simplicity that contemporary engineering is determined to accomplish. Moreover, the convergence of sensor technologies and new actuation methodologies illustrates the significance of ongoing research and competition in advancing the field of autonomous systems. Such activities as robotic competitions for autonomous toy cars have not only provided environments for testing and validating ideas about algorithms but also encouraged interdisciplinary teamwork (Berger et al., 2014). Such settings provide a rich soil for experimentation, leading to the enhancement of both hardware and software design in ways that reduce development risk materially. These activities place emphasis upon the need for bridging theoretical underpinnings with practical application, ultimately leading to advancements in vehicle technologies that promise a safer, more efficient mobility future. The integration of these considerations provides a firm foundation for the continued evolution of sensors, actuators, and perception systems that will determine the course of automation for the years to come.

Sensors integration is very important in automating robotic platforms and autonomous vehicles, allowing for accurate perception of the world. Out of all the sensors used, vision systems are important to recognize and interact with objects, and LIDAR provides spatial maps in exquisite detail by measuring distances with laser light. Inertial measurement units (IMUs) and global positioning systems (GPS) provide crucial information regarding orientation and positioning, respectively. All of these sensors combined offer a wide range of data points required for autonomous operation decision-making mechanisms. The choice of the sensors greatly relies on the overall system performance and adaptability to different operating conditions, putting much focus on how sensor technologies are selected based on activity and environment. Data sensing and signal processing are the foundational elements in using sensor data in perception systems. Raw data collected with multiple sensors typically contain noise and require sophisticated processing to convert them into meaningful information. Signal processing techniques enhance data quality and usability via filtering, normalization, and transformation processes. Moreover, high-fidelity data acquisition ensures measurements to be actual representations of environmental conditions. The efficiency of these processes has direct ramifications on the dependability of the systems outputs; therefore, these processes must be optimized in order to achieve robust performance in safety-critical applications. Therefore, knowledge of the principles on which signal processing and data acquisition are based is essential for successful development of high-end sensor systems. Sensor fusion theory seems to be a strong way of enhancing environmental perception where data from various sensors is merged to provide an integrated perception of the environment. Sensor fusion leverages the strength of individual sensors to reduce constraints and decrease uncertainty associated with separate data sources. For instance, combining data from vision and LIDAR sensors can create rich environmental maps incorporating color and depth and improving navigation and decision-making capability. Also, advancements in machine learning algorithms have augmented sensor fusion techniques such that systems can learn over time and improve performance. The resultant layered perception enhances situational awareness and aids the development of additional autonomous and smart systems, demonstrating the ongoing innovation and necessity of sensor fusion within modern vehicle automation and robotic systems.

REFERENCES

1. Abidi, S. H., Cianchetti, M., Falotico, E., Laschi, C., et al. (2019). A bistable soft gripper with mechanically embedded sensing and actuation for fast closed-loop grasping. https://core.ac.uk/download/543788444.pdf (accessed on 10th September 2025).
2. Ahmed, M. A., Gallardo, J. L., Zúñiga, M., Pedraza, M. A., Carvajal, G., Jara, N., & Carvajal, R. (2022). LoRa-based IoT platform for remote monitoring of large-scale agriculture farms in Chile. *Sensors, 22*, 2824. https://doi.org/10.3390/s22082824 (accessed on 10th September 2025).
3. Aiken, E. W., Denery, D. G., Eshow, M. M., Hindson, W., et al. (2025). Rotorcraft in-flight simulation research at NASA Ames Research Center: A review of the 1980s and plans for the 1990s. https://core.ac.uk/download/pdf/42811995.pdf (accessed on 10th September 2025).
4. Archimède, B., & Desforges, X. (2006). Multi-agent framework based on smart sensors/actuators for machine tools control and monitoring. https://core.ac.uk/download/12041379.pdf (accessed on 10th September 2025).
5. Askarian, M., Benítez Iglesias, R., Graells Sobré, M., Zarghami, A., et al. (2016). Data-based fault detection in chemical processes: Managing records with operator intervention and uncertain labels. https://core.ac.uk/download/81571354.pdf (accessed on 10th September 2025).
6. Ay, N., Martius, G., & Zahedi, K. (2013). Linear combination of one-step predictive information with an external reward in an episodic policy gradient setting: A critical analysis. https://core.ac.uk/download/pdf/82833004.pdf (accessed on 10th September 2025).
7. Bacciu, D., Bahr, R., Bröring, A., Cavallo, F., et al. (2017). Internet of robotic things: Converging sensing/actuating, hypoconnectivity, artificial intelligence and IoT platforms. https://core.ac.uk/download/84045908.pdf (accessed on 10th September 2025).
8. Badii, A., Glackin, C., Khan, A., Leu, F., et al. (2013). CORBYS cognitive control architecture for robotic follower. https://core.ac.uk/download/29850548.pdf (accessed on 10th September 2025).
9. Beck, S. M., Breckenridge, R. A., Clark, L. G., Collier, R., et al. (2025). Technology for the future: In-space technology experiments program, part 2. https://core.ac.uk/download/pdf/42817270.pdf (accessed on 10th September 2025).
10. Berger, C. (2014). From a competition for self-driving miniature cars to a standardized experimental platform: Concept, models, architecture, and evaluation. http://arxiv.org/abs/1406.7768 (accessed on 10th September 2025).

11. Brochu, P., Jing, Y., Kalajian, M., Kargar, M., et al. (2013). Tactile feedback display with spatial and temporal resolutions. https://core.ac.uk/download/323078717.pdf (accessed on 10th September 2025).
12. Carlin, C. M., Lambregts, A. A., & Ray, J. K. (2025). High-speed civil transport flight- and propulsion-control technological issues. https://core.ac.uk/download/pdf/42813729.pdf (accessed on 10th September 2025).
13. Chang, J., & Fiorini, P. (2025). A procedure concept for local reflex control of grasping. https://core.ac.uk/download/pdf/42821752.pdf (accessed on 10th September 2025).
14. Copeland, R., Crespi, N., Rehmani, M. H., Saleem, Y., et al. (2019). Internet of Things-aided smart grid: Technologies, architectures, applications, prototypes, and future research directions. http://arxiv.org/abs/1704.08977 (accessed on 10th September 2025).
15. Creese, S., De Roure, D., Hall, W., Madaan, A., et al. (2018). A storm in an IoT cup: The emergence of cyber-physical social machines. http://arxiv.org/abs/1809.05904 (accessed on 10th September 2025).
16. Crowder, R., & Dubey, V. (2006). A dynamic tactile sensor on photo elastic effect.
17. D'Andréa-Novel, B., Fliess, M., Gruyer, D., Menhour, L., et al. (2015). A new model-free design for vehicle control and its validation through an advanced simulation platform. http://arxiv.org/abs/1503.07026 (accessed on 10th September 2025).
18. Dautenhahn, K., Dragone, M., & Saunders, J. (2015). On the integration of adaptive and interactive robotic smart spaces. https://core.ac.uk/download/29853781.pdf (accessed on 10th September 2025).
19. Dipippo, S., Labini, G. S., & Pasquariello, G. (2025). Information management in an integrated space telerobot. https://core.ac.uk/download/pdf/42821745.pdf (accessed on 10th September 2025).
20. Duda, K. R., Hansberry, M., Jacobs, S. E., Middleton, A., et al. (2025). Variable vector countermeasure suit (V2Suit) for space habitation and exploration. https://core.ac.uk/download/pdf/189597222.pdf (accessed on 10th September 2025).
21. Eberle, H., Hayashi, Y., Jing, X., Madgwick, S., et al. (2019). Formulation of a new gradient descent MARG orientation algorithm: Case study on robot teleoperation. https://core.ac.uk/download/210582747.pdf (accessed on 10th September 2025).
22. Fluitman, J. (1996). Microsystems technology: Objectives. https://core.ac.uk/download/pdf/11453403.pdf (accessed on 10th September 2025).

23. Glackin, C., Greaves, M., Polani, D., Ristić-Durrant, T., et al. (2014). Information-theoretic measures as a generic approach to human-robot interaction: Application in CORBYS project. https://core.ac.uk/download/29850402.pdf (accessed on 10th September 2025).

24. Gordon, T. J., & Lidberg, M. (2015). Automated driving and autonomous functions on road vehicles. https://core.ac.uk/download/29177272.pdf (accessed on 10th September 2025).

25. Grissom, W. A., & Rafla, N. I. (2025). Ground robotic hand applications for the space program study (GRASP). https://core.ac.uk/download/pdf/42805521.pdf (accessed on 10th September 2025).

26. Lachmayer, F., & Čyras, V. (2012). Multisensory legal machines and legal act production. https://core.ac.uk/download/14524802.pdf (accessed on 10th September 2025).

27. Liñán Cembrano, G., Maldonado López, R., Roca Moreno, E., Rodríguez Vázquez, Á., et al. (2005). Tactile on-chip pre-processing with techniques from artificial retinas. https://core.ac.uk/download/286565282.pdf (accessed on 10th September 2025).

28. Maurer, M., Nolte, M., Rose, M., Stolte, T., et al. (2017). Model predictive control-based trajectory generation for autonomous vehicles–An architectural approach. http://arxiv.org/abs/1708.02518 (accessed on 10th September 2025).

29. Nyagudi, N. M. (2013). Post-Westgate SWAT: C4ISTAR architectural framework for autonomous network integrated multifaceted warfighting solutions Version 1.0: A peer-reviewed monograph. http://arxiv.org/abs/1401.6379 (accessed on 10th September 2025).

30. Shaer, O., & Hornecker, E. (2009). Tangible user interfaces: Past, present and future directions. https://core.ac.uk/download/9033674.pdf (accessed on 10th September 2025).

31. Vermesan, O., John, R., Pype, P., Daalderop, G. H. O., Kriegel, K., Mitić, G., Lorentz, V., et al. (2021). Automotive intelligence embedded in electric connected autonomous and shared vehicles technology for sustainable green mobility. *Frontiers in Future Transportation, 2*. https://doi.org/10.3389/ffutr.2021.688482 (accessed on 10th September 2025).

32. Yang, G. Z., Bellingham, J., Dupont, P. E., Fischer, P., Floridi, L., Full, R. J., Jacobstein, N., et al. (2018). The grand challenges of *Science Robotics*. *Science Robotics, 3*. https://doi.org/10.1126/scirobotics.aar7650 (accessed on 10th September 2025).

33. Yurtsever, E., Lambert, J., Carballo, A., & Takeda, K. (2020). A survey of autonomous driving: Common practices and emerging technologies. *IEEE Access, 8*, 58443–58469. https://doi.org/10.1109/access.2020.2983149 (accessed on 10th September 2025).

34. Zhang, C., Yang, C., Xu, J., Liu, S., Liu, T., Zhai, X., Xie, X., et al. (2017). A review of smart materials in tactile actuators for information delivery. http://arxiv.org/abs/1708.07077 (accessed on 10th September 2025).

CHAPTER 4

CONTROL ARCHITECTURES FOR AUTONOMOUS SYSTEMS

Contents

Introduction 123
4.1. Open-Loop Control Systems 127
4.2. Closed-Loop Control Systems 131
4.3. Hierarchical Control Architectures 136
4.4. Distributed Control Architectures 141
4.5. Feedback Mechanisms in Intelligent Systems 147
4.6. Comparison of Control Architectures 152
Summary 158
References 160

INTRODUCTION

The developing environment of autonomous systems is supported by heterogeneous control architectures that determine how these agents perceive and react to their worlds. Familiarity with the underlying differences between open-loop and closed-loop control systems is essential; the former acts without feedback, making it easier but less responsive, whereas the latter includes feedback to improve decision-making and system robustness. Such richness makes it possible for closed-loop systems to adjust in real time, adapting to prevailing environmental conditions. Emerging technologies have emerged the importance of robust feedback control in smart systems, bringing responsiveness and dependability in uncertain environments. Hierarchical and distributed control architecture are two prominent paradigms in autonomous system control. Hierarchical control systems, with a top-down approach, optimize decision-making efficiency by breaking tasks into multiple layers of command; this approach can streamline operations but becomes limited by bottlenecks in information transfer. Distributed control architectures, in contrast, offer a decentralized approach where individual agents cooperate to steer tasks, potentially generating higher resiliency and flexibility under variant conditions.

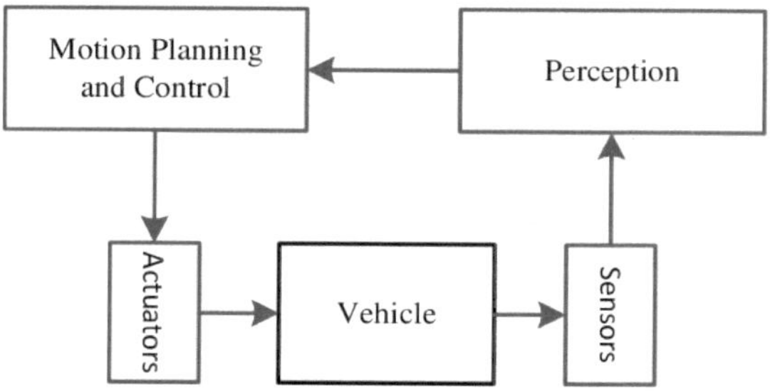

Figure 4.1. *High-level diagram of the control system architecture of an autonomous vehicle.*

Source: https://www.researchgate.net/profile/Constantin-Ilas/publication/261280271/figure/fig1/AS:668937104670720@1536498442051/High-level-diagram-of-the-control-system-architecture-of-an-autonomous-robot.png.

The selective distinction between the two systems depends upon the inherent needs of the application since each possesses inherent strengths and weaknesses that need to be balanced against the overall goals of the larger system. One of the essential components in the development of autonomous systems is the dynamic interplay among feedback loops that facilitate continuous learning and adaptation. As artificial intelligence continues to explode, new architectures have gravitated toward integrating intelligent decision-making in autonomous systems. Postulates of Perceptual Control Theory (PCT) are becoming increasingly influential in the structure of these systems, focusing on the necessity to align behavior with simultaneous environmental feedback. This intersection is simple, as it broadens the ability of intelligent agents to simulate behavior in a significant manner to allow them to respond suitably to advanced human-robot interactions. The ongoing synthesis of sophisticated feedback mechanisms will undoubtedly continue influencing the course of autonomous systems, bringing about a paradigm shift towards more intelligent and responsive actions in an extensive variety of applications (Figure 4.1).

Definition of Autonomous Systems

The definition of autonomous systems entails a range of technologies that can independently operate with limited intervention from humans. These systems possess different control architectures to perform in complex worlds and perform tasks independently. With developments in technology, the perception of autonomy has evolved as artificial intelligence, machine learning, and robotics have improved. The ability of such systems to perceive their surroundings, process information, and make choices is the secret to their operational efficiency. In particular, attention to feedback mechanisms during autonomous system development is an important aspect in ensuring their flexibility and reliability in operation. Control architectures are organizational structures in autonomous systems, defining how such systems process information and interact with outer worlds. Open-loop control systems are those that make no use of feedback but follow pre-defined trajectories, which can be functional in plain operations but are vulnerable to errors based on unforeseen variables. Closed-loop control systems, however, employ feedback measures, continuously adjusting from real-time data to enhance performance as well as decision-making capacity. Understanding the distinction between the two types of control is essential in the design of autonomous systems able to adapt and respond to changing conditions across a wide variety of applications from autonomous cars to robot-operated manufacturing processes. The new designs for autonomous systems incorporate hierarchical and distributed control architectures, both of which have some special advantages and challenges. Hierarchical control designs organize decision-making structures in a tiered arrangement to facilitate complex tasks to be divided into chunks that are easy to manage, hence ensuring smooth operations. Distributed control systems, on the other hand, provide independence to individual agents to interact with each other and decide on their own, ensuring strength and flexibility in dynamic environments. Application of such frameworks typically incorporates principles borrowed from the General Systems Theory (GST), according to research aimed at consolidating control approaches in cognitive systems. Implementation of integrated control mechanisms can significantly enhance the operational efficiency of autonomous systems and satisfy demands for real-time processing and adaptive response in complex scenarios, as seen in technological developments pioneered by organizations like NASA (Beck et al., 2025) and recommendations from research into cognitive architectures (Hernandez et al., 2008).

Importance of Control Architectures

The significance of control architectures in autonomous systems is more than technical in nature; it defines the way such systems interact with and exist within the environments they operate within. Control architectures determine how systems gather information and react to their environments, impacting performance, reliability, and adaptability. In the area of intelligent systems, architectures uniting control paradigms in multiplicity, such as open-loop and closed-loop control, are essential to improving feasible decision-making within dynamic environments. The evolution towards advanced architectures, which incorporate feedback loops, enhances the capacity of systems to learn and adapt over time, thereby promoting efficiency and accuracy in activity execution. This is the focus of recent work that aims to develop communicative behavior in robots, which relies on a robust understanding of perceptual control theory. This trend is indicative of the necessity of highly structured control architectures for facilitating intelligent interaction with people and environments (Arkin et al., 2020). Moreover, the contrast between hierarchical and distributed control paradigms is central in determining the efficiency and scalability of autonomous systems. Hierarchical control structures facilitate centralized decision-making, where top-level commands are decomposed into manageable tasks at lower levels, thus facilitating coordinated behavior of complicated systems. Distributed control, in contrast, allows autonomous systems to operate more independently, making them more resilient to single points of failure and facilitating problem-solving flexibility.

Each architecture has its own advantages; for instance, in autonomous vehicles, hierarchical architectures can help demystify navigation protocols, and distributed systems can efficiently manage real-time feeds of data from large numbers of sensors. There is a need to combine these paradigms, especially in sectors like aerospace, where new automated technologies are needed to explore new horizons such as space technology (Beck et al., 2025). Finally, feedback mechanisms are also of significance in maximizing autonomous systems' performance by allowing them to make decisions based on current information. Feedback mechanisms allow systems to learn and improve progressively through interactions with their worlds, thus enhancing decision-making. This iterative learning pattern is quickly becoming essential as systems try to master more sophisticated functionalities, e.g., human-robot collaboration in tasks requiring communicative behavior. As suggested, a strong architecture founded on perceptual control principles can lead to inventions in intelligent communicative agents, with implication that feedback is not just an operational necessity but a key element within adaptive control structures. Lastly, embedding feedback mechanisms within various control architectures is a move towards increasingly responsive, agile, and smart autonomous systems, an astronomical leap in engineering as well as AI applications (Arkin et al., 2020).

4.1. OPEN-LOOP CONTROL SYSTEMS

Open-loop control systems constitute a fundamental form in the context of autonomous control arrangements, characterized by a lack of feedback. The control actions in open-loop systems are pre-set and executed without monitoring the output to adjust it. This arrangement often leads to systems that are not so sensitive to variations in the environment at the current time, a severe shortcoming in dynamic environments. For instance, in cases involving rapid adaptation, the absence of feedback can cause suboptimal performance or inability to achieve intended outcomes. Thus, while open-loop systems may operate superbly under stable conditions, their rigidity emphasizes the need for integration of feedback systems with an ability to adaptively optimize system responses and boost overall effectiveness as noted by Liu et al. (2025). Despite their limitation, open-loop systems possess a number of benefits when employed in certain applications, particularly in those processes whose environment is predictable and well-organized. Manufacturing sectors prefer employing open-loop techniques in straightforward operations, such as conveyor belts, where complex feedback is unnecessary. With the identification of precise control inputs, these systems can be quite efficient without the use of complex feedback loops. Also, the comparative ease of open-loop control can also yield simplicity of implementation and fault-finding. This formulation is compatible with sequential decision-making paradigms used in distributed control systems, where the deterministic nature of open-loop systems can facilitate transparency in operation and reduce levels of computation, as seen in examples like those given by Liu et al. (2025). In the shifting landscape of intelligent systems, the presence of open-loop and closed-loop control highlights basic issues to be addressed in future development. While the inherent simplicity of the former is apt for some uses, incorporation of feedback mechanisms is increasingly crucial for achieving flexibility in complex, variable settings. Use of AI-driven frameworks, such as those referenced by Elkael et al. (2025), implies a new direction with the utilization of smart agents to interpret and learn from real-time data feeds. This breakthrough approach might revolutionize open-loop structures so that they can function as part of an extended adaptive ecosystem. As researchers move toward architectures that borrow elements from both control paradigms, the path toward adaptive, robust autonomous systems is more visible, calling for balanced approaches to design that leverage the strengths of both open and closed-loop controls (Figure 4.2).

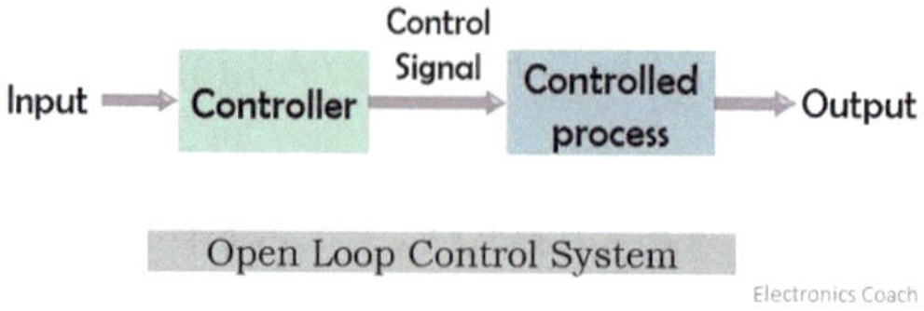

Figure 4.2. *Open-loop control system.*

Source: https://electronicscoach.com/wp-content/uploads/2019/11/open-loop-control-system-1.jpg.

4.1.1. Characteristics of Open-Loop Systems

Open-loop systems are characterized by the way they operate without using feedback in an attempt to change their output based on the actual performance or environmental conditions. Since open-loop systems do not use feedback, they are extremely simple and uncomplicated and operate only on programmed guidelines for system operation. In autonomous systems, these types of systems can be employed in cases where environmental spontaneity is minimal or deterministic. For instance, washing machines or factory conveyor belts illustrate open-loop control, in which processes are executed on the basis of a pre-programmed algorithm with no accommodation for variations of input or output. Although the straightforward approach can work well in clearly defined processes, it limits responsiveness to changing conditions, a determining factor in dynamic circumstances typical of intelligent systems (Albayrak et al., 2018). The lack of feedback also means open-loop systems do not monitor their performance and therefore are likely to have errors or inefficiency when an unplanned change is made. This deficiency can lead to fatigue of the system, which is a requirement for conscious design considerations, particularly in systems where system reliability is critical. In complex autonomous systems, for example, multi-robot systems under emergent behaviors, the absence of feedback mechanisms can significantly disable adaptability and coordinated action to feedback from the environment. As it has been observed in recent studies, the integration of self-organization principles typical for swarm robotics can enhance the effectiveness of open-loop systems by enabling them to adapt behavior to some extent, though without explicit feedback (Andrawos et al., 2019). This research aims to find out how the inherent liabilities of open-loop systems are reinforced by the promotion of emergent behavior complementary to the static demands of traditional open-loop designs. Moreover, the process of designing open-loop control systems entails careful consideration of how and under what circumstances they will be operating, with their performance assessed in terms of desired outcomes rather than responsive modification. Consequently, when building the system, the engineers must examine the expected operating conditions in detail, creating algorithms that will perform optimally within those conditions. While the openness of open-loop systems presents an advantage in implementation, this also calls for extensive testing in order to ensure that there are no traps in operating the system. The exploration of transitioning to more complex models, such as hybrid systems with feedback loops, is an exciting possibility for optimizing the performance of the system. Welcoming such innovations has the potential to improve system robustness and flexibility and eventually close the gap between open-loop predictability and the flexibility in closed-loop systems (Albayrak et al., 2018; Andrawos et al., 2019).

4.1.2. Advantages of Open-Loop Control

One of the strengths of open-loop control in the context of autonomous systems control architectures is simplicity and ease of implementation, which is one of the primary advantages. Open-loop systems do not employ feedback mechanisms, and thus design and control action implementation are readily possible. As such, they are extremely useful in situations where system dynamics are understood and can be predicted. When conditions within the environment are static, open-loop control is able to regulate operations with full effect without the need for continuous feedback or adjustment. This simplicity reduces computational requirements and costs of employing intricate feedback mechanisms, making it a desirable solution for some processes such as simple automatic routines or mundane jobs. Consequently, this can result in steady performance in conditions where parameters remain unchanged, and stability in system operations is provided. The other significant advantage of open-loop control is the response time. Since these systems do not have the slow feedback loops typical of closed-loop systems, they can issue commands with minimal delay. The near-instant execution of commands comes in particularly handy in applications that require real-time performance, such as in certain robotics or automated manufacturing applications. For instance, in production lines where machines must operate at high speed and in rhythmic sequence, open-loop control systems can maintain operations without delay due to incessant analysis of output errors. This rapid response can boost productivity and efficiency levels so that systems can meet high demand without sacrificing operational integrity and performance. Furthermore, the cost-effectiveness of employing open-loop control systems cannot be underestimated. Since these systems tend to have fewer sensors and less sophisticated infrastructure than closed-loop systems, they can significantly lower initial capital expenditures for companies.

This cost-effectiveness offers a chance to smaller business or lower-resource companies to adopt automation technologies and thereby increase productivity across different industries. In addition, as the conditions of working change, adjusting open-loop systems might be simpler than adjusting closed-loop systems since they are less complicated. This field aligns with the increasing need for low-barrier technology, as simpler control architectures can be more accessible and less daunting for new entrants into the market. Accordingly, while open-loop control may not be usable in all areas, the advantages of open-loop control make it an attractive option for a wide variety of autonomous applications across a wide array of markets.

4.1.3. Applications of Open-Loop Control in Autonomous Systems

The applications of open-loop control systems in autonomous systems typically take place in contexts where precise timing and sequences of action are essential, but feedback mechanisms are impracticable or do not exist. An example is the application of open-loop control in robotic arm movement in assembly processes, in which pre-programmed motion works to propel machinery through a predetermined sequence to perform specific functions. This strategy provides more efficiency and predictability, particularly on mass-production lines under relatively stable environments. The application of precisely defined command sequences is the best indication of open-loop system operating efficacies and the simplicity that negates the necessity for dynamic adjustments, thereby minimizing complexity. The constant improvement and application of open-loop control therefore provide important operating advantages, guaranteeing reliability and performance in automatic operations. In educational research on open-loop control systems, their set of features makes knowledge into their utilization. The design of open-loop systems is an intrinsic construct in macro-control structures, suggesting different knowledge engagement levels among groups of individuals. This is particularly relevant when referring to knowledge transfer initiatives in organizations that employ independent systems. The distinct generational traits bring about differences in knowledge uptake and operational practices, creating challenges in the successful adoption of open-loop control methodologies. Handling the complexities is important for successful future development and application of such control systems in intelligent systems. As organizations increasingly rely on automation, creation of an impression of open-loop systems through different mechanisms of knowledge transfer becomes important for both the current and future generations. In addition, the frontier of open-loop control and neuromechanics is of significant interest for autonomous system innovation. Open-loop control systems, although typically limited by their incapacity to react to evolving situations, can leverage understanding gained from neuromechanics, which examines dynamic interactions among physical structures and task requirements. As suggested, specifying the balance among under- and over-determined mechanics offers a backdrop to optimize the autonomous system design efficiency. It invites the exploration of maximum performance versus established behavior so that engineers can design sophisticated models. Through the application of these results, industries and universities can pursue sophisticated training programs that will yield a better-informed study of the uses and consequences of open-loop control, thereby driving innovation for autonomous systems (Bernardin et al., 2020; Santello et al., 2017) (Figure 4.3).

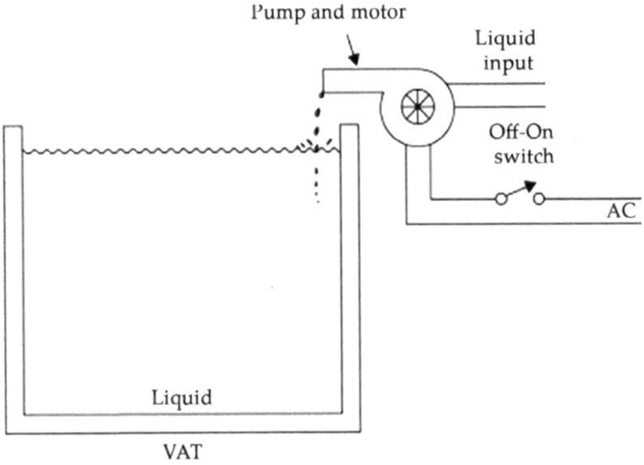

Figure 4.3. *Open-loop control system for filling a vat with liquid via a pump.*

Source: https://ars.els-cdn.com/content/image/3-s2.0-B9780128116418000126-f12-02-9780128116418.jpg.

4.2. CLOSED-LOOP CONTROL SYSTEMS

Closed-loop control systems play a central role in autonomous system creation and use through response adaptation based on feedback loops. In comparison with open-loop systems that function exclusively based on pre-determined commands without any adjustment based on outside conditions, closed-loop systems dynamically monitor performance and change actions in real time. Closed-loop systems' adaptability is crucial in dynamic situations where variables can change unexpectedly. Implementation of the feedback loops in closed-loop systems allows the independent agents not only to execute functions but also to quantify their performance. Neuromechanics on the rise brings out the fact that concrete real-world interactions dictate behavior, illustrating how effective performance is achieved by closed-loop systems by adapting to emergent conditions and not necessarily rigid pre-programmed rules (Santello et al., 2017). Closed-loop systems are also demonstrated in their application to intelligent systems that require robust communicative behaviors, i.e., human-robot interaction. With research improving, the problem of natural interaction between human beings and robots shows the need for complex feedback systems with the ability to process and react to subtle human cues. Using concepts from Perceptual Control Theory (PCT), closed-loop control systems can be designed to prefer needs-directed communication, and therefore address how they might make responses simulated in a human-like fashion. The models have the potential to bridge the gap between high-level rule-based programming and more sophisticated, embodied architectures that engage with actual physical worlds (Arkin et al., 2020).

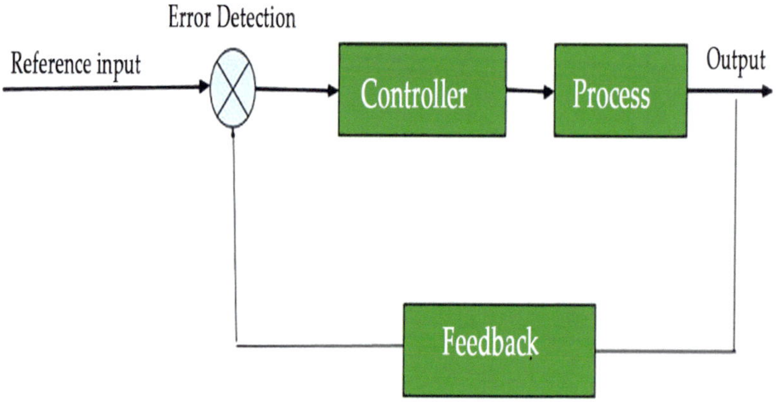

Figure 4.4. *Closed-loop control system.*

Source: https://media.geeksforgeeks.org/wp-content/uploads/20231109204652/closed-loop-control-system-660.jpg.

Thus, closed-loop control systems play a growing role not only in the accomplishment of tasks but also in enabling relational dynamics among autonomous systems and their human counterparts. In addition, the changing environment in robotics mirrors the theoretical advances within the area of closed-loop control systems. With scientists acknowledging the convergence of biological and robotic strategies toward grasping, the realization of shared but fundamentally disparate mechanisms raises questions regarding control architecture design. The difference between under-determined and over-determined mechanics, as well as the limitations of balancing learning and adaptation, demonstrate the character of the intricacies present in the design of effective closed-loop systems. The greater understanding of these dynamics enables engineers to optimize the performance of autonomous systems to effectively juggle a variety of activities while learning to accomplish unexpected obstacles in their environments (Santello et al., 2017). The continued examination of these intricacies further establishes the place of closed-loop control systems in bringing about the lofty vision of highly autonomous intelligent agents (Figure 4.4).

4.2.1. Characteristics of Closed-Loop Systems

Closed-loop system features are intrinsic to their functionality in autonomous systems and distinguish them from open-loop systems. Closed-loop systems employ feedback, so they can continually monitor output and adjust input to meet this, thereby optimizing performance and stability. Self-correction is particularly beneficial in dynamic systems, where shifting conditions in the outside environment require real-time adjustments. For example, an autonomous vehicle employs sensors to quantify its location along the road so that it can adjust its trajectory instantly. This correlation of input and output not only contributes to accuracy but also to the rigidity of the control system architecture so that systems can operate as intended under a range of operation conditions. The last feature of closed systems is the ability to incorporate learning and adaptation. As opposed to open-loop systems which execute preprogrammed orders, closed-loop configurations are capable of adapting their behavior based on historical data and gained outcomes. Such real-time learning capability is most essential for intelligent systems which must act in unstructured and ambiguous surroundings. For instance, manipulator robots can learn from previous experiences and modify their methods such that they become increasingly efficient over time. Neuromechanics theory describes the intricate relationship between physical structure, task requirements, and neural control processes involved in such adaptive systems. Such a perspective is compatible with contemporary efforts at unification of biological and engineering paradigms, emphasizing convergences while acknowledging the distinctive engineering challenges involved in the design and implementation of adaptive robot systems (Santello et al., 2017). In addition, the hierarchical organization of closed-loop systems provides for greater coordination and delegation between their structure. Layering of control within closed-loop systems can easily oversee various aspects of feedback and processing from the lower-level sensors and actuators to higher cognitive levels that support decision-making and planning. Distributed control in this manner ensures that complicated tasks are controlled as a whole, optimizing resource usage and responsiveness. In the context of space exploration technology, the importance of this hierarchical paradigm is apparent, as it becomes possible to perform complex operations that require precision under extreme conditions. The associated technology needs of such missions underscore the need for advanced automation and robotics, in which closed-loop devices are capable of constantly adapting in accordance with challenges unique to the in-space environment (Beck et al., 2025). Hence, understanding such characteristics of closed-loop systems is very important for control architecture development in autonomous systems.

4.2.2. Advantages of Closed-Loop Control

The advantages of closed-loop control systems are particularly glaring as far as enhancing the performance and resilience of autonomous systems are concerned. Compared to open-loop systems, in which case no feedback exists, closed-loop systems dynamically adjust their operations according to real-time monitoring of conditions of output. This capability facilitates improved accuracy and efficiency because the system has the opportunity to make adjustments for any such deviation from desired performance. For instance, in such systems that require maximum accuracy, such as in robotics or autonomous transportation, closed-loop control ensures minimal error and hence more reliable outcomes. This is also indicated by the incorporation of neuromechanics in robots to ensure the emphasis placed on feedback when achieving maximum performance despite inherent mechanical complexities and natural environment variations, according to Santello et al. (2017). Besides, closed-loop control offers disturbance and uncertainty robustness, something that is critical to autonomous systems operating in uncertain environments. By incorporating feedback mechanisms, such systems can automatically adapt to variations in the environment, e.g., load fluctuations, unexpected barriers, or variable operating conditions. This adaptability is at the core in production environments, particularly with emerging technologies like Industrial Additive Manufacturing, where prompt response to current data is critical for sustainability and productivity. Therefore, closed-loop control can effectively manage interactions among various variables impacting production, allowing for maximum resource allocation and system performance, as addressed by Eyers et al. (2017). Furthermore, closed-loop control systems have a significance that extends to improved performance; they also give learning and adaptation over time. This is particularly important in intelligent systems, where it is possible to learn from past performance and enhance working strategies to make a potent impact in the long run. By creating feedback loops, closed-loop systems are able to develop a deeper sense of their working environment, allowing them to make prudent adjustments that optimize functionality. This learning process narrows down the limitations of pre-programmed responses typical in open-loop systems, paving the way for more complex and self-maintained operations. Closed-loop control is thus essential in industries where continuous improvement and innovation are costly, aligning with the dynamic requirement of the technological era in modern-day autonomous systems (Figure 4.5).

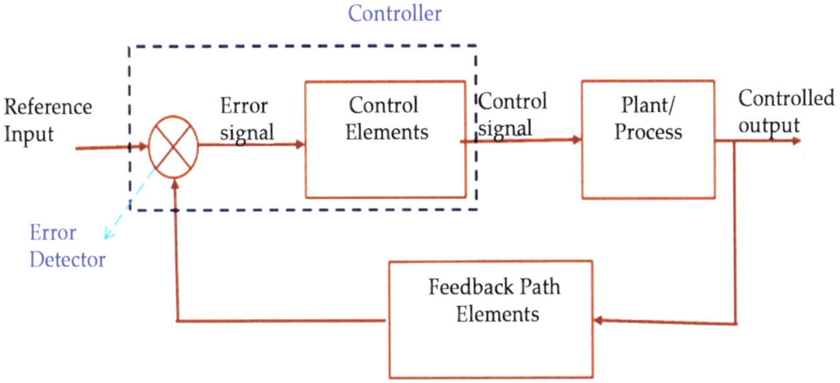

Figure 4.5. *Closed-loop control system, the one in which the output is fed back and compared with the desired output (reference input). If there is a deviation, the controller generates the necessary control signal to alter the conditions of the plant or process in order to achieve the desired output (controlled output).*

Source: https://www.quora.com/What-are-the-advantages-and-disadvantages-of-a-closed-loop-control-system.

4.2.3. Applications of Closed-Loop Control in Autonomous Systems

Closed-loop control in use within autonomous systems is most important in ensuring accuracy and reliability within the operation of complex procedures. In robotic applications, closed-loop control facilitates real-time feedback mechanisms that make continuous tunings of performance according to environmental factors. This ability to adapt is critical in contexts where external influences may change uncontrollably, such as in autonomous navigation. As noted in current developments, integration of hybrid algorithms can lead to significant improvement in cooperative behavior between the subsystems with an overall enhancement of robotic platform performance. The architecture currently in development that addresses decoupled building blocks enables the systems to be well-performing even during adverse conditions like in robotic competitions. By abstracting implementations from the hardware, the development enables processes to be streamlined and design innovation, eventually giving rise to autonomous systems across a range of applications for use in industry as well as research (Kloss et al., 2007). Besides, the use of closed-loop control is not limited to robots but can be extended to a range of automated systems, including industrial processes and smart technologies. Closed-loop control systems are employed to control desired operating levels, ensuring that systems respond in the right way to changing inputs. In industry, for instance, feedback mechanisms help regulate such parameters as temperature and pressure, optimizing efficiency and quality in production. Applications of machine vision and remote process control, as mentioned in recent studies, demonstrate the broad application of closed-loop systems in complex industrial designs. This adaptability is useful as it provides faultless integration into various contexts, conforming to the principles of Service Oriented Architecture (SOA) that promote platform independence and ease. Such approaches enhance autonomous system dependability, particularly in time criticality applications (Kloss et al., 2007). The merger of biological and robotic manipulation and grasp provides a fascinating perspective for closed-loop control applications. While biological systems are controlled by neuromechanical laws, robotic systems are controlled by artificially engineered feedback processes to execute their corresponding functions. Identifying the similarities and differences of these two paradigms opens the ground for further evolution of autonomous technologies. Learning from nature can impart lessons on new forms of control that are more flexible and resilient. For instance, understanding how living systems employ sensory feedback to refine motor actions can potentially translate to robotic control structures that are more leaning and adaptive rather than rigidly implemented. Closing these gaps allows researchers to produce more effective imitations of the nuanced response patterns of living systems, rendering robotic systems more efficient and more adaptable for real-world applications.

4.3. HIERARCHICAL CONTROL ARCHITECTURES

Hierarchical control structures are the pivotal idea in the control of autonomous systems since they facilitate structured decision-making and task achievement. The structures are designed to imitate structuring of complex systems, allocating tasks and responsibilities in layers that can be managed with ease. Strategic goals are determined at the pinnacle of the hierarchical framework, and tactical and operational details are carried out at lower levels. Such a structure not only facilitates efficient resource management but also ensures system robustness via local failure. Stratification of the control ensures specialized processes within each tier are simpler to implement, and the overall mission planning and operation complexity is minimized. Hierarchical control benefits are best seen in applications demanding adaptability and rapid response, such as in autonomous space missions where real-time decision-making is crucial to the success of the mission (Beck et al., 2025). Feedback mechanisms added in hierarchical control structures also make them more effective, facilitating real-time adjustment to a dynamic environment. Feedback mechanisms enable lower levels of the hierarchy to respond to disturbance and changes, passing implicitly required adjustments to higher levels, where they can reassess goal priority and resource allocation. Such collaboration between hierarchical organization and feedback elements brings about an integrated working framework to better manage uncertainties in intelligent systems.

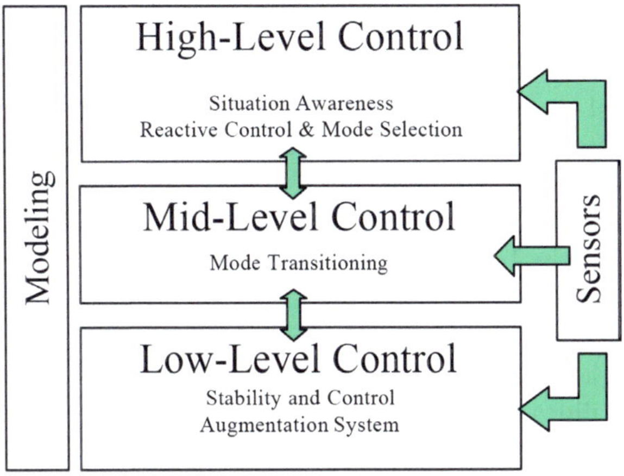

Figure 4.6. *Hierarchical control architecture.*

Source: https://www.researchgate.net/profile/Suresh-Kannan-7/publication/3869717/figure/fig1/AS:670012377407496@1536754807325/Hierarchical-Control-Architecture.png.

For instance, for robotic manipulations, understanding the contextual differences between biological and robotic systems enhances the use of feedback controls, whose performance is optimized through iterative increments. An understanding of the interaction between mechanistic actions and feedback enables researchers and engineers to optimize the design of control architectures further, blurring the line between robotic and biological abilities (Santello et al., 2017). Furthermore, hierarchical control structures have adapted to reflect ongoing advancements in technology and theoretical design. As autonomous systems become more complex, particularly as artificial intelligence and machine learning enter the mix, older hierarchical paradigms must be updated to provide a more sophisticated picture of interactions between system parts. Under-actuated and over-actuated conditions pose difficulties to current designs, and reassessments of those systems are in order to optimize performance. By the implementation of neuromechanical principles and making a link between biological and robotic systems, researchers are able to advance the design of architecture for such systems. Hierarchical control architectures' ability to not just execute tasks effectively but also facilitate learning and development within intelligent systems is advanced with this new approach, paving the way for more research on autonomous technologies (Beck et al., 2025; Santello et al., 2017) (Figure 4.6).

4.3.1. Definition and Structure of Hierarchical Control

The concept of hierarchical control is at the core of autonomous system development and operation. This form structures the control processes in a manner of strata or levels such that each level has clearly defined roles and responsibilities. The most prominent strength of this form is its capacity to handle complexity by fragmenting large tasks into smaller-sized sub-tasks, which are manageable. This decomposition not only improves efficiency but also enhances the versatility of systems in dynamic environments. A good example is robotics, where hierarchical control allows distinct levels of decision-making, ranging from high-level strategic planning to low-level motor control, such that diverse functionalities collaborate. This strategy is consistent with the overall principles of Perceptual Control Theory (PCT), which contends that behavior is a means of controlling perception, validating the central role of perceptions at different levels of control in which systems regulate their environments (Arkin et al., 2020). Moreover, hierarchical control complemented with feedback processes is also fundamental in maintaining the stability and accuracy of autonomous systems. Hierarchical architectures' embedded feedback loops allow continuous monitoring and adjustment of behavior based on real-time data. Through this procedure, these systems can correct themselves and react to new situations, becoming more resilient to uncertainty and disturbance. In the context of automated vehicle or robot use, for instance, feedback processes ensure that sensory data informs decision-making up the hierarchy so that responses to unforeseen obstacles or environmental change occur in a timely manner. The interdependence and hierarchy of feedback is obviously a pointer to the necessity of an integrated system of control, whereby all levels of control communicate with each other efficiently, leading to improved performance and overall system dependability (Beck et al., 2025). In addition, hierarchical control is extremely important when considering future autonomous system trends in several industries. As technology advances further, the boundaries of artificial intelligence and machine learning will continue to widen, and incorporating these systems into a hierarchical framework will continue to allow more sophisticated interactions and decision-making processes. Intelligent communicative agents are one of the best examples of how such frameworks are required in order to provide proper response with situational awareness and user needs. The continued search and formulation of hierarchical control principles will be crucial in the development of designs that allow for smooth and natural interaction between humans and machines. Therefore, the understanding of the definition and structure of hierarchical control not only provides insight into the modern applications but also forms the basis for future developments in autonomous system architectures (Arkin et al., 2020) (Figure 4.7).

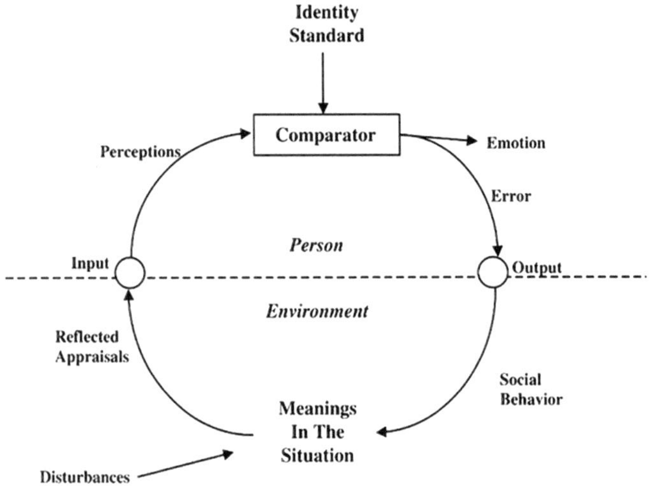

Figure 4.7. *Perceptual control theory.*

Source: https://www.researchgate.net/profile/Janice-Anderson-7/publication/290443921/figure/fig1/AS:616387974266895@1523969753451/Perceptual-Control-Theory-Components-of-a-feedback-loop-associated-with-the-theory.png.

4.3.2. Benefits of Hierarchical Control Systems

Hierarchical control systems offer a structured approach to countering the increasing complexity of autonomous systems. Multiple hierarchical control tasks at various levels are offered by these systems, allowing effective processing and decision-making across various operational environments. The topmost hierarchies are able to handle high-level strategic planning, with lower hierarchies handling individual tasks. There are specialized algorithms that are geared towards various levels of control, and they enhance the overall efficiency and responsiveness of the system. For instance, in robotic applications, high levels may define generic movement routines in response to environmental input, while low levels handle limb movement micro-adjustments. As hierarchical, this process not only optimizes efficiency but also the flexibility of the system in response to varying environments. Significantly, without such an integrated framework as that provided through hierarchical control, the control of advanced autonomous behaviors would be excessively burdensome and likely ineffective within systems that must adapt in real-time to user commands or environmental factors (Arkin et al., 2020). A significant advantage of hierarchical control systems is that they have very effective error management capabilities. In advanced autonomous systems, error detection and correction at the right level of hierarchy will prevent tiny troubles turning into monstrous failures. For example, when an autonomous car runs into a hurdle, the supervisory high-level system translates high-level information in order to deduce a correct plan, while low-level control modulates physical action in real time, i.e., steering and braking. This two-tiered system mitigates the risk of system failure, so that even if there is unexpected failure, the system remains functional and effective. Hierarchical control architecture also improves fault tolerance. Separating functions into various levels enables the system to attain overall performance even if a part of it is affected by a fault, demonstrating how this structure diligently balances operation efficiency with fault tolerance (Santello et al., 2017). Also, hierarchical control systems are better in scalability, and thus they are particularly ideal for large autonomous systems. Along with new tasks and challenges arising, the systems can be easily scaled by introducing new levels of control without necessitating designing frameworks from the ground up. This modularity allows accommodation of new technologies and methodologies as they arise, thereby future proofing the system against the rapid evolution of artificial intelligence and robotics. For instance, in new domains like shared autonomous vehicles and drone fleets, additional layers of hierarchy can be introduced to support collaborative behavior and communication protocols. This is not only useful for preserving operational effectiveness but also for the innovation within control strategy. Continuous development of hierarchical systems underscores their crucial function in the evolution of autonomous technologies towards more complex applications that are both effective and manageable in practical contexts (Arkin et al., 2020).

4.3.3. Examples of Hierarchical Control in Practice

Hierarchical control systems play a vital role in the control of advanced autonomous systems in order to facilitate organized decision-making at various levels. In real life, an interesting example is visible in an urban traffic control system wherein control is structured in tiers. City planners establish general traffic regulations at the highest level, and middle-levels may utilize real-time information for the management of traffic signals and optimization of flow. On the ground level, there are locally reacting autonomous navigation units embedded in each car. This multi-level system enables high-level commands to be executed accurately at the operational level, showcasing the strength of hierarchical control. These systems not just improve traffic efficiency but are also vital to safety mechanisms. This orchestration of hierarchical control proves the effectiveness of hierarchical structures in various disciplines, their adaptability and responsiveness to dynamic environments. Another compelling instance of hierarchical control can be found in aerospace technologies, indeed in unmanned aerial vehicles (UAVs). The systems utilize a multi-layered approach to control the complex interplay between navigation, communication, and mission planning. At the highest level, the mission planner specifies high-level goals, which are then broken down into particular sub-goals for autonomous flight control systems. The intermediate control levels manage altitude, speed, and tuning navigation on feedback from real-time environmental sensors. The hierarchical structure allows strong response to changing conditions while it keeps it in alignment with the overall mission. With hierarchical control, UAVs can execute sophisticated tasks, such as surveillance or reconnaissance, with excellent reliability and autonomy, hence justifying the existence of this paradigm in high-risk situations. Hierarchical control structures have their use beyond the traditional field, as is observed from communications network design. The transition from 5G to subsequent 6G networks reflects the requirement for hierarchical systems to address the future complexity of telecommunications. For this reason, control layers must effectively tradeoff between user requirements, network provision, and higher-level regulatory frameworks established by standardization organizations. Every hierarchical level address specific aspect of network performance—e.g., resource management, connection establishment, and security protocols—finding a balance between operational efficiency and user experience. As identified in recent publications, overcoming the challenges of these intricate networks demands a clearly defined structuring of roles in the control hierarchy to optimize functionality and minimize risks. The hierarchical strategy highlights the applicability of hierarchical control systems in evolving technological landscapes, showcasing their capacity for resilience against mounting pressures and complexity in the telecommunication sector (Cheng-Wang et al., 2023; Wang et al., 2022).

4.4. DISTRIBUTED CONTROL ARCHITECTURES

In the context of distributed control structures in autonomous systems, one must understand their fundamental purpose of maximizing efficiency and robustness of operation. Distributed control is distinct from hierarchical control structures because these compound decision-making functions. Distributed control emphasizes cooperative interaction among multiple autonomous agents. Decentralization allows for flexible task distribution and coping with dynamic environments. When these systems communicate, they are able to share data and distribute tasks of processing, resulting in increased redundancy and fault tolerance. These are the type of attributes that pay off in complicated missions, like going to space, where systems are needed to function in unknown conditions (Beck et al., 2025).

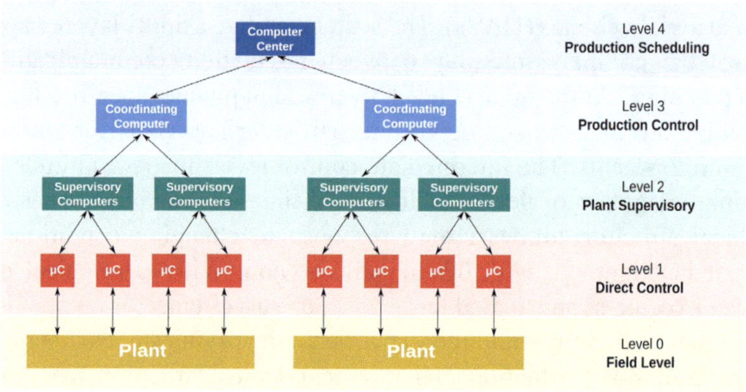

Figure 4.8. *Distributed control systems.*

Source: https://lh7-rt.googleusercontent.com/docsz/AD_4nXe6vXup91K9mGCaftbfgm6o60dBSU6TFclVG5Iaa-awNFhNB-1J4lilRZKx4ky6fdkcJpLaoRf-CX-7zp9cLdtmPvD8gtRIYrlmjXZU_XhC0BanxoIfHXCV3eT5diEzdXoBsAD-o2w?key=V3q1Vs9YfSutoUcbty7sljCo.

The flexibility that is embedded in distributed control architectures allows them to react very quickly to changes, promoting a more resilient system design as a whole that can utilize the complexity of autonomous operation in a positive way. Distributed control architectures' other key aspect is their compatibility with feedback mechanisms that can help improve decision-making processes. Feedback loops in autonomous systems allow agents to adapt their behaviors in response to real-time data from their environment, hence making it possible for adaptive learning and continuous refinement. This is crucial in applications that involve interacting with humans or other autonomous agents since it makes it possible to gain a better understanding of intention and context. The recent advances in Artificial Intelligence have transformed the lives of these systems, increasingly developing from rigid rule-based operation towards more dynamic, embodied structure (Arkin et al., 2020). Through the convergence of complex feedback mechanisms and distributed control, autonomous systems can better perceive and understand their surroundings, leading to improved performance in cooperative tasks. The relationship between distributed control structures and open-loop and closed-loop systems is also something to explore. Open-loop systems lack feedback but operate based on pre-established commands that are non-adaptive. Closed-loop systems, however, operate with the use of feedback in influencing actions based on realized outcomes, allowing for a more adaptive response. Distributed control frameworks predominantly exist in a closed-loop framework, where local feedback from all agents is utilized to inform overall decision-making. This process creates an atmosphere where information in real-time enhances cooperative operations, ultimately leading to improved results in complex contexts. Hence, the use of distributed control structures in autonomous systems not only enhances efficiency but also creates a wiser and adaptive framework that can accommodate the rising demands of modern applications in diverse environments (Beck et al., 2025) (Figure 4.8).

4.4.1. Definition and Structure of Distributed Control

Distributed control theory is at the heart of understanding new control structures in autonomous systems. This contrasts with centralized control, where a single entity has total control over the operation, to distributed control, where several agents that converse with one another for a common purpose are involved. Each agent behaves autonomously, making decisions based on local knowledge and sending messages to other agents to achieve maximum global performance. This architecture adds robustness, as the failure of any agent will not paralyze the system in general. Moreover, distributed control systems are extremely suitable for complex environments in need of real-time decision-making. The adaptability and scalability of this architecture make it more and more suitable in applications ranging from robotics to networked systems, paving the way for more adaptive responses to changing circumstances. One of the defining features of distributed control is its reliance upon multi-agent systems (MAS) such that each agent possesses its own beliefs, desires, and intentions (BDI). This paradigm enables agents to reason locally upon their conditions and make decisions consistent with the global objectives, which maximizes the system efficiency overall. MAS research has a tendency to focus on applying these principles to project management scheduling issues and demonstrate how distributed control can address the complexity and uncertainty of contemporary tasks. For instance, having regard to Time-Constrained Project Scheduling (TCPSP) and Resource-Constrained Project Scheduling (RCPSP), researchers illustrate how the BDI model promotes decision-making by agents and maximizes the utilization of shared resources within diverse contexts (Constanța et al., 2025).

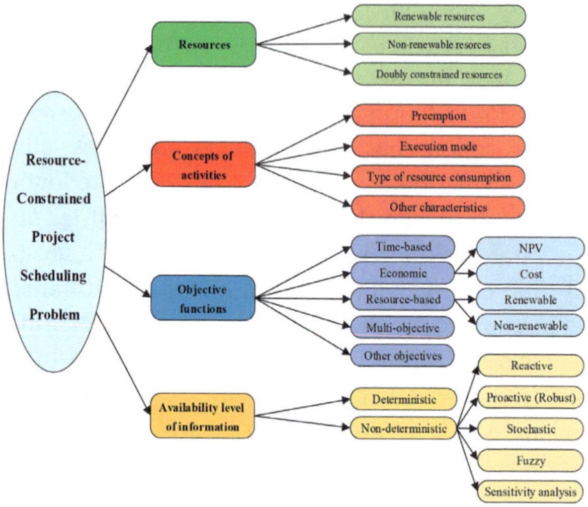

Figure 4.9. *Classification of resource-constrained project scheduling problems.*

Source: https://www.researchgate.net/publication/322515659/figure/fig1/AS:583230462349312@1516064386207/Classification-of-resource-constrained-project-scheduling-problems.png.

This development highlights the importance of distributed control in facilitating intelligent action in autonomous systems. Distributed control also has another important application in its utilization with automation technologies, particularly in high-risk domains such as space exploration. In this regard, AI-based automation systems are designed in order to maximize crew safety and efficiency by reducing extravehicular activities (EVA) and promoting overall operational autonomy. The application of distributed control systems allows the systems to monitor complex environments and respond best to issues in real-time. For example, the application of teleoperation and robotics enhances the capability to repair and maintain equipment, thereby making space missions more efficient and safer (Firschein et al., 2025). Through the use of distributed control, such sophisticated systems can perform complex operations, ultimately aiding the agility and robustness needed in the wake of unforeseen adversity in space ventures (Figure 4.9).

4.4.2. Benefits of Distributed Control Systems

In the context of control architectures of autonomous systems, distributed control systems have tremendous advantages that achieve increased operational efficacy and flexibility. They decentralize decision-making processes as one of the foremost benefits. Unlike centrally controlled systems where a master controller controls the entire operation, distributed systems allow individual components or nodes to operate autonomously. This autonomy enables each part of the system to act instantaneously based on local conditions, leading to quicker response times and less susceptibility to single points of failure. Distributed behavior allows for scalability; with added nodes, the system can execute larger workloads and complexities without performance loss. Such features have enhanced applications in numerous areas, including robotics and telecommunications, where responsiveness in real time is critical to system functioning. Moreover, distributed control systems enable resilience through enhanced fault tolerance. In a centralized system, a failure of a main controller could collapse the entire operation. However, distributed systems are designed to be fault-tolerant in that they can withstand node failures by allowing other components to continue to operate without interruption. Such ruggedness becomes especially critical in applications like the Internet of Things (IoT), where devices must contend with uncertain networks and changing operating conditions all the time. As Cristianini et al. (2025) suggests, what results from the interaction between technical and human factors in such a system is an organized response mechanism that may have complex behaviors, allowing the networked environment as a whole to be adaptive and resilient. So long as distributed control architectures guarantee that the failure of one component will not bring down the system, they are fundamental to the stability of autonomous systems. In addition, the addition of feedback mechanisms to distributed control systems greatly enhances their performance and responsiveness. Feedback loops allow nodes to be continuously assessing their performance and adjusting their behavior in real-time, according to sensor feedback and environmental input. This is fundamental to the capacity of systems to learn and refine their behaviors over time, a concern highly applicable to dynamic and complex environments. As the self-organization literature has highlighted, these adaptive behaviors have been applied extremely well in deployments like wireless sensor networks to showcase the ability of distributed systems to self-organize and enhance network performance under varying conditions (Aliu et al., 2013). Therefore, the advantages of distributed control systems not only consist of structural robustness and expandability but also rely on utilizing feedback mechanisms to strive for continuous improvement, further proving their vital function in autonomous systems today.

4.4.3. Examples of Distributed Control in Practice

The distributed control principle is demonstrated in vehicle formations where every vehicle forms part of a big unit but possesses autonomous control systems. In distributed formation, the motion of every vehicle is controlled by its velocity and relative position to a small number of neighborhood vehicles. This connectivity is accounted for by an information graph enabling control and communication without the presence of a central controller. This control law is mathematically modeled using partial differential equations (PDEs) to describe the position error dynamics of the formation. Such models suggest that with an increase in vehicles, the stability margin, which is crucial in maintaining formation integrity, reduces to zero; in other words, while formations increase, it is more challenging to stabilize them employing optimal control policies and requires innovative design methodologies (Barooah et al., 2010). Distributed control is another salient example in the employment of deep learning algorithms in robots.

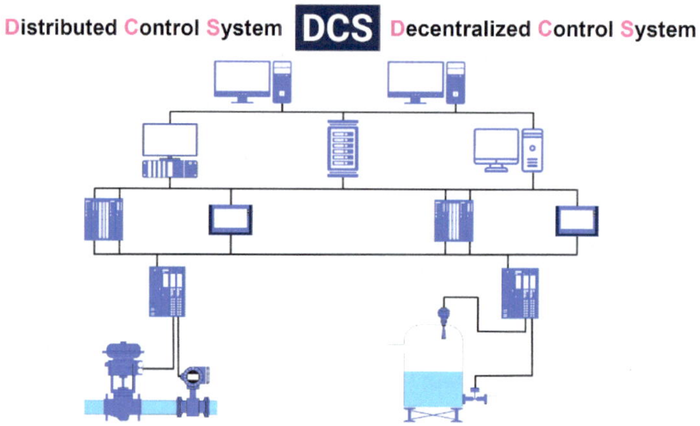

Figure 4.10. *Understanding DCS in industrial automation.*

Source: https://cdn.prod.website-files.com/63dea6cb95e58cb38bb98cbd/66e8e146ac600c85b87f8d4a_AD_4nXcHsKC-LckDAkfK9BfgvLLBsy-j-gAFMgDG6jZZhtlf-H4nNROzbVb-DyiweKfLx9g8xlaWtxJ9TGxE8HoDMGHQqYGOh3kTvOsVAdy-BN_OAYH_uozjE_TgIWT8qrInV4DbnjCLUcqgp1sgJT2Hq_y48DtPXjWRXOK-MZ-0kqJ.png.

The past 10 years have witnessed tremendous advances in deep artificial neural networks, and it has unlocked the gates of robotics to a new era wherein these systems would be able to learn and adapt from their surroundings. One can see a recent corpus of literature through which distributed learning techniques can be employed to enhance the autonomy and the productivity of the robot operations where different agents are able to share knowledge and adjust their behavior accordingly. This distributed approach encourages flexibility and robustness as the agents are able to converge on best policies independently while communicating with their peers. These innovations not only emphasize the potential of artificial intelligence enhancement in robotics but also the intrinsic challenges, including guaranteeing successful synchrony of learning activities across different agents (Gashler et al., 2017). Lastly, distributed control application to large-scale systems, such as smart grid technology, shows how adaptable the architecture is. In smart grids, numerous appliances and generators communicate with one another and coordinate with each other to most efficiently allocate energy distribution and utilization. Every one of these devices has a distributed control system that works on local information, enhancing efficiency without the need for a centralized command and control setup. This decentralized ability provides dynamic response to changes in demand and supply, thus increasing system reliability and sustainability. The distributed control structure enhances the stability of grids since it accommodates renewable sources of energy that are inherently variable. Moreover, the successful incorporation of such systems necessitates mutual coordination where individual agents must maintain their local goals while facilitating higher-level system goals. Thus, the development of distributed control in such applications highlights its prospect to solve intricate problems in autonomous and intelligent systems (Figure 4.10).

4.5. FEEDBACK MECHANISMS IN INTELLIGENT SYSTEMS

The inclusion of feedback mechanisms in intelligent systems plays a significant role in the development of autonomous systems that can dynamically interact with their environments. Such mechanisms allow for systems to adapt their behavior based on discrepancies between desired outcomes and actual-world performance. Intelligent robotic agents, for instance, utilize feedback loops to structure their actions so that they adapt to shifting conditions as well as guarantee operational efficiency. This type of adaptability is essential, especially in applications where human-robot interaction is important. Although artificial intelligence has moved beyond rule-based paradigms to more embodied architectures, there is still a great gap towards achieving strong communicative ability in such intelligent agents, as highlighted by recent research into feedback as a control strategy. This gap may be closed to enhance the effectiveness of feedback mechanisms to yield complex interactions and a better understanding of autonomy within robotic systems (Arkin et al., 2020). When examining feedback mechanisms, one must realize the difference between closed-loop and open-loop control systems.

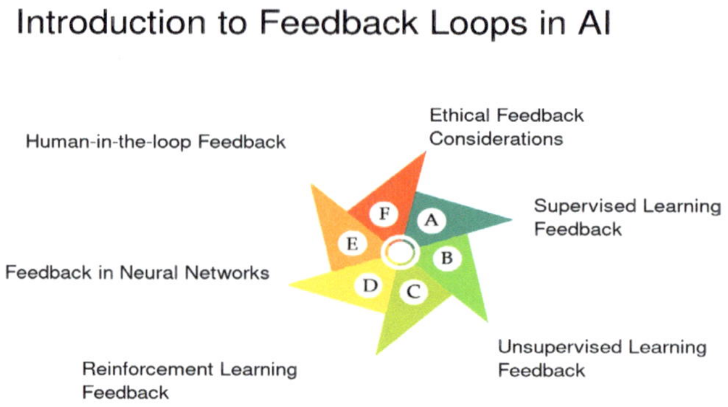

Figure 4.11. *Feedback in artificial intelligence.*

Source: https://fastercapital.co/i/Feedback-loops--Feedback-in-Artificial-Intelligence--Learning-Machines--Feedback-in-Artificial-Intelligence--Introduction-to-Feedback-Loops-in-AI.webp.

Closed-loop control systems operate on feedback for constant adjustment of operation based on their performance, defining a responsive and adaptive model of interaction. Conversely, open-loop systems lack such feedback control and therefore are less responsive and deterministic in response to environmental change. An understanding of these distinctions is key to the advancement of autonomous system architecture, particularly for control hierarchy distribution. For instance, hierarchical control frameworks would likely employ feedback loops that are capable of filtering and processing signals at various levels of the system to facilitate rich decision-making potential and improved operations. Merging multiple feedback channels can improve responsiveness and robustness to make feedback mechanisms central to intelligent system architectural design (Beck et al., 2025). Research in intelligent communicative behavior in autonomous systems shows that there exists a need for a general theory that effectively integrates feedback mechanisms with perceptual control theories. Theoretical framework postulates that behavior is anchored on perceptual control and that simulating behavior plays a fundamental role in developing communicative agents. Their use can facilitate system-level design choices based on feedback mechanisms that are not only reactive but predictive as well, thus improving the overall quality of interactions. By enhancing the importance of feedback in not only defining the action of automated systems but also their communicative skills, researchers can develop the field further to great extents. This emergence cannot be dissociated from the quest to create smart machines that can operate independently with effective communication in complex environments (Arkin et al., 2020) (Figure 4.11).

4.5.1. Types of Feedback Mechanisms

Feedback mechanisms are important components in autonomous systems in order that they are able to learn and respond to changing environments in an efficient manner. In general, feedback mechanisms are either negative or positive feedback. Negative feedback mechanisms aim to maintain stability in a system by opposing changes, so the output does not exceed prescribed limits. For example, in robot movement control, feedback may reverse or alter the speed of a robot when deviating from course. Positive feedback, however, builds on changes, tending to lead to exponential growth in system response. This type of feedback appears in systems that need to overcome strong hindrances or induce extreme changes but, by itself, can lead to instability if not controlled accurately. Thus, understanding these mechanisms is imperative for creating dependable control architectures in autonomous systems. Feedback mechanisms in intelligent systems are tasked with enabling self-regulation and adaptability. A practical application can be seen in the healthcare industry, namely in Intensive Care Units (ICUs), where data retrieval is critical for making timely decision-making. The COSARA architecture illustrates the mechanism by which feedback systems, employing autonomic control loops, enhance system performance through dynamic monitoring and correction of data queries. The reactive, deliberative, and reflective loops provide a model facilitating systems to handle data streams independently, with the end result being minimal intervention by humans and optimal efficiency and dependability in the domain of critical care. As underlined by research, the activation of these feedback mechanisms significantly improves query execution time to a very large degree, thus improving patient data visualization for healthcare providers (Turck et al., 2013). These applications indicate the revolutionary power of feedback mechanisms in creating more intelligent and more efficient systems. Also, feedback processes can be explained even further within the context of Perceptual Control Theory (PCT), which posits that behavior is really all a matter of perception control. In this scheme, feedback processes involve intricate interaction among sensory inputs and behavioral outputs, which effectively allows an intelligent agent to guide itself in its world based on perceived discrepancies between its desired state and its current state. The extension of PCT to the design of communicative agents is a shift towards systems that not only react but also actively engage in meaningful exchanges by translating perceptual information into terms that result in the modification of their actions (Arkin et al., 2020). This view stresses that systems should be designed with robust feedback mechanisms to enable not just reactive reactions, but deliberative processes that enhance communication and adaptability in ambiguous situations. As autonomous systems progress, the integration of feedback loops will be crucial in order to achieve higher-level interactions with users and other systems (Figure 4.12).

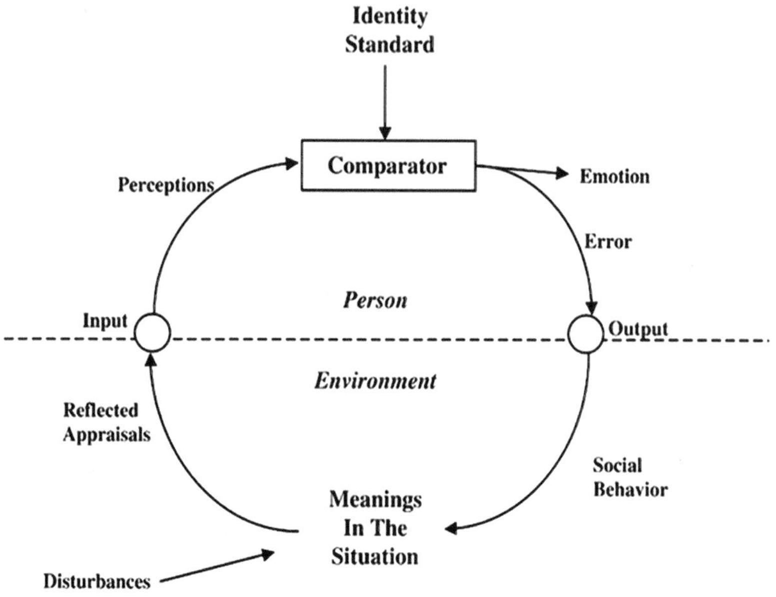

Figure 4.12. *Perceptual control theory.*

Source: https://www.researchgate.net/profile/Janice-Anderson-7/publication/290443921/figure/fig1/AS:616387974266895@1523969753451/Perceptual-Control-Theory-Components-of-a-feedback-loop-associated-with-the-theory.png.

4.5.2. Role of Feedback in System Performance

Feedback constitutes a critical method of enhancing the performance of autonomous systems, especially where closed-loop control systems are used. Feedback systems in this case provide instantaneous feedback regarding the output relative to levels of desired performance, hence enabling the system to respond in kind. This repeated process enables proactive solutions to be made that will effectively minimize errors and optimize system operation. For instance, when an autonomous vehicle detects a discrepancy with its intended route, it can apply feedback to re-map its navigation choices, improving accuracy as well as security. Feedback-based approaches are utilized in numerous areas, including aerospace and medicine, where immediate adjustments are vital to achieving good outcomes. Without such mechanisms, operation and performance of these sophisticated systems would be drastically reduced, highlighting the focal role played by feedback in system operation and control. Incorporating feedback into hierarchical and distributed control systems enhances flexibility in autonomous systems. Hierarchical systems are commonly based on multiple control levels, with each level controlling and giving feedback to the lower levels. This multilayer approach enables intelligent information processing that enables the system to respond appropriately to new information and changing conditions. Distributed control, on the other hand, is concerned with local feedback loops that reinforce coordination among interdependent subsystems. These arrangements are resilient and efficient because subsystems can respond to variability autonomously while staying aligned with system goals as a whole. A good case in point is feedback application in health information systems, where real-time inflow of data requires an adaptive mechanism of control. Dynamic process adaptation through feedback here can lead to significant performance gains in data retrieval and, by extension, healthcare provision as noted by Turck et al. (2013). The effectiveness of feedback mechanisms is also illustrated through their use in the field of autonomous robots, where systems need to repeatedly function within unstructured environments. Within this space, feedback lays the groundwork for motor action optimization and learning, allowing robots to make real-time decisions based on sensor data. Experiments have shown that feedback control greatly enhances the efficiency of robotic task execution since systems can learn from the past and adjust their behavior accordingly. Additionally, according to Beck et al. (2025), utilization of feedback loop-based technologies becomes vital for continuous performance enhancement in a large number of engineering disciplines. Through regular application of feedback towards adaptive learning, stand-alone systems not only optimize the performance of tasks but also calibrate their operating systems so that they continue to thrive amidst novel challenges. This incessant optimization is evidence of the critical function played by feedback in optimizing system performance across many different applications.

4.5.3. Case Studies of Feedback Mechanisms in Autonomous Systems

An analysis of feedback systems in autonomous systems offers interesting lessons regarding the successful design of control structures to maximize performance. Case studies illustrate the real-world applications of closed-loop and open-loop control techniques under different environments. Closed-loop systems utilize continuous feedback to adjust their activities, ensuring responsiveness under real-time conditions. For instance, autonomous vehicles rely on sensory data to adjust speed and direction according to the dynamic changes of their surroundings. Open-loop systems, however, do not have this looped feedback, which may result in less adaptive performance but is cheaper and simpler to apply. Every case study emphasizes how intricate and unusual are the challenges engineers have to achieve such control mechanisms in equilibrium, with the need for robust architectures that integrate effective feedback processes to enable responsive and intelligent autonomy in autonomous systems. In feedback mechanisms, the distinction between hierarchical and distributed control is of extreme importance. Hierarchical control architectures generally consist of a centralized configuration wherein decisions are made at different levels of abstraction to enable complexity management and resource allocation. Distributed control, on the other hand, is designed to support several autonomous agents that interact with each other and their environments, hence enabling resilience and flexibility. Feedback mechanisms are important in both architectures and impact the way systems communicate and adjust to real-time data. For example, in robot grasping tasks, a self-acting robot uses tactile sensor feedback to adjust its grip force as a function of altering object properties. This is an example of the convergence of biological inspiration in robot design, which reveals shared principles for application in research and functional application. Ultimately, the exploration of these case studies bridges the gap between ideas from various disciplines, making successful paths for creating smart control systems in autonomous environments. One of the fundamental aspects of intelligent system feedback mechanisms is that they learn and change, as is seen in robotic manipulation case studies. Neuromechanics theory provides a framework for how physical interaction and mechanical requirements intersect with control strategies. Through a sensory-integrative feedback loop, robotic systems can adjust movement based on past experience and shifting surroundings. For instance, through real-time feedback from vision and touch sensors, robots can learn their manipulation skill through a set of trials, embodying the philosophy of learning through interaction. Such a unity not only emphasizes the imperative of responsivity in autonomous action as embodied in the principles of Perceptual Control Theory (PCT), but it also obliterates traditional paradigms dichotomizing biological organisms from engineered solutions. As formulated in recent research, these interdisciplinary results pave the path towards developing more adaptive and intelligent feedback mechanisms that can perform high-order operations within dynamically changing environments (Arkin et al., 2020; Santello et al., 2017).

4.6. COMPARISON OF CONTROL ARCHITECTURES

The examination of autonomous system control structures determines that there exists a basic distinction between open-loop and closed-loop control techniques. Open-loop control does not involve feedback, taking preordained commands in a straightforward step-by-step fashion, which could limit flexibility to real-time environmental changes. Closed-loop control employs feedback systems through which the system may adjust its responses based on actual performance outcomes. This reactive capacity provides stability and reliability, especially in complex and changing environments where novel challenges can arise. Consequently, closed-loop systems are generally desirable in applications with higher safety and precision. Feedback input allows for continuous learning and adaptation, showing how well these systems handle complexity and uncertainty and thus embody a gigantic leap in control architecture design. Hierarchical and distributed control structures offer varying forms of control for autonomous systems each with its own advantages and drawbacks.

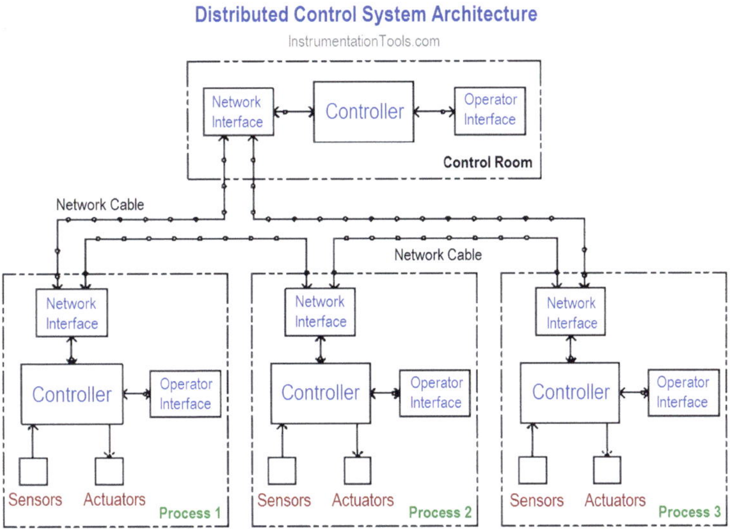

Figure 4.13. *Different types of control systems architecture.*

Source: https://media.licdn.com/dms/image/v2/C4E12AQGUzbG8Hb0t_g/article-cover_image-shrink_720_1280/article-cover_image-shrink_720_1280/0/1592041602927?e=2147483647&v=beta&t=_8IoJUivVHfzsd9lqvrQ8WlLmRCD3scBjrO7Ol9Sdgs.

Hierarchical control traditionally uses a top-down hierarchical framework, where higher-level decision-making modules direct lower-level entities. This organization can facilitate simplicity and ease the coordination of system components. However, it can become a bottleneck in complicated scenarios that require rapid responses since the higher level may decelerate the processing of vital tasks. In contrast, distributed control promotes autonomy by providing separate units with the power to act based upon local information and interaction in their world. Decentralization can become more flexible and robust, embracing a broader range of conditions and tasks. It is important to have knowledge of these differences while designing systems that integrate the strengths of each architecture with a reduction in their inherent limitations. The advent of advanced technologies has propelled the development of sophisticated control structures, such as ICAROUS-2, an excellent example of a hybrid control structure in fusing distributed algorithms for unmanned systems. This novel structure provides support for a variety of functions, including geofencing, path planning, and autonomous decision-making, ensuring mission effectiveness and operational safety. Formal verification methods added another layer of reliability in curtailing the risk of failure in real-time operations. These advancements underscore the key role of control architectures to push the autonomy and capability of robotic systems. While research continues to expand, implications of integrating deep learning into such architectures continue to be important, offering novel frontiers for complex interactions and decision-making capabilities, as indicated by the growing body of literature on deep learning in robotics (Gashler et al., 2017). These probes put a focus on the ongoing adjustment of control methods to make autonomous systems responsive and effective in challenging environments (Balachandran et al., 2025) (Figure 4.13).

4.6.1. Open-Loop versus Closed-Loop Control

The distinction between closed-loop and open-loop control systems is at the heart of control structures in autonomous systems. Open-loop control operates through a predetermined set of instructions without the use of feedback, making it simpler to implement but less responsive to dynamic environmental conditions. It typically occurs in applications where environmental conditions are constant and predictable. Examples of open-loop systems are washing machine clocks or toasters where input provides the only parameter for control, and maximum operation can be achieved without constant recalibration. However, due to its lack of feedback devices, open-loop systems are less responsive to disturbances, and thus generate inefficiencies and inaccuracies in dynamic systems. It is important to know this limitation in order to determine the appropriate setting wherein one can apply open-loop control methods. Closed-loop systems, however, possess feedback loops in order to allow them to adjust their output based on the level of performance in relation to an ideal set point. This flexibility is essential in autonomous systems that move in uncertain environments and where circumstances continually change rapidly. An example would be a sensor-equipped robotic arm, which can continuously check its position and adjust its movement to accurately perform tasks regardless of external forces or obstacles. Through operating feedback loops, closed-loop systems are very precise and dependable in the long term, adapting their performance based on real-time input. This control design has special relevance in complex applications, where the ability to maintain performance despite variance is essential. Incorporating feedback into these systems enables them to act in response effectively before the event occurs rather than acting in response. With the evolution of the autonomous systems field, the integration of open-loop and closed-loop control paradigms results in more sophisticated architectures that enhance their capabilities in real-world use. Closed-loop systems dominate domains that require accuracy and flexibility, but niches still remain where open-loop approaches might gain dominance, particularly in cases that require speed over accuracy. Additionally, the intermixing of hierarchical and distributed control mechanisms allows for the optimization of both modes in line with operational objectives. For example, an autonomous vehicle may employ closed-loop control for navigation but leverage open-loop processes in terms of making adjustments to speed when driving on predictable sections of the road. This intermixture of control techniques hence leads to the further evolution of intelligent systems to further support dynamic tasks and minimize waste in diverse conditions, as shown by Arkin et al. (2020); Beck et al. (2025).

4.6.2. Hierarchical versus Distributed Control

Hierarchical and distributed control structures differ significantly in the design and operation of autonomous systems. Hierarchical control is a top-down approach where decisions are based on a central source, which oversees and directs lower-level processes. This structure is easy to develop with separate authority and enables the integration of compound tasks through assigned layers, which can contribute to efficiency and ensure harmonious decision-making within systems. In situations where worldwide system performance is a necessity, hierarchical structures are advantageous due to the streamlined communication and control flow. Conversely, this centralized control also renders hierarchical systems vulnerable to points of failure, thus impacting their resilience in a changing world. Distributed control schemes, however, favor decentralization, with individual parts or agents able to act autonomously while colluding to interact with their world. This method is geared to improve resilience because if one agent fails, it does not have to bring down the entire system. Distributed control systems excel in applications requiring flexibility and adaptability because they can easily respond to real-time variations without necessarily looking for commands from a master controller. The integration of intelligent computing concepts into distributed systems further complements their ability by allowing them to execute higher-order cognitive capabilities that facilitate learning and adaptation. Emergence of technologies like biofeedback mechanisms proves the potential for distributed control systems to evolve to higher levels of machine-environment interaction as indicated by current developments in such areas of brain organoid computing (Smirnova et al., 2023) and intelligent computing (Zhu et al., 2023). Choice between hierarchical and distributed control is extremely situation-dependent and tends to significantly influence autonomous system potency. Complexity of the system, the context of operation, and specific task requirements must decide this option. Although hierarchical control may be suitable in situations where coordination is central, distributed control offers critical benefits in less stable or more interactive environments. Future development and research efforts need to be focused on merging the strengths of each architecture into a potentially hybrid model that can take advantage of hierarchical systems' structured decision-making functions in combination with distributed systems' adaptive flexibility. Synergetic models like these have the potential to create more robust and powerful autonomous systems, which would be more useful and effective across a wide variety of applications (Figure 4.14).

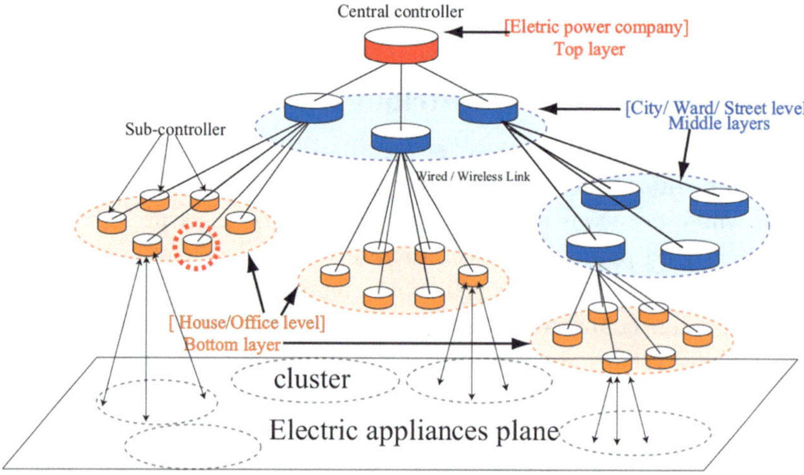

Figure 4.14. *Hierarchical distributed control network.*

Source: https://www.researchgate.net/publication/236878986/figure/fig3/AS:614266524663833@1523463960712/Hierarchical-distributed-control-network.png.

4.6.3. Impact of Control Architecture on System Efficiency

The analysis of control architectures illustrates a fascinating interplay between system performance and design selection within autonomous systems. The most significant difference here is between open-loop and closed-loop control architectures. Open-loop systems use pre-set instructions with no regard to prevailing output conditions, typically leading to inefficiencies, especially when used in dynamic environments. Closed-loop systems, however, possess the capability of embedding feedback mechanisms such that system adjustments are possible in real-time based on output performance. This flexibility is the source of resilience and performance ability, enabling systems to achieve unintended variables, thereby leading to efficiency as a whole. In systems when performing complex tasks, the intrinsic benefits of closed-loop feedback are realized, supporting the argument that control architecture choice directly impacts the performance outcomes of autonomous systems and, in turn, guides the design toward optimized operational performance capabilities (Beck et al., 2025). Hierarchical and distributed paradigms of control present a further dimension along which systems can be measured in terms of efficiency. Hierarchical design of structure layers control tasks to give structured decision-making and optimized communication. This can speed response in systems with centralized coordination to optimize the efficiency of information processing. The inordinately large bottlenecks to communication channels, however, can hinder performance in dynamic environments. On the other hand, distributed control systems provide actors in a network with autonomous capabilities that permit more localized response and decision-making. The autonomous capability enables greater adaptability and efficiency when sophisticated problems are addressed by agents collectively. The coexistence of such control models indicates that a well-organized architecture not only optimizes task performance but also enhances resource utilization, touching the overall efficiency of autonomous systems (Chung et al., 2018). Furthermore, their implementation of feedback mechanisms emphasizes their critical role of enhancing system performance. Feedback mechanisms are the cornerstone of responsiveness in hierarchical and distributed control systems. They enable an endless loop of monitoring and adjustment, which is pivotal in maintaining optimum levels of performance. For instance, in air swarm systems, feedback allows individual drones to communicate and coordinate with one another, bridging the gap between localized decisions to mission goals. This distributed intelligence manifests in the form of higher efficiency of operation, for example, in current advances in swarm robotics, where algorithms for task allocation and communication give rise to significant performance gains. Feedback loops are thus an essential design factor, as they enable continuous adaptation that maximizes the systems functionality over time, affirming the thesis that control architecture significantly impacts efficiency in autonomous complex systems (Chung et al., 2018).

SUMMARY

The study of control architectures of autonomous systems exhibits a dynamic relationship between open-loop and closed-loop systems and hierarchical versus distributed control structures. This strongly integrated framework has far-reaching implications for intelligent system design and behavior. Open-loop systems, with their predictable yet rigid nature, offer critical advantages in circumstances where conditions are stable and predictable. Conversely, closed-loop systems introduce adaptability functionality with feedback loops that enhance performance under uncertain conditions. This dynamic interaction underscores the significance of selecting an appropriate control structure against the needs and context variables of each application, from self-driving cars to robots working within complex environments. As the field progresses, it becomes increasingly evident that hierarchical and distributed control models play a key role in the efficient functioning of autonomous systems. Hierarchical frameworks allow decision-making at various levels, allowing optimal allocation of resources as well as coordination among subsystems. Distributed control models, however, allow agents to act autonomously based on local knowledge, ensuring resilience as well as adaptability within complex networks. These architectures are extremely important for enhancing system robustness, addressing scalability challenges, and facilitating overall operational efficiency in an abstract sense. Coupled with feedback structures, these models point towards the future of intelligent systems in terms of greater autonomy and elegance, suggesting a revolutionary trajectory of applications in the future across a variety of industries. Combined with high-performance computing technologies, control architectures in the future have a very promising area of exploration in the field of autonomous systems. The arrival of organoid intelligence signals a potential shift to employing biological computing along with artificial systems, thus increasing the cognitive power of autonomous agents (Smirnova et al., 2023). In addition, the fusion of intelligent computing techniques redraws conventional paradigms, opening doors for perceptive and cognitive intelligence in operation models (Zhu et al., 2023). This merging suggests a new paradigm in which smart systems can learn, adapt, and enhance performance on an ongoing basis in real-world scenarios. Thus, the convergence of these various control architectures with latest computing advancements will certainly open up new avenues, creating more advanced and responsive autonomous systems for the future ahead. In exploring control structures within autonomous systems, there is a basic distinction between open-loop and closed-loop control. Open-loop systems do not have feedback and simply perform commands on the basis of pre-set instructions, which can limit flexibility within dynamically changing environments. Closed-loop systems employ feedback mechanisms to adjust responses according to current information, enhancing responsiveness and precision. This adaptability is indispensable, especially under complex interaction with human operators or changing external conditions. As highlighted in controversies surrounding smart autonomous machines, e.g., robots, application of feedback mechanisms informs adaptive solutions to perform complex activities while remaining safe and dependable in operations (Arkin et al., 2020). Hence, the acknowledgment of these innate differences in control approaches is essential for the design of autonomous systems that can effectively work in actual-world environments. Architectural frameworks of control systems are generally explained as either hierarchical or distributed, both accommodating various operation necessities and complexities. Hierarchical control systems promote ordering by articulating commands in a hierarchical framework, with the top levels setting overall goals and the lower levels addressing performing detailed tasks. The system can be able to simplify decision-making and minimize mistakes by making functions clearly defined in the architecture. Distributed control systems, however, favor a decentralized arrangement where agents or modules are assigned autonomy to make decisions based on local information. This can foster greater resilience and flexibility, especially in contexts such as swarm robotics and cooperative actions among different autonomous entities. The comparative effectiveness of these architectures will largely be mission goals and environment-dependent, requiring a developed sense of their respective strengths (Beck et al., 2025). One of the most significant characteristics of control architectures is the integration of feedback mechanisms that enhance the intelligence and responsiveness of autonomous systems. Feedback loops allow for continuous assessment and real-time system behavior adaptation, which is most critical in applications where high interaction and adaptability are required. Intelligent systems are advantaged by such an ability through the synthesis of sensory information to guide decision-making procedures, which in turn results in more precise and efficient performances of tasks. For example, the tenets of Perceptual Control Theory (PCT) can guide the formulation of architectures with the focus on needs-led behavior as the means to enhance the idea that knowing and anticipating outcomes is key to success. This is paralleled with the theoretical shift from conventional rule-based paradigms to more sophisticated, body-based frameworks that can effectively run in complicated environments, significantly enhancing the operational ability of autonomous systems (Arkin et al., 2020).

REFERENCES

1. Albayrak, S., Hrabia, C. E., & Lützenberger, M. (2018). Towards adaptive multi-robot systems: Self-organization and self-adaptation. https://core.ac.uk/download/222444390.pdf (accessed on 10th September 2025).
2. Aliu, O. G., Evans, B., Imran, A., Imran, M. A., et al. (2013). A survey of self-organization in future cellular networks. https://core.ac.uk/download/74362769.pdf (accessed on 10th September 2025).
3. Andrawos, M., Aßmann, U., Bass, L., Behara, D. G. K., Betts, T., Bianco, P., Bryant, D., et al. (2019). Microservice transition and its granularity problem: A systematic mapping study. https://core.ac.uk/download/426749569.pdf (accessed on 10th September 2025).
4. Arkin, R., Bickhard, M., Bourbon, D., Braitenberg, V., Bridle, J., Brooks, R., et al. (2020). PCT and beyond: Toward a computational framework for 'intelligent' communicative systems.
5. Balachandran, S., Consiglio, M. C., Feliu, M. A., Muñoz, C., et al. (2025). Independent configurable architecture for reliable operation of unmanned systems with distributed onboard services. https://core.ac.uk/download/pdf/286392855.pdf (accessed on 10th September 2025).
6. Barooah, P., Hao, H., & Mehta, P. G. (2010). Stability margin scaling laws for distributed formation control as a function of network structure. https://core.ac.uk/download/357713994.pdf (accessed on 10th September 2025).
7. Beck, S. M., Breckenridge, R. A., Clark, L. G., Collier, R., et al. (2025). Technology for the future: In-space technology experiments program, part 2. https://core.ac.uk/download/pdf/42817270.pdf (accessed on 10th September 2025).
8. Bernardin, E., & Leybourne, S. (2020). Project knowledge into project practice: Generational issues in the knowledge management process. https://open.bu.edu/bitstream/2144/39065/1/IRNOP%202018%20paper%20-Leybourne%20%26%20Bernardin%20-%20FINAL.pdf (accessed on 10th September 2025).
9. Bodea, C. N., Badea, I. R., & Mogoș, R. I. (2025). A multi-agent system with application in project scheduling. https://core.ac.uk/download/pdf/6331286.pdf (accessed on 10th September 2025).
10. Chung, S. J., Dames, P., Kumar, V., Paranjape, R., et al. (2018). A survey on aerial swarm robotics. https://core.ac.uk/download/211379963.pdf (accessed on 10th September 2025).
11. Cristianini, N., Ladyman, J., & Scantamburlo, T. (2025). Social machinery and intelligence. https://core.ac.uk/download/223236092.pdf (accessed on 10th September 2025).

12. De Turck, F., Decruyenaere, J., Latré, S., Steurbaut, K., et al. (2013). Autonomic care platform for optimizing query performance. https://core.ac.uk/download/55763037.pdf (accessed on 10th September 2025).
13. Drgoňa, J., Arroyo, J., Cupeiro Figueroa, I., Blum, D., Arendt, K., Kim, D., Perarnau Ollé, E., et al. (2020). All you need to know about model predictive control for buildings. *Annual Reviews in Control, 50*, 190–232. https://doi.org/10.1016/j.arcontrol.2020.09.001 (accessed on 10th September 2025).
14. Elkael, M., D'Oro, S., Bonati, L., Polese, M., Lee, Y., Furueda, K., & Melodia, T. (2025). AgentRAN: An agentic AI architecture for autonomous control of open 6G networks. https://www.semanticscholar.org/paper/c6a1db97eb88a635d8b13d2696b362d1a6bd6f09 (accessed on 10th September 2025).
15. Eyers, D. (2017). Control architectures for industrial additive manufacturing systems. https://core.ac.uk/download/78077319.pdf (accessed on 10th September 2025).
16. Firschein, O., Georgeff, M. P., Kautz, W. H., Levitt, T. S., et al. (2025). NASA space station automation: AI-based technology review. https://core.ac.uk/download/pdf/42844682.pdf (accessed on 10th September 2025).
17. Gashler, M. S., & Pierson, H. A. (2017). Deep learning in robotics: A review of recent research. http://arxiv.org/abs/1707.07217 (accessed on 10th September 2025).
18. Hernandez, C., Lopez, I., & Sanz, R. (2008). Consciousness in cognitive architectures: A principled analysis of RCS, Soar and ACT-R. https://core.ac.uk/download/pdf/87106.pdf (accessed on 10th September 2025).
19. Kloss, G. K. (2007). Performance evaluation of a distributed integrative architecture for robotics. https://core.ac.uk/download/148639212.pdf (accessed on 10th September 2025).
20. Liu, H., Moreu, L. S., Andersen, T. S., Puche, V. V., & Fumagalli, M. (2025). Stinger robot: A self-bracing robotic platform for autonomous drilling in confined underground environments. https://www.semanticscholar.org/paper/8aac945c0fe6dba6190f2838622f8878ffd6d85a (accessed on 10th September 2025).
21. Park, S., & Kim, Y. G. (2022). A metaverse: Taxonomy, components, applications, and open challenges. *IEEE Access, 10*, 4209–4251. https://doi.org/10.1109/access.2021.3140175 (accessed on 10th September 2025).
22. Santello, M., & Valero-Cuevas, F. J. (2017). On neuromechanical approaches for the study of biological grasp and manipulation. http://

arxiv.org/abs/1710.08557 (accessed on 10th September 2025).

23. Smirnova, L., Caffo, B., Gracias, D. H., Huang, Q., Morales Pantoja, I. E., Tang, B. H., Zack, D. J., et al. (2023). Organoid intelligence (OI): The new frontier in biocomputing and intelligence-in-a-dish. *Frontiers in Science, 1*. https://doi.org/10.3389/fsci.2023.1017235 (accessed on 10th September 2025).
24. Wang, C. X., You, X., Gao, X., Zhu, X., Li, Z., Zhang, C., Wang, H., et al. (2023). On the road to 6G: Visions, requirements, key technologies, and testbeds. *IEEE Communications Surveys & Tutorials, 25*, 905–974. https://doi.org/10.1109/comst.2023.3249835 (accessed on 10th September 2025).
25. Wang, Y., Su, Z., Zhang, N., Xing, R., Liu, D., Luan, T. H., & Shen, X. (2022). A survey on metaverse: Fundamentals, security, and privacy. *IEEE Communications Surveys & Tutorials, 25*, 319–352. https://doi.org/10.1109/comst.2022.3202047 (accessed on 10th September 2025).
26. Zhu, S., Yu, T., Xu, T., Chen, H., Dustdar, S., Gigan, S., Gündüz, D., et al. (2023). Intelligent computing: The latest advances, challenges, and future. *Intelligent Computing, 2*. https://doi.org/10.34133/icomputing.0006 (accessed on 10th September 2025).

CHAPTER 5

MACHINE LEARNING FOR EMBEDDED INTELLIGENCE

Contents

Introduction 163

5.1. Historical Context 169

5.2. Key Concepts in Machine Learning 174

5.3. Applications of Machine Learning in Embedded Systems 178

5.4. Challenges in Implementing Machine Learning for Embedded Intelligence 183

5.5. Tools and Frameworks 189

5.6. Future Trends in Machine Learning for Embedded Intelligence 194

Summary 198

References 201

INTRODUCTION

The advent of machine learning (ML) has transformed numerous domains, especially in the area of embedded systems where the intelligence is made a part of ordinary appliances. Embedded intelligence refers to the capability of the devices to execute the processing of data and make decisions without human involvement, depending on algorithms that are experienced in nature. As technology progresses, computing systems become smaller and communication technologies become more sophisticated, paving the way for advanced applications that can perceive their environment. Such technologies enable machines not only to perform pre-programmed instructions but modify their actions based on real-time feedback, a paradigm shift necessary for applications today. These improvements announce an era of intelligent gadgets, which will call for a thorough understanding of the role that part machine learning plays in optimization of performance and enhanced user experience. A chief aspect of embedded intelligence is learning how to draw from various sources of data when making informed decisions.

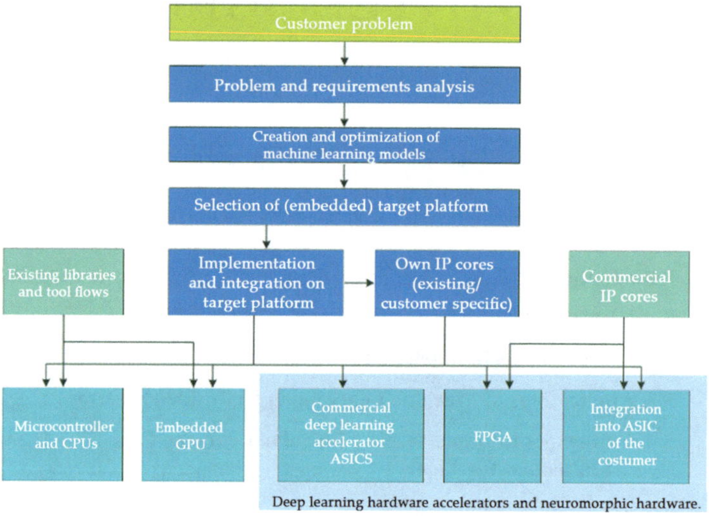

Figure 5.1. *Embedded machine learning.*

Source: *https://www.iis.fraunhofer.de/en/ff/kom/ai/embedded-ml/jcr:content/contentPar/sectioncomponent_1057262409/sectionParsys/textwithasset/imageComponent/image.img.4col.large.png/1740037389331/Embedded-ML-Flow-en.png.*

For instance, automotive technology has cars garnering vast quantities of data from sensors, which ultimately will enable the ability to predict driver intent and behavior. This integration of data both from internal measurements and environmental surroundings involves both the measures and context information, which needs advanced computational techniques for processing effectively. The progress in fuzzy computational modeling and other methods of computational intelligence (CI) is at the core of such processing. By processing these various streams of information, systems can deliver timely and relevant feedback, significantly improving safety and economy of driving conditions. Recent research has highlighted the value of calendar information as forward-looking inputs to driver purpose, as proof of the new methods being adopted in such a field (Abtahi et al., 2018). Moreover, the intersection of Big Data and computational intelligence is crucial to developing solutions that address real-world problems, particularly in terms of smart cities. As cities increasingly rely on intelligent systems to manage urban infrastructure, the efficiency of ML in handling gargantuan amounts of data cannot be overestimated. Effective data modeling methods such as the Hierarchical Spatial-Temporal State Machine (HSTSM) embody the potential of machine learning deployments in enabling enhanced decision support for cities. The convergence of this technology has far-reaching implications that inform policy, enhance public services, and optimize resource utilization. As the demand for intelligent infrastructure grows, the legal and ethical aspects that encompass data use as well as privacy concerns will be paramount to manage for the realization of sustainable smart city solutions (Doctor et al., 2020) (Figure 5.1).

Definition of Machine Learning

The term machine learning refers to a subdomain of artificial intelligence that deals with the ability of systems to learn from data, recognize patterns, and make decisions with limited supervision. It is founded on algorithms mainly that enable computers to study large amounts of information and improve over time through new knowledge acquisition. In embedded intelligence, machine learning is the vital tool for intelligent devices to react and adapt to changing circumstances. The ability to respond is central to applications ranging from autonomous vehicles to home automation systems, where real-time processing of data and decision-making are critical. As industries are now more readily embracing advanced technologies, the necessity of understanding machine learning principles is of utmost significance in encouraging innovation and maintaining competitive advantage through the evolving force of technology. Information technology and operation technology coming together through the use of machine learning is particularly of note as it relates to the fourth industrial revolution, referred to as Industry 4.0. This period is characterized by the emergence of smart factories and interconnected systems that utilize data intelligence to optimize processes. Machine learning plays a central role in this transformation by allowing complex tasks to be automated as well as predictive maintenance, thus enhancing overall operational efficiency. For instance, the Industry 4.0 Technologies Laboratory at Catalonia's Technical University is an initiative intended to catalyze engineering education innovation around these digital technologies (Prieto et al., 2019). The proactive development of machine learning capabilities forms a subset of the broader objectives for innovation and competitiveness in the modern industrial environment. Besides, applications of machine learning techniques such as low-rank decomposition (LRD) show evidence of the broad impact of this technology in data analysis and modeling. LRD has been found to have large potential in visual information reconstruction and modeling in scenarios that are plagued by issues such as noise, occlusion, and misalignment (Chen et al., 2017). With the addition of manifold priors to LRD, scientists have been able to increase the performance of the technique and, more importantly, extract important features from complex data sets. This shows that machine learning is not only automation but enables systems to handle real-world data imperfections and make better choices. Thus, as machine learning technology evolves, new frontiers open for embedded intelligence applications that basically revolutionize how systems exchange information with their surroundings and render them more efficient in dealing with the problems of the real world (Figure 5.2).

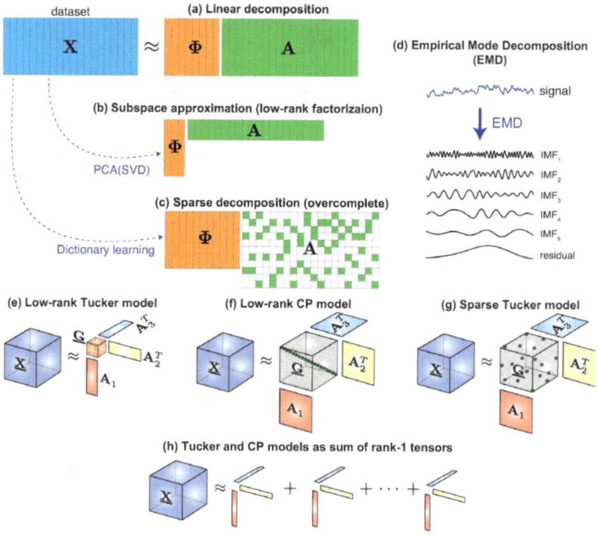

Figure 5.2. *Decomposition methods for machine learning with small, incomplete, or noisy datasets.*

Source: https://www.mdpi.com/applsci/applsci-10-08481/article_deploy/html/images/applsci-10-08481-g001.png.

Overview of Embedded Intelligence

Embedded intelligence is the integration of sophisticated computing capacity in appliances that are not computers in themselves but can perform sophisticated tasks independently. The model is particularly pertinent in the automotive industry, where predictive driver intent is increasingly important. The invention of miniaturized computerizing and ever-present connectivity has created an abundance of information that can be utilized by the in-car systems to enhance the user experience and safety. Advanced systems are able to interpret data from an array of sources, including e-calendars and georeferencing systems, which assists with deciphering driver behavior and intent. The involved complexity resulting from processing and tapping such enormous and diversified data calls for ongoing research in selecting a suitable set of computational techniques to ensure optimal performance, emphasizing embedded intelligence power and the necessity in modern vehicles (Abtahi et al., 2018). With the advancement of embedded intelligence, new avenues are laid open to enhance decision-making in numerous fields. These advances in Computational Intelligence (CI) and Big Data processing have significant implications not just for single vehicles but for traffic management and city planning. For instance, by filtering data and analyzing the surroundings to which they respond, such as traffic levels and driver timetables, vehicles can be optimized to drive more efficiently. This allows not just for an improved driving experience but for potentially reducing congestion and emissions on a broader scale. However, these developments also compel a re-examination of existing ethical and legal frameworks under which the use of self-learning cars is governed, suggesting that there is a need for stringent regulations ensuring safety and accountability in intelligent vehicle systems (Abtahi et al., 2018). Furthermore, the landscape of embedded intelligence demands a careful evaluation of existing variable selection methods and analysis, particularly with respect to the constraints of computational resources and time-constrained regimes. Research projects are investigating methods like the EventTracker Sensitivity Analysis, which illustrates how time-sensitive information can be used to inform variable selection in complicated systems. This methodology emphasizes the need to detect inputs that have a significant influence on system outputs, particularly when it is a matter of prioritizing operational safety and efficiency. By combining real-time monitoring with sophisticated data processing, embedded intelligence technologies can greatly improve decision-making in many applications, from autonomous transport vehicles to factory floors, and thus shape the future world of intelligent systems.

Importance of the Intersection of Machine Learning and Embedded Systems

The interaction between embedded systems and machine learning is increasingly important in a variety of applications, powering innovation in fields like medicine, automotive engineering, and home automation. Embedded systems, with specialized function within a larger mechanical or electrical system, generally must cope with the stress of low processing power and memory. The addition of machine learning algorithms, nevertheless, equips such systems with the ability to process data and learn from it in real-time, hence enhancing their decision-making power.

For instance, in the medical sector, wearable technology with embedded systems integrated in them can monitor patient vitals and apply machine learning to predict health complications or alert caregivers to abnormalities, hence improving patient outcomes and operational efficiency. These advancements underscore the core position that machine learning takes in the transformation of traditional embedded systems into intelligent entities endowed with adaptive actions to dynamic environments. Furthermore, the complementarity between machine learning and embedded systems provides new avenues for more sophisticated autonomous devices. Technologies such as drones and self-driving cars rely heavily on embedded systems to function optimally in real-world environments.

By applying machine learning algorithms, the vehicles are able to handle vast amounts of sensor information to determine patterns and make decisions in real-time. Not only does this enhance the reliability of operations but also makes safety standards better through the capability of the cars to react to dynamic road conditions and obstacles. With the continuous advancement of such technologies, it is important that engineers and developers ensure maximization of the integration of machine learning into embedded systems in order to realize efficiency, security, and ethical considerations such as privacy and adherence to regulations in consideration of innovation in this area. Integration of machine learning into embedded systems also has ethical considerations that cannot be overlooked. As more devices are embedded to gather and process individual data, privacy, data security, and machine learning biases are high on the agenda. For instance, the use of AI in education technology raises questions regarding learner agency and ethical application of data-driven insights for enhanced learning outcomes. Mitigating these ethical challenges is key to establishing trust among stakeholders and end-users who increasingly rely on smart systems daily. While researchers continue to develop frameworks for machine learning ethical use, they emphasize the need to adopt transparency, responsibility, and fairness in system development and deployment. Ultimately, the successful embedding of machine learning and embedded systems depends not only on technology advancement but also on a humble approach to ethical issues, making their benefits available without compromising public values (Nguyen et al., 2022; Yogesh et al., 2023).

5.1. HISTORICAL CONTEXT

The historical context of machine learning for embedded intelligence rests upon the rapid growth of computing technologies and their application in various disciplines, including the automotive industry. As automobiles migrated away from mechanical systems to highly advanced electronic architectures, embedding computational intelligence became prominent. The existence of computer innovations that were miniaturized provided the capabilities to incorporate sophisticated algorithms into cars for augmenting their ability to process data in real time. Modern cars today rely on various data sources, ranging from sensors to external environmental inputs, providing opportunities to predict driver behavior and augment safety, efficiency, and experience. It is concerned with the need to employ robust computational techniques to manage the volume and variety of data inherent in such systems, as explored by Abtahi et al. (2018), which highlights challenges and solutions proposed to leverage driver intent prediction effectively. Furthermore, the introduction of Big Data has also revolutionized the application of machine learning techniques in smart systems. When urban areas evolved into increasingly integrated systems, the collection and processing of massive datasets became integral to their function. Big Data analysis along with computational intelligence offers tools for the development of smart city infrastructures that utilize resources more efficiently and improve life quality. The roots of these advancements lie in the need for increased urban planning and management, where data-driven insights were a major factor. Insights gained from the processing of massive data enable cities to react optimally to systemic problems, as highlighted by Doctor et al. (2020). This converging of technologies is a reflection of the central role that machine learning plays in developing intelligent solutions that react to emerging needs in real-time and adapt to the changing needs of today's society. Lastly, the socio-ethical implications of using machine learning on embedded intelligence cannot be overlooked. As smart technologies become more integrated into daily life, privacy concerns, security, and ethical usage arise. Historical examples, such as the evolution of surveillance tools, illustrate the dangers in continually collecting and analyzing personal data through intelligent systems. As discussed by Abtahi et al. (2018) and Doctor et al. (2020), the deployment of advanced computational models in the public space, for example, in cars and cities, requires proper attention to schemes of regulation to protect the rights of people while tapping the potential of innovation. Finding the balance between these requirements and the potential for wider societal benefit is critical, as stakeholders navigate the complexity of implementing machine learning solutions that work but also are equitable and equitable in application.

5.1.1. Evolution of Machine Learning Techniques

The evolution of machine learning techniques has largely been influenced by advances in computing power and data availability. Machine learning was originally constrained by limited processing power, encouraging more straightforward techniques with focus on linear regression and basic classification models. But with the advent of big data, there has been a revolution in machine learning techniques with sophisticated models that are able to deal with complex data. For instance, techniques such as neural networks and ensemble methods have been established, and they have become central to applications in various domains, including systems of embedded intelligence. Such advancements enable real-time handling of data so that learning and adaptive automatic systems can be developed, improving their ability to make decisions in intelligent applications within cities. Big data's impact on machine learning applications illustrates the ubiquitous interplay of algorithmic innovation and technology, all of which continue to drive the intelligent systems of tomorrow (Doctor et al., 2020). Machine learning's integration into compiler optimization has also illustrated the evolution of these techniques within specific, high-stakes domains. Scientists have sought to use machine learning since the mid-1990s in order to tackle challenging compiler optimization problems, such as optimization selection and phase-ordering. The complexity of modern-day compilers has created a need for more subtle techniques, as previous techniques of optimization are no longer capable of keeping pace with the expanding horizon of optimization with new implementations and frameworks. Learning-based techniques have proven highly effective in this regard, enabling the compiler to select optimally on the basis of choice and determine the most optimal sequence of implementation. As a result, these algorithms not just boost performance but also learn dynamically to cope with various programming environments, which reflects the diversity and transferability of machine learning in various sectors (Ashouri et al., 2018). The integration of machine learning within embedded systems reflects precisely the ongoing development that takes place within this field. As more intelligence gets embedded into devices, the expectations on machine learning techniques rise by the day. This is not just about basic algorithmic breakthrough but also about streamlining systems to operate within restricted environments. With the application of new machine learning techniques, such as reinforcement learning and real-time data processing, embedded intelligence systems can now better process input data and respond with instant results. This streamlining is a reflection of a larger paradigm shift in which machines no longer simply execute based on pre-programmed rules but also learn and evolve through experiential learning. Thus, the trajectory of machine learning methods is a testament to their lead role in driving the evolution of embedded intelligence, toward increasingly intelligent, autonomous systems capable of survival in real-world contexts (Figure 5.3).

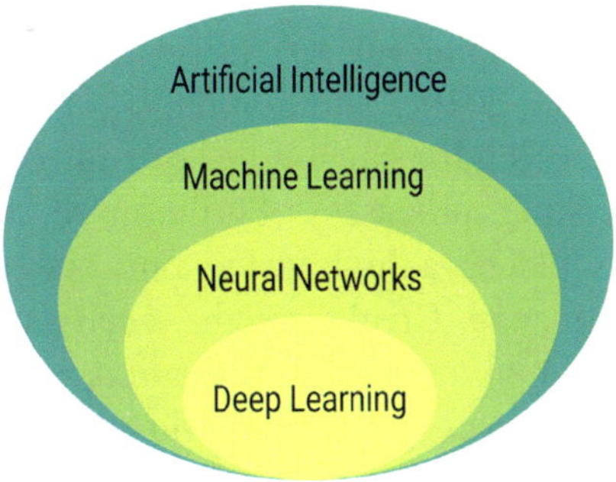

Figure 5.3. *Machine learning methods.*

Source: https://images.deepai.org/news-links/9e9c413e4f6e46098efa4d1a11851ddf/machine-learning-methods-explained-1848075285.jpg.

5.1.2. Development of Embedded Systems

The building of embedded systems has increasingly become more essential in recent years, particularly with their embedding in machine learning to provide enhanced functionality and autonomous capabilities. Embedded systems, being specialized computing systems with the ability to perform specialized functions within larger systems, have currently begun to be utilized across industries like automotive, healthcare, and industrial automation. This combination of machine learning technology allows these systems to learn from data, adapt to changes in the environment, and make decisions in real-time. For instance, in agriculture, scientists are currently working on intelligent farming technologies that use autonomous navigation, assisted by advanced embedded systems, to optimize farm practices and raise productivity (Aghi et al., 2020). The marriage of machine learning algorithms and embedded systems enables one to execute complex operations with little human intervention, a truly groundbreaking advancement in how such technologies operate. With each advance towards Industry 4.0, more and more intelligent embedded systems are needed than ever before. Not only do they pull and process vast amounts of data, but they enable one to respond that quickly as well, further magnifying the efficiency of operation. The systems evolved around embedded systems must now incorporate elements like connectivity, security, and responsiveness to enable efficient operation in networked environments. Education facilities play a pivotal role in this revolution through the development of a workforce that has expertise in both machine learning and embedded system design. Initiatives like the Industry 4.0 Technologies Laboratory at the Technical University of Catalonia are a clear indication of how the gap between industry and academia could be bridged by developing an ecosystem that promotes innovation and competences (Prieto et al., 2019). Through the learning process, such learners are contributing to enhancing the capabilities of embedded systems that enable intelligent decision-making processes. Furthermore, the ongoing evolution of embedded systems places a high value on cost-efficient solutions with minimal loss of performance. Breakthroughs, such as the proposed low-cost local motion planner for autonomous navigation, demonstrate the promise of embedded systems to execute effectively in constraint-rich environments. With the help of cutting-edge computer vision technology through RGB-D cameras and two-layer control algorithms, such systems not only offer precise mapping and localization but also exhibit resilience under various conditions of operation (Aghi et al., 2020). Resilience is crucial as it enables embedded systems to learn and adapt over time based on their experiences. As we continue deeper into the meeting of embedded systems and machine learning, the future of developing more intelligent, autonomous technologies starts to take shape, ultimately realigning the approach of various industries toward automation and optimization.

5.1.3. Milestones in the Integration of Machine Learning and Embedded Intelligence

The union of embedded intelligence and machine learning (ML) has witnessed a series of exemplary milestones that characterize the evolution of the two technologies. Among the developments has been the introduction of intelligent data interfaces that facilitate real-time analytics. The technology is specifically applicable in complex project situations, like those within NASA, where fast decision-making and information sharing are a priority. Optimizing data access, these systems allow engineers and project managers to make judgments and address risks more efficiently. The possibilities are vast; for example, scheduled design reviews can be turned into on-demand, dynamic analyses, enhancing teamwork and eliminating time limits. The application of ML methods into such frameworks is successfully a milestone for project management, further enhancing the potential of embedded intelligence in high-risk environments (Evans et al., 2025). A milestone in cognitive robotics is reached when machine learning is employed to enhance embodied cognitive agents' capabilities, including humanoid robots. Progress in this area has been on how such agents learn to acquire knowledge and encode actions and language needed to interact with the environment and also with people. The research is intended to learn complex sensorimotor skills and social interaction dynamics, which are crucial for adaptive learning. By resolving points like the combination of action and language representations, researchers are establishing routes to better cognitive systems that are capable of executing tasks independently and adjusting to new conditions. The path that comes through such milestones is not just designed to inform cognitive robotics research but also to make realistic promises of achievement in the combining of machine learning with embedded intelligence that performs well in real-world scenarios (Belpaeme et al., 2010). Last but not least, the inclusion of the analysis of real-time data in embedded systems shows the transformative capability of machine learning in numerous applications. Through the use of ML, such systems can offer actionable intelligence and inform decision-making in ways not currently possible. As project management becomes more data-centric, the capability to tap into and interpret enormous amounts of information in real-time is vital. The potential is far-reaching, allowing teams to accomplish ambitious projects efficiently with less risk. As embedded intelligence continues to evolve, the necessity for sound frameworks to merge data and analysis will grow all the more. Making sure that the insights from these technological advances are useful and germane to project objectives will be paramount to the future of machine learning implementations in embedded systems (Evans et al., 2025) (Figure 5.4).

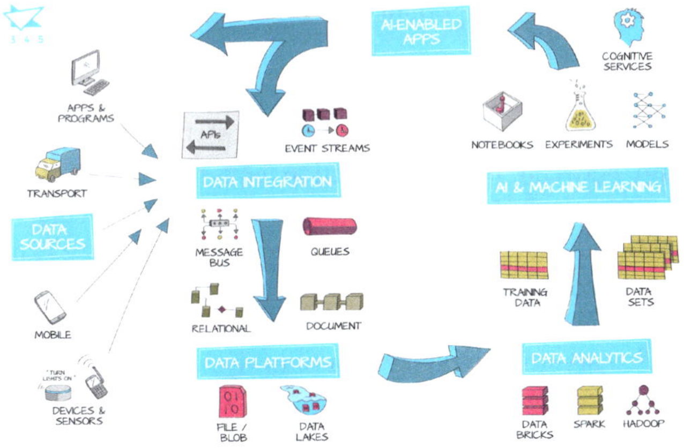

Figure 5.4. *Integrating data: AI and machine learning.*

Source: https://345.technology/wp-content/uploads/2020/08/B170-345-Link-Between-Integ-and-Ai-01.jpg.

5.2. KEY CONCEPTS IN MACHINE LEARNING

In machine learning, supervised learning, unsupervised learning, and reinforcement learning are key concepts that are the pillars of improving embedded intelligence comprehension. Supervised learning is characterized by the utilization of labeled data, where algorithms learn and make prediction from input features. This is important in applications where historical data is present to actually train models to recognize patterns and predict. Conversely, the focus of unsupervised learning is on the detection of hidden patterns in unclassified data. It becomes relevant in cases such as clustering and association when unlabeled data is the only thing available. These learning paradigms together indicate the flexibility of machine learning and prove it to be an essential component for intelligent embedded systems that process huge volumes of data real-time, which go on to enable innovation across various sectors. Having some knowledge about feature selection and extraction is helpful as it helps improve machine learning models, particularly in embedded intelligence architecture. Feature extraction is the process of transforming raw data into model-compatible forms such that appropriate characteristics responsible for prediction can be directly discovered. In embedded systems, since computation resources are typically constrained, appropriate feature selection is all the more crucial to make the models efficient and prevent overfitting. Techniques such as Principal Component Analysis (PCA) and mutual information are often utilized to retain the most important information while eliminating redundant or irrelevant data.

This not only enhances model performance but also minimizes the risks of high dimensionality of the dataset typical in real-world issues. Therefore, the relationship between feature extraction and feature selection is critical in developing robust machine learning architectures particular to effective embedded intelligence. Also, the impact of big data on machine learning cannot be overstated, particularly as it is concerned with the design of smart systems and technologies. The blend of big data and machine learning is key in the development of smart cities and new economic systems. As seen in innovation, the application of computational intelligence approaches to big data analysis involves challenges and opportunities calling for innovative solutions to enhance efficiency. Applications in smart cities like traffic management, air and noise pollution monitoring, and law enforcement are great illustrations of how big data and machine learning unite to create operational efficiencies and intelligent decision-making processes. By using models that can process large datasets, such as those based on algorithms developed by Doctor et al. (2020), embedded systems can deal with real-time situations, optimizing functions within cities and benefiting the quality of life for citizens. Knowledge of these connections is a major part of the theoretical and applied domain of machine learning in embedded intelligence (Figure 5.5).

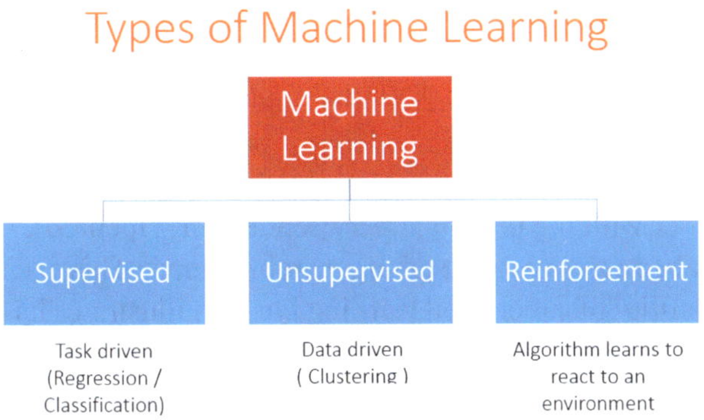

Figure 5.5. *Machine learning types.*

Source: https://www.analyticsvidhya.com/wp-content/uploads/2015/06/machine-learning-types.png.

5.2.1. Supervised Learning

Supervised learning is one of the building blocks of machine learning, traditionally speaking, especially in its uses within embedded intelligence. Supervised learning permits algorithms to learn from labeled data, whereby the relationship of the input features to their corresponding output is established well enough. Using these data, supervised learning algorithms are able to identify patterns and make predictions on new data. This trait is especially crucial for embedded systems, which operate most of the time in dynamic environments that require adaptive and accurate decision-making processes. For instance, in wireless sensor networks, supervised learning techniques enhance the system's ability to detect and respond to environmental alterations, eventually enhancing resource utilization and extending the operational lifespan of the network, as outlined by Alsheikh et al. (2014). In embedded intelligence, supervised learning may be used to solve real-world issues in various applications, like image recognition, natural language processing, and predictive maintenance. The power of supervised learning models rests on the ability to generalize far beyond the training set, a requirement that is essential in real-life situations where conditions may change unpredictably. In consequence, advances in supervised learning are strongly relevant to fields like healthcare, transportation, and smart manufacturing, where accurate prediction means resource allocation and improved outcomes. In addition, as embedded systems grow increasingly sophisticated, the selection of effective supervised learning algorithms that are relevant to specific tasks has become critical to optimizing system performance. The evolving landscape of embedded intelligence also emphasizes the relationship among supervised learning, advanced compiler optimization techniques. They have tried to identify how machine learning can make compilers more effective by addressing underlying issues like optimization selection and phase-ordering, as highlighted by Ashouri et al. (2018). Compilers can make better choices about which optimizations to use and in what order with the help of supervised learning methods, thus achieving better performance across a range of target architectures. This exchange not only shows the versatility of supervised learning but also illustrates its dominant role in pushing the boundaries of embedded intelligence. As further advancements are made in machine learning methods, their application to embedded systems will be unavoidable in order to tackle new challenges and realize the full potential of smart applications.

5.2.2. Unsupervised Learning

Unsupervised learning is a key characteristic of machine learning, particularly embedded intelligence. Unlike supervised learning, which trains models using labeled data, unsupervised learning learns from unlabeled data in order to discover patterns and relationships without explicit guidance. This independent search enables intrinsic structures to be discovered in large data sets, making clustering, dimensionality reduction, and anomaly detection possible. The ability of unsupervised learning techniques to adapt and respond to changing conditions makes them particularly useful in dynamic situations, where pre-existing labels are unavailable or might rapidly become so. This property resonates well with the requirements of embedded systems that often deal with huge volumes of unstructured data and need complex analysis in order to improve decision-making ability and operational effectiveness. One of the most common applications of unsupervised learning is in feature selection, one of the most important steps in machine learning model optimization. Feature selection reduces data dimensionality by choosing and retaining the most significant features to train models. Studies demonstrate that applying strong feature selection techniques can significantly alleviate the curse of dimensionality that normally weakens model performance and interpretability in machine learning. The Feature Selection Library (FSLib) provides an extensive collection of feature ranking and selection algorithms that are utilized to identify the most suitable features, hence enhancing data understanding and speeding up the learning process (Roffo et al., 2018). The combination of these algorithms allows embedded intelligence systems to extract the maximum from their data processing capacity such that only valuable information guides decision-making. Furthermore, unsupervised learning is very applicable in the case of wireless sensor networks (WSNs) since the ability to adapt dynamically to changing environments is critical. WSNs generally follow changing conditions that change on the basis of external factors or system design specifications. Utilizing unsupervised learning techniques, WSNs can analyze patterns of data in real-time and respond adaptively, optimizing resource utilization and extending the service life of the networks. With various machine learning approaches, designers can address typical challenges like power efficiency and data reliability and ultimately achieve improved network performance (Alsheikh et al., 2014). As embedded intelligence continues to evolve, the incorporation of unsupervised learning into WSNs is guaranteed to offer new solutions and considerably improve the responsiveness and efficiency of such systems (Figure 5.6).

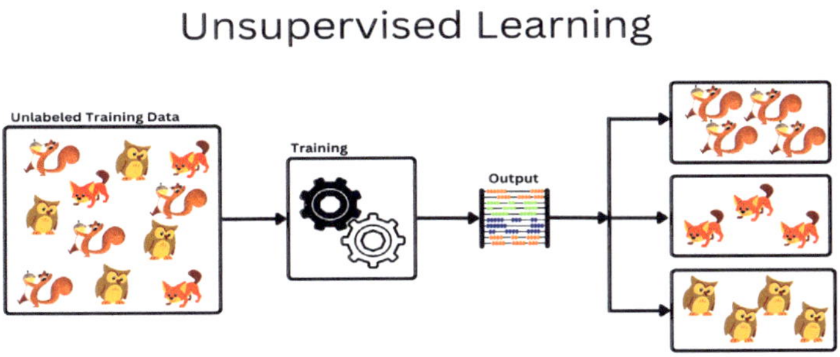

Figure 5.6. *Supervised versus unsupervised learning.*

Source: https://datamapu.com/images/20231017_supervised_unsupervised/unsupervised.gif.

5.2.3. Reinforcement Learning

The RL research has broad applicability to the creation of embedded intelligence across different domains. RL is basically a paradigm in which an agent learns to take a decision by interacting with an environment and getting feedback in terms of rewards or penalties. The iterative method enables the agent to learn how to optimize its actions towards achieving long-term objectives, a characteristic that makes it most suited to dynamic situations. Given that embedded systems operate in the fast-response environments that change rapidly, the strength of RL enables such systems to learn without necessarily full redefinition of their architectures. This is a crucial aspect of embedded intelligence, where operational flexibility translates into enhanced functionality and longer system life, able to resolve the concerns of real-world applications (Alsheikh et al., 2014). Further, RL's relevance extends to some domains such as autonomous driving, where traditional supervised learning approaches fail due to the multidimensionality and ambiguity of driving situations. As has been observed through studies in deep reinforcement learning, the integration of complex neural network models has helped systems deal with complex sensory data and adapt to the consequences of their actions. For instance, through the incorporation of Recurrent Neural Networks, vehicles are able to attain situational awareness in the case of uncertainty by ignoring irrelevant information through the application of attention models. Along with ushering in more robust decision-making, this new method also concurs with the realities of embedded systems where computing power is not necessarily given in sizeable amounts. Simulation outcomes, such as the TORCS 3D automobile racing simulator, illustrate the potential of RL in supporting autonomous navigation in complicated environments (Abdou et al., 2017). Application of reinforcement learning in embedded systems in the real world illustrates the monumental advances in machine learning for intelligence augmentation. With RL's help, software developers can create intelligent, more resource-efficient systems that can learn from their use experiences in real-time. Such ability to adapt and change leads to improved use of resources since learning agents can tune their operations according to feedback, thereby conserving energy and extending the lifespan of sensor networks. According to high rates of literature reporting in the last 10 years, machine learning methods like RL have addressed common issues found in contexts ranging from wireless sensor networks to driverless cars and evaluated their robustness and performance. The incorporation of such methodologies within embedded intelligence will necessarily set the course for technological innovation, solidifying RL as the foundation upon which future innovations in smart systems will be built (Alsheikh et al., 2014).

5.3. APPLICATIONS OF MACHINE LEARNING IN EMBEDDED SYSTEMS

The integration of machine learning in embedded systems has revolutionized numerous sectors, particularly in medicine. Using mobile phones equipped with inertial sensors, significant advancements have been made to monitor and evaluate the mobility of older adults. For instance, the Timed-Up and Go test to assess mobility and balance can now be successfully conducted using smartphones that contain embedded accelerometers and gyroscopes. This approach not only enhances the validity of health assessments but also offers instant feedback, thereby improving preventive care interventions. The varied parameters from these movement sensors give a detailed analysis of patient movement, allowing timely interventions for patients with conditions such as Parkinsons disease and orthopedic disorders, emphasizing the relevance of embedded machine learning in optimizing the quality of life of elderly populations (Garcia et al., 2020). In car environments, machine learning in embedded systems uses vast amounts of data from many different sources to enhance driving safety and efficiency.

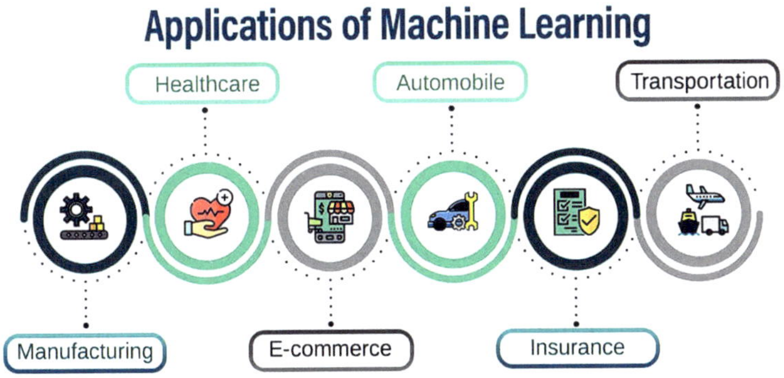

Figure 5.7. *Machine learning applications.*

Source: https://external-content.duckduckgo.com/iu/?u=https%3A%2F%2Ftse3.mm.bing.net%2Fth%2Fid%2FOIP.5_rmyx-FZWv56u-IsIknfgHaDo%3Fr%3D0%26pid%3DApi&f=1&ipt=6c0871066c4589b3aa083aec32177f3fc274dfa6d5e364b80fd800d200965973.

Modern vehicles integrate advanced computer technologies with the capabilities to integrate sensor data as well as data sources from the outside, such as e-calendars and geo-referencing networks. Predictions of driver intention based on these input data represent a promising innovation to make it possible for cars to react to user needs and behavior. For example, with the analysis of location data integrated into contextual signals, machine learning algorithms can provide anticipatory guidance and make driving safer. Furthermore, application of advanced computational intelligence techniques in cars raises ethical questions and legal issues, particularly regarding user privacy and data protection (Abtahi et al., 2018). This combined focus on direct practical applications and broader social impact indicates the transformative potential of machine learning application within embedded automotive systems. The general impact of machine learning on embedded systems extends beyond the automotive and health sectors to be applied in smart home devices and automation industries. Embedded systems used in smart homes leverage machine learning algorithms in order to optimize energy consumption, enhance security, and ease the user's life. For instance, smart thermostats analyze usage patterns in order to create optimized heating and cooling schedules, and security cameras utilize machine learning for threat detection in real-time. In manufacturing settings, machine learning maximizes predictive maintenance such that machines are able to warn operators of probable failures beforehand, minimizing downtime and maintenance expenses. Through providing such systems with smart decision-making abilities, businesses are in a position to achieve higher efficiency, safety, and user experience, indicating the vast scope of applications of machine learning across various embedded systems (Abtahi et al., 2018; Garcia et al., 2020) (Figure 5.7).

5.3.1. Smart Home Devices

The integration of smart appliances in homes is a giant leap for the technology of embedded intelligence, with machine learning used to enhance the user experience and efficacy. From intelligent thermostats to smart lighting systems, these appliances use real-time processing of data to make functionalities adjust based on user behavior and environmental factors. Machine learning methods enable them to learn from experience, thereby increasing predictive power with time. This not only makes home maintenance easier but is also useful for energy saving, as appliances are capable of maximizing their performance, potentially reducing bills and damaging the environment less. With the incorporation of intelligence in common items, the idea of a smart home goes beyond mere automation, to an interactive environment that automatically adapts to its inhabitants' requirements. However, the incorporation of machine learning technology with smart home appliances also presents significant challenges that must be overcome. The very large amounts of information gathered by such sensors necessitate robust data analytics platforms that are able to extract beneficial insights without infringing on user privacy and security. Big Data analytics, also being discussed in the context of smart cities, is equally relevant here, where such sheer scale and variety of information require advanced computational intelligence techniques to generate actionable feedback. Moreover, the efficiency of machine learning algorithms is directly proportional to the input data quality. Therefore, ensuring a smooth unification of various home appliances and consistency in data quality is absolutely essential. All these matters have to be handled with utmost caution in order to realize the maximum potential of smart home technology while ensuring users' rights. Besides personal convenience, the social implications of smart home technology extend beyond individual level; it represents a change in the norms of daily life and urbanization. As the prevalence of smart homes increases, they are part of the new trend of smart cities where interconnected devices enhance city infrastructure and resource utilization. For instance, data from sensors within a smart home are utilized to inform urban planning by allowing insight into existing energy consumption patterns and environmental conditions. Intelligent interaction between home appliances and urban systems as a whole illustrates one of the most promising machine learning-computational intelligence synergies. However, it is essential to enter this convergence with regard to ethical issues and policy frameworks to inhibit misuse of data. It will be crucial for smart home technology developments in the future to balance social responsibility with innovation.

5.3.2. Wearable Technology

The innovation of wearable technology is a big step towards human-machine interaction and individual health monitoring. With continued progress in smartwatches and fitness trackers, the fact that they can integrate machine learning algorithms makes them offer personalized suggestions and feedback based on individual behavior patterns. Application of embedded sensors in handhelds, particularly in omnipresent smartphones, highlights the potential of such devices to monitor health-related parameters effectively. Specifically, it has been proven through studies that the use of inertial sensors like accelerometers and gyroscopes not only allows the recording of important information but also facilitates the identification of complexities related to movement and performance while undergoing tests like the Timed-Up and Go test, which results in improved diagnostics and intervention procedures according to the necessities of older adults with varying pathological states (Garcia et al., 2020). Further, new applications of machine learning in wearable technology extend beyond basic health monitoring to enhance safety, particularly for vulnerable populations such as the elderly. Wearable sensors that sense falls are now essential in addressing the public health concerns of falls, which have dangerous implications among older people. Through the use of recurrent neural networks, these systems are intended to process accelerometer signals efficiently in real time, detecting fall events and also potential threats with high accuracy. The ability to process temporal signals in low-power microcontrollers makes it possible for users to be constantly monitored and without needing high battery life, hence making real-time feedback not only possible but also consistent. These models' established sensitivity and specificity underpin the efficacy of integrating leading-edge machine learning techniques into wearables, ultimately enabling independence and safety for older adults within their daily settings (Balcells et al., 2019). In addition to individual health monitoring and safety applications, wearable technology facilitates population-level data-driven insights into physical activity and health behaviors. This matters to public health practice and research because aggregating wearable data across devices can inform interventions to achieve optimal population improvement in health. Machine learning embedded in analysis pipelines enables detection of patterns and associations that may not otherwise be evident. This capability not only enhances the effectiveness of health interventions in targeted population groups but also facilitates more engaged healthcare. As wearables become better in parallel with advancements in machine learning algorithms, there will be opportunities for improving health outcomes through more integrated and targeted approaches to be sure, and it will be a key component of future health policy in an increasingly ageing population (Garcia et al., 2020) (Figure 5.8).

Figure 5.8. *Wearable technology.*

Source: https://www.biolab.vic.edu.au/wp-content/uploads/2020/06/Wearable-Technology-1.png.

5.3.3. Autonomous Vehicles

Autonomous cars represent a critical intersection of advanced technologies and machine learning systems, illustrating how intelligence can be combined in order to revolutionize transport. Designed to travel in varied environments on their own, these vehicles are heavily reliant on algorithms that have been trained from huge volumes of data collected by sensors, cameras, and real-time interaction with the environment. The integration of machine learning enables autonomous cars to enhance decision-making, maximizing their potential to properly interpret real-world situations. As artificial intelligence continues to advance, the potential for unmatched efficiency and safety on highways becomes increasingly enticing day by day. However, in order to attain these benefits, concerted efforts are required to reconcile the resulting safety and ethical problems, hence rendering research into solid AI structures imperative for the assurance of the technology's continued utility to society (Dewey et al., 2015). Reinforcement learning, one of the most exciting machine learning algorithms being used with autonomous vehicles, is one of the most important departures from typical supervised learning through direct system interaction with the environment. Compared to more conventional methods that rely on pre-tagged data, reinforcement learning allows vehicles to enhance their skills based on feedback regarding their actions, which induces a form of trial and error closer to the manner human beings learn. This is particularly beneficial when addressing the complex situations involved in driving, with infinite variables—other vehicles, pedestrians, and varying states of the road, for instance—to be considered simultaneously. Despite the challenge of adapting this emerging paradigm in reinforcement learning to automotive systems, researchers are making great strides towards significant implementation (Abdou et al., 2017). Additionally, machine learning model architecture for autonomous vehicles increasingly concentrates on efficiency, particularly when it comes to their deployment in embedded systems. As such vehicles often operate in real-time, computation efficiency becomes a building block concern. Present developments in deep reinforcement learning apply recurrent neural networks and attention models to make environmental data processing simpler so that it enhances the ability of vehicles to manage complex driving conditions. With focus on reliable information and compensation for computational cost, these developments facilitate realistic application of sophisticated machine learning algorithms to vehicle systems. With the evolution of self-driving vehicles in the future, the emphasis on inserting intelligence in their operating systems showcases the emphasis on performance as much as safety on future mobility platforms (Abdou et al., 2017).

5.4. CHALLENGES IN IMPLEMENTING MACHINE LEARNING FOR EMBEDDED INTELLIGENCE

Machine learning (ML) integration into embedded systems has drawn considerable attention with the potential of simplifying decision-making in real-time applications. When implemented, however, there are different issues that are at play. One of the stand-out ones is that little computational capacity is available in embedded devices, which in most instances makes complex ML algorithms impossible to implement. These machines tend to have to operate on models not just efficient in nature but also comfortable operating in sparse resource constraints. Therefore, researchers must balance the correctness of ML models with the requirements of saving power and memory, with the possible outcome of diminished performance in high-critical applications. The efficient use of ML in these systems necessitates novel optimization techniques to bypass the inherent hardware constraints without sacrificing the functionality required for health monitoring and autonomous applications. Data gathering and preprocessing present another significant challenge necessary to create effective ML models.

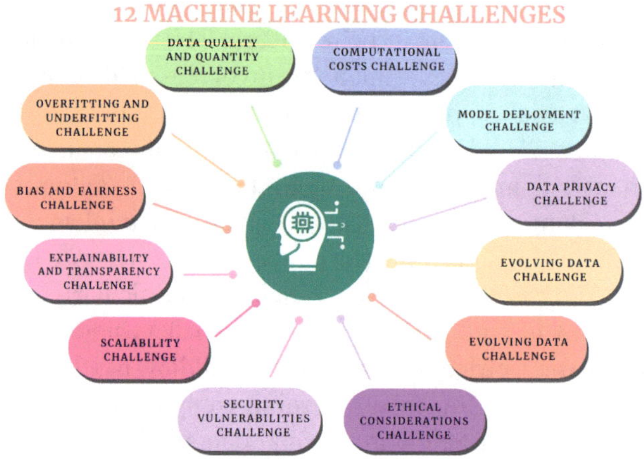

Figure 5.9. *Machine learning challenges and solutions.*

Source: https://external-content.duckduckgo.com/iu/?u=https%3A%2F%2Fmtplcdn.s3.amazonaws.com%2Flive%2Fbigdatacentric%2Fwp-content%2Fuploads%2F2024%2F05%2F05083841%2F12-Machine-Learning-Challenges-1.jpg&f=1&nofb=1&ipt=89e1e5b2b1e043f3a0c3cb98e34e08fb7bf7504f987acc961c1099c9d5e4f7e4.

Embedded systems generally operate in dynamic conditions, which lead to variation in data being collected, i.e., noise and inconsistent signal quality. This variation has major impacts on the robustness and reliability of the machine learning systems employed in such a setting. For instance, in identifying the functioning of older adults through methods like the Timed-Up and Go test, the reliability of the inertial sensors that are part of mobile devices becomes the focal point. It is important to apply advanced signal processing techniques to enhance data quality prior to model training, as in systems with accelerometers and gyroscopes (Garcia et al., 2020). In case data preprocessing is inadequate, the machine learning models cannot generalize, resulting in degraded operational performance. In addition, ethical considerations and compliance with legislation are main challenges facing the incorporation of machine learning into embedded intelligence. Since ML systems increasingly influence sensitive decision-making domains, such as healthcare and security, the right consent and data privacy become required yet demanding facets of implementation. Sectors undergoing change towards Industry 4.0 are particularly impacted, as they must grapple with shifting systems of accountability and transparency towards data use (Prieto et al., 2019). Such a complex setting calls for deep understanding of legal frameworks as well as ethical aspects that are specific to machine learning technologies. Developers can, by so doing, guarantee that embedded systems using machine learning do not just optimize performance but uphold standards of ethics that society demands (Figure 5.9).

5.4.1. Resource Constraints

In embedded intelligence, resource limits pose tremendous challenges to be overcome before being able to leverage the full capability of machine learning. Embedded systems used by industries wide and far, including automotive and healthcare, are normally under severe computation power, memory capacity, and energy consumption constraints. These limitations render it difficult to implement sophisticated algorithms common in machine learning, as these typically require significant processing resources and memory. These limitations necessitate the re-examination of classic machine learning mechanisms to ensure that they are able to operate effectively within the limits of operation in embedded systems and comply with real-time performance levels required in critical applications in such fields (Shi et al., 2023). Estimates for offsetting these requirements are significant to ensure smooth integration of machine learning in the field of embedded systems. Resource constraint handling within embedded systems comprises innovative methods that optimize machine learning efficiency in terms of restricted resources. One approach is to design light models specifically suited for low-power, low-memory environments. Techniques such as model pruning, quantization, and knowledge distillation are utilized to reduce the size and complexity of machine learning models so that they can be supported by the limitations of embedded hardware (Shamsa et al., 2022). Furthermore, the utilization of algorithms aware of the resources is in a way that computationally and energy-expensive parts of machine learning can be carried out without compromising computational and energy constraints typical of embedded systems. This optimization work not only enhances model execution performance but also provides better sustainability for devices that are embedded through minimized usage of energy, hence extending their lifetime in the field. Furthermore, dynamic adjustment schemes are responsible for dealing with varying patterns of resource constraints within embedded systems. With changing workloads and operating conditions, machine learning-enabled systems must react quickly to coping with changes in order to guarantee performance levels. This demands the implementation of self-aware resource management strategies that use online learning mechanisms in order to learn about real-time situations while balancing contradictory priorities such as energy efficiency, performance requirement, and user preferences. Through the implementation of full-resource management techniques that employ machine learning for both offline modeling and online control, system designers can better respond to the complex and usually conflicting demands on embedded systems (Shamsa et al., 2022). Consequently, the addition of these intelligent resource management methods is a big advance in the implementation of machine learning in resource-constrained environments so that there are more function and flexibility in embedded intelligence systems.

5.4.2. Data Privacy and Security

In the domain of embedded intelligence, the security and privacy of data are of utmost concern due to the delicate information processed and stored in the systems. Since machine learning models govern the operational decisions of several technologies, it becomes necessary to have the data as private as possible and shielded against the wrong users. Conventional security solutions tend to fail when dealing with the specific challenges of next-generation cyber-physical systems (NG-ICPSs) in which the interconnected nature of cyber and physical layers introduces new vulnerabilities. In fact, the heterogeneity of NG-ICPSs, precipitated by the integration of machine-embedded sensors and advanced cyber-infrastructure, calls for sophisticated security frameworks that are resilient enough to counteract emerging security threats. Heightened sensitivity to these issues highlights the imperative for innovative solutions that leverage the potential of artificial intelligence to create adaptive and resilient security measures capable of keeping pace with evolving threats, setting the stage for better data privacy and system resilience in embedded intelligence applications. Machine learning's contribution to enhancing data privacy cannot be overstated, especially when addressing the menace of the Internet of Things (IoT). With the increasing number of devices being integrated, the collection and transmission of confidential data, the risk of security intrusions has risen. The landscape demands Intrusion Detection Systems that will respond to detecting and shielding IoT networks. Random Forest and Deep Neural Networks are becoming major technology here with their potential to identify abnormal behavior and discover intrusion attempts in real-time. Through such cutting-edge methods, network administrators can foretell potential threats, negating risks beforehand. Such a preemptive strategy not only fortifies security measures but also assists in building more public confidence in the application of IoT technologies, subsequently aiding in privacy and integrity programs in embedded intelligent systems. Further, the incorporation of AI-based methods, especially in the authentication procedures of NG-ICPSs, is a major leap forward in data protection. Conventional authentication methods tend to be inefficient and non-scalable because of the intricate verification processes, which compromise effectiveness in increasingly changing digital landscapes. However, machine learning-based authentication mechanisms can simplify verification and improve security at the same time. By recognizing patterns and responding to various datasets, these systems can learn to deal with new threats and user behaviors, providing privacy threats with an adaptive countermeasure. Such advancements not only satisfy existing security needs for embedded intelligence applications but also pave the way for future breakthroughs. With the continued development of AI, ML, and deep learning, their application towards data privacy will be sure to lead to sophisticated systems capable of addressing the intricate problems within the machine learning for embedded intelligence space, capturing the fine balance between innovation and security in the era of technology (Adil et al., 2025; Varaprasad et al., 2024) (Figure 5.10).

Figure 5.10. *Data security versus data privacy.*

Source: https://www.stealthlabs.com/wp-content/uploads/2020/10/whats-the-core-difference-between-data-security-and-data-privacy.jpg.

5.4.3. Algorithm Complexity

Algorithm complexity is also a crucial factor in the efficiency and usability of machine learning models, especially in embedded intelligence systems. Algorithm complexity relates to the amount of computation resources, such as time and memory, which are required to process data and produce outputs. In embedded systems, operation is typically under restriction due to limited processing and power usage. Optimal selection of algorithm complexity becomes very important here. By knowing the complexity of algorithms at a deeper level, researchers and developers can make an efficient decision so that they employ models with optimal performance rates without overwhelming the restricted resources of embedded devices. As machine learning becomes more advanced, being able to support ever-increasing complexity algorithms must involve a balance of measuring their complexity so that they can remain relevant for practical implementations in embedded environments. Moreover, the relationship between algorithm accuracy and complexity cannot be overstated in inductive concept learning. Such learning involves classifying inputs based on learned models, and algorithm complexity can greatly influence its efficiency in classification.

For instance, more advanced algorithms may require longer periods of training and more extensive databases to achieve acceptable accuracy. However, they may be likely to cause higher costs in terms of misclassification errors whose effects can reach far beyond theoretical confines into practicality. Researchers can be able to acquire more cost-effective models for learning by examining the various types of costs associated with the complexity of algorithms. Operating through such cost factors raises the awareness of the interplay between complexity, precision, and procedural effectiveness in machine learning systems (Turney et al., 2000).

As Big Data becomes increasingly linked with machine learning, algorithm complexity implications extend beyond computational considerations. Embedded intelligence systems are likely to make use of large amounts of data as part of the decision-making process, and this boosts the issue challenged by algorithm complexity. Big Data analytics is able to leverage effective computational intelligence techniques that limit these complexities by enabling improved data processing and modeling. For example, in smart city contexts, Big Data algorithms must balance complexity with real-time processing to enable effective integration into urban infrastructures. The use of innovative methods, such as the Hierarchical Spatial-Temporal State Machine (HSTSM), indicates the extent to which advanced algorithmic methods facilitate the intelligent use of data as they deal with complexity. As researchers search for the various uses of Big Data and machine learning, it becomes more necessary to recognize the supreme importance of algorithm complexity in designing intelligent and dynamic solutions that can deal with the evolving needs of modern society (Doctor et al., 2020).

5.5. TOOLS AND FRAMEWORKS

Embedded intelligence relies heavily on making the right tool and frameworks choices in a bid to facilitate the deployment of machine learning models in resource-constrained settings. The rapid advancement that has been witnessed in machine learning, particularly with Convolutional Neural Networks (CNNs), has led to an ample number of software frameworks for the purpose of enhancing performance on numerous hardware architectures. While traditional CPUs and GPUs dominate, the addition of reconfigurable hardware like Field Programmable Gate Arrays (FPGAs) has emerged as the dominant alternative. FPGAs give a balanced control between performance and power consumption that makes them perfectly suited for real-time application in embedded systems. A review of existing CNN-to-FPGA toolflows emphasizes the importance of architectural choices and design space exploration methods, thus demonstrating the diversity of options that are available to engineers for creating power-efficient machine learning models for embedded intelligence (Bouganis et al., 2018). Besides hardware, the widespread adoption of open-source machine learning (ML) software is increasingly dominating the terrain of embedded intelligence. Software such as TensorFlow, DeepLearning4j, and H_2O is at the forefront of development, providing flexible functionality that caters to various application requirements. A comparison of the software reveals their distinct strengths and potential vulnerabilities, particularly when operating in systems with low computational capability. Surprisingly, hyperparameter optimization in such systems can be drastic for model performance, as seen in MNIST tests with improper selection of parameters causing extreme runtime hikes without corresponding increases in accuracy. Engineers thus need to fine-tune such critical parameters to ensure efficient deployment of machine learning algorithms on embedded devices (Alienin et al., 2017). The convergence of hardware innovation with good software architectures is allowing unparalleled growth in the field of embedded intelligence. With the leverage of the capabilities of FPGAs and adaptable machine learning architectures, professionals can develop systems that not only perform complex operations but also react to the dynamic nature of real-world environments. Constant innovation in such tools necessitates constant investigation into means to project machine learning algorithms onto target hardware configurations efficiently. It also necessitates focus on the importance of comprehensive evaluation methodologies considering application-specific factors. With embedded intelligence spreading increasingly across numerous industries, collaborative research in building frameworks and tools will be the key to unlocking the capabilities of machine learning in achieving the needs of next-generation intelligent systems (Alienin et al., 2017; Bouganis et al., 2018) (Figure 5.11).

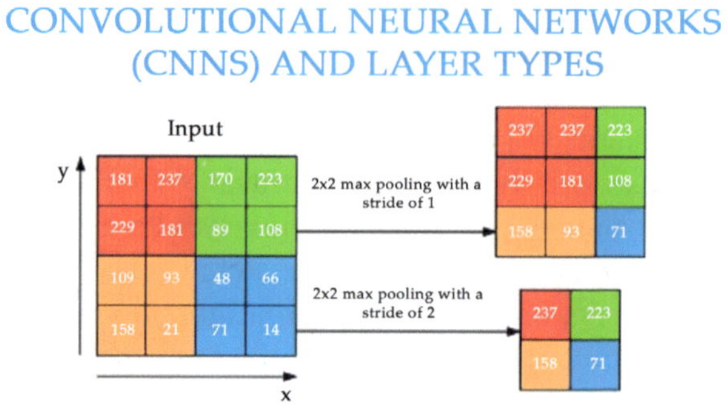

Figure 5.11. *Convolutional neural networks (CNNs) and layer types.*

Source: https://pyimagesearch.com/wp-content/uploads/2021/05/Convolutional-Neural-Networks-CNNs-and-Layer-Types.png.

5.5.1. TensorFlow Lite

In the rapidly changing world of machine learning, TensorFlow Lite has been a game-changing platform for deploying highly efficient ML models onto embedded hardware. Designed for mobile and edge platforms, TensorFlow Lite enables developers to convert standard TensorFlow models into an optimized and light-weight form that can be used with less latency and lower power consumption. This allows for real-time inference on low-computation devices like smartphones, wearables, and IoT devices. Consequently, the distribution of ML applications across industries is easier today because organizations can now put smart functionality in consumer devices directly. Overall, TensorFlow Lite is an important breakthrough in allowing machine learning for more applications to be processed efficiently on devices. The significance of TensorFlow Lite could be underscored by the increased need for machine learning systems that are able to bridge across a range of hardware, from embedded solutions to data-center configurations. This emerging need is pointed out by the increase in the number of ML inference systems, where over 100 firms are engaged in developing specialty ML chips for use across different applications. The capabilities of such systems bridge across an unparalleled range, with performance and power consumption differing widely. The TensorFlow Lite integration enables this diversity by providing a universal and adaptable platform for executing models between hardware architectures. Through the unification of varying levels of processing capability and application need, TensorFlow Lite not only broadens the scope of machine learning but also responds to the fundamental demand for standardization and replicability in this evolving universe of ML applications (Anderson et al., 2020). Further, TensorFlow Lite aligns with the trend of leveraging machine learning for novel health care solutions, particularly mobile health. For instance, intelligent robots that will perform clinical interventions require efficient data processing to heighten interaction with humans. TensorFlow Lite enables such applications by providing efficient model deployment that satisfies key requirements of privacy and performance. The platform can facilitate data exchange and collaboration among multiple robotic units in different healthcare settings, thereby enhancing their learning process without infringing on patient privacy. As illustrated through efforts in developing platforms like RoboChain, where machine learning is coupled with secure data sharing methods, TensorFlow Lite is an essential component toward revolutionizing healthcare interactions and thereby improving patient care (Ferrer et al., 2018) (Figure 5.12).

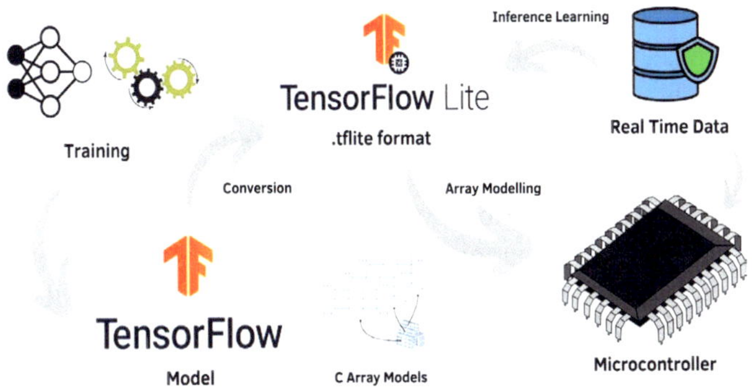

Figure 5.12. *TensorFlow lite.*

Source: https://static1.makeuseofimages.com/wordpress/wp-content/uploads/2021/05/General-Work-Flow-for-TensorFlow-Lite-Micro.jpg.

5.5.2. PyTorch Mobile

PyTorch Mobile is a cutting-edge way of deploying machine learning models on embedded platforms. As there is a growing demand for cognitive applications within mobiles, frameworks facilitating this integration are of high importance. PyTorch Mobile leverages the PyTorch ecosystem's inherent flexibility since it optimizes the models for deploying them in resource-restricted environments. This ensures that sophisticated deep learning applications can be run effectively on devices with constrained computational capacity and battery life. PyTorch Mobile not only has cross-platform support but also provides the user with a deep experience through a lightweight and efficient solution without compromising performance. Such support is essential in the context of the nature of machine learning applications that span across multiple domains ranging from mobile health monitoring to real-time visual recognition tasks in autonomous driving. The simplicity of deploying models releases developers to steer innovation toward smarter, more powerful mobile technologies. Secondly, the application of PyTorch Mobile aligns with broader industry trends highlighting the rapid evolution of machine learning inference frameworks. Indeed, ubiquitous heterogeneity of ML hardware architectures establishes a demand for those platforms that may adapt dynamically across various computation platforms. As noted in literature, firms are developing multiple ML inference chips and in the process, abandoning a knotted ecosystem of embedded solutions on various platforms (Anderson et al., 2020). PyTorch Mobile counters this challenge through a robust set of tools that allow developers to tailor their models to suit specific hardware setups, thereby enhancing performance and efficiency. Increased adoption of standardized benchmarking like MLPerf Inference provides indispensable insights in obtaining similar performance metrics across different systems, further making flexible frameworks like PyTorch Mobile indispensable in preserving consistency and reproducibility. Finally, real-world applications of models implemented via PyTorch Mobile provide evidence of the strength in enhancing embedded intelligence. As a case in point, recent advances such as M2U-Net architecture demonstrate significant improvements in retinal vessel segmentation and can be employed on mobile devices for real-time use (Jalali et al., 2020). By annihilating the processing requirements without compromising on accuracy, these apps pave the way for innovative applications in healthcare, which can further be used in biometric authentication systems and disease detection in early stages. The fact that PyTorch Mobile has the capability to execute such sophisticated models on ARM-based embedded platforms says a lot about its role in transforming the development and usage of intelligent applications. As the limits of mobile computing are pushed further and further, PyTorch Mobile is a vital tool in the search for intelligent and efficient embedded systems and eventually in the creation of machine learning towards future-proofed user experiences in everyday technology.

5.5.3. Edge Impulse

The release of Edge Impulse is a significant advancement of the integration of machine learning technologies into embedded intelligence systems. The platform is intended to simplify machine learning model development and deployment for resource-constrained environments, such as microcontrollers and edge devices. By facilitating an end-to-end process—data acquisition to model training and deployment—Edge Impulse makes it possible for developers to unlock the power of artificial intelligence without needing to possess advanced expertise in the field. This machine learning democratization paves the way for a wide variety of applications ranging from industrial automation to environmental sensing, which lead to enhanced operational efficiency and innovative solutions across many domains. Its key to success is its ability to perform rapid processing of high-level data types, enabling the embedded systems to perform predictive analytics in real time, a critical consideration for reliability and safety in embedded use. Edge Impulse's foundation architecture does not end with model deployment; it extends to high-level machine learning paradigms to optimize system performance. With the aid of deep neural networks and other algorithms, Edge Impulse enjoys great flexibility in tailoring models to specific tasks and therefore maximizing the utilization of resources.

Figure 5.13. *Object classification with edge impulse using raspberry.*

Source: https://external-content.duckduckgo.com/iu/?u=https%3A%2F%2Fstatic.cytron.io%2Fimage%2Ftutorial%2Fedge-impulse-with-raspberry-pi-4%2Fpresentation1.jpg&f=1&nofb=1&ipt=2964d6d42a552450dd85325a2cd49f2d17dad904c13194b8dac423e615a452e8.

The platforms' approach aligns with recommendations by Hong et al. (1994), where the emphasis is placed on learning system integration in applications for adaptation and ongoing improvement. By being able to integrate various learning techniques, including symbolic and neural network learning, Edge Impulse allows for more flexibility under shifting conditions. This is particularly beneficial where information is continuously being collected and models need to be revised to accommodate new information, enhancing the systems resilience and performance in real-world application. Additionally, the incorporation of sparse computation techniques complements the abilities of Edge Impulse. As described by Kepner et al. (2017), deep neural network performance can be greatly improved through the application of sparse matrix operations, a fundamental component of Edge Impulses architecture. By minimizing memory consumption and computation associated with dense matrix computation, these techniques enable complex machine learning models to execute at an optimal level within the limitations of embedded hardware. This power-efficiency strategic focus ensures that Edge Impulse can provide high-performance analytics without compromising on power efficiency, a critical need in embedded systems. Ultimately, the intersection of ease-of-use, flexibility, and computational efficiency puts Edge Impulse firmly at the forefront of the machine learning for embedded intelligence arena, paving the way for groundbreaking applications across various industries (Figure 5.13).

5.6. FUTURE TRENDS IN MACHINE LEARNING FOR EMBEDDED INTELLIGENCE

The future of embedded intelligence in machine learning (ML) is going to witness monumental advancements through ongoing improvements in computation power and effectiveness of algorithms. With increasingly powerful devices in the pipeline, the embedded intelligence platforms are expected to process complex data without any glitch at the source. This saves latency and enables better real-time decisions, which is crucial for autonomous vehicle and IoT appliance use. Further, edge computing technologies will likely quicken this trend, allowing machine learning algorithms to execute without relying on centralized server clouds and network environments to be less vulnerable to network outages. As foreseen in the contemporary debate regarding Big Data's role, integration of ML with embedded intelligence will enable intelligent urban environments, with smarter transport and resource management technologies for smart cities, as outlined in the intelligent transport case study with new data modeling approaches (Doctor et al., 2020). Another crucial nascent trend is machine learning with embedded systems' increased potential for personalization. The next-generation ML algorithms are expected to learn user preferences over time, hence providing a more tailored user experience across various devices such as wearables, smart home devices, and personal assistants. With advancing machine learning models, they will increasingly utilize data-driven insights to facilitate frictionless interactions, acting on the distinct user behavior in real-time. This level of personalization not only enhances user satisfaction but also drives efficiency in energy consumption and performance improvement, as dictated by the principles witnessed at the intersection of Big Data and computational intelligence. Integrating advanced ML techniques into embedded systems will thus pave the way for more responsive environments, highlighting the extent of transforming everyday objects into data-driven intelligent objects (Sandford et al., 2009). Also, the examination of policy rules and ethical aspects becomes imperative as embedded intelligence machine learning continues to expand. This aspect will dictate the way of technology use in the future while ensuring that embedded systems are not only efficient but also ethical. The specter of data misuse, privacy breaches, and algorithmic bias necessitates a systemically based model of government and regulation in ML usage. As highlighted in the discussions on Big Data value and safeguard, a robust framework will be crucial in balancing innovation and ethical accountability. Policymakers and technologists will need to collaborate to establish parameters that will ensure protection for users as they promote technological advancement. By solving such dilemmas beforehand, the future of embedded intelligence machine learning can thrive in a culture that values both progress and ethical integrity (Doctor et al., 2020) (Figure 5.14).

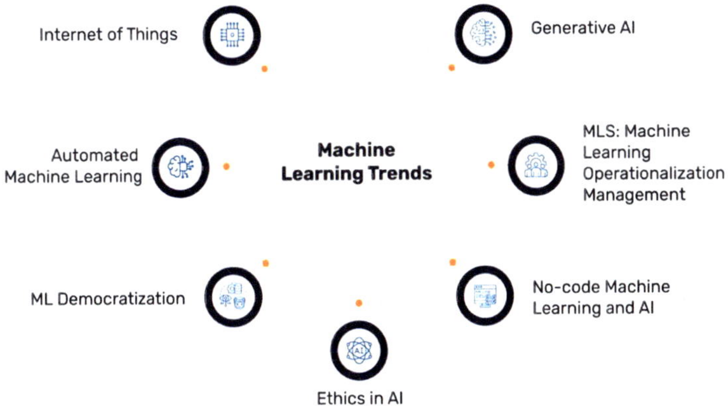

Figure 5.14. *Machine learning trends.*

Source: https://ewzduhvhjkj.exactdn.com/wp-content/uploads/2022/09/02200254/Machine-Learning-1024x572.jpg?strip=all&lossy=1&ssl=1.

5.6.1. Increased Adoption of AI at the Edge

Expanding edge AI adoption represents a revolution within the domain of embedded intelligence, allowing devices to process locally rather than relying solely on centralized cloud infrastructure. A number of drivers are making this trend increase, from the need for real-time analytics in applications such as the Industrial Internet of Things (IIoT) and smart manufacturing. As firms strive to achieve operational efficiencies and reduce latency, edge AI systems can facilitate these by enabling data to be processed close to where it is generated. By enabling real-time decision-making, edge AI directly influences enhanced responsiveness in high-stakes applications, therefore enhancing innovation and competitiveness in sectors relying on instant data-driven information, for example, innovations covered in Industry 4.0 initiatives globally (Ali et al., 2017; Prieto et al., 2019). Besides, advancements in low-power, high-efficiency processors have made it possible for machine learning algorithms to be utilized effectively in devices of all edge environments. This technology has made it possible for embedded systems to perform complex calculations with minimal energy expenditure, thereby prolonging the lifecycle and allowing for operations. As a result, industries such as automotive, manufacturing, and healthcare are increasingly adopting AI in their edge devices to simplify operations and improve user experience. The arrival of specialized edge AI hardware augments software advancements with a robust platform for machine learning and data analysis applications that previously were confined to traditional computing platforms. This innovation is a textbook example of the intersection of hardware and software in realizing the maximum potential of embedded intelligence (Ali et al., 2017; Prieto et al., 2019). Using AI at the edge also creates learning and development challenges that universities are only beginning to address. As Industry 4.0 reshapes the technology environment, universities and engineering education schools must adapt their curricula to equip the next generation of professionals to meet the challenges of this new field. Initiatives like Industry 4.0 Technologies Laboratory at the Technical University of Catalonia are illustrations of shifting the education model with the focus being on increasing capacities in the context of edge computing and embedded intelligence (Prieto et al., 2019). By fostering an active learning environment that includes hands-on application of AI technologies, such initiatives seek to develop a workforce that can understand the interdependencies of complicated systems, thereby fueling the innovation required for a digitized globe. The resulting innovation in education represents the larger societal effect of the broadening edge-based AI adoption, demonstrating the dire necessity of the conjunction of technology and knowledge creation.

5.6.2. Advances in Hardware for Machine Learning

Hardware development for machine learning is essential for advancing embedded intelligence systems. Machine learning has traditionally demanded significant computational power, with CPUs and GPUs being the workhorse for computation. Application-specific integrated circuits (ASICs) and field-programmable gate arrays (FPGAs) have revolutionized this tradition. These hardware technologies enable the acceleration of machine learning workloads with lower power consumption and latency while maintaining high performance levels. This capability is particularly important in embedded systems where efficiency is critical. With particular hardware design for machine learning software, developers are able to create optimized solutions that best fit the unique constraints and requirements of such systems in order to improve real-time decision-making and overall operational effectiveness in applications from autonomous vehicles to smart sensors in urban environments (Stein et al., 1999). Such futuristic hardware solutions for machine learning integrated in embedded systems also enhance the handling of big data, which is the hallmark of smart city initiatives. The integration of powerful computational intelligence techniques with state-of-the-art hardware enables analysis and processing of huge volumes of data generated by city technology and thereby facilitates urban management as well as efficient resource allocation. For instance, the Hierarchical Spatial-Temporal State Machine (HSTSM) is a cutting-edge data model that leverages biological inspiration to increase the processing capabilities at the hardware level. This facilitates smart city intelligent transportation solutions, presenting an example of how advanced hardware can enable successful applications by ensuring seamless data flow and real-time processing. Hence, not only do these advancements improve computational effectiveness but also allow cities to be more responsive and resilient (Doctor et al., 2020). Furthermore, these hardware advances also challenge a re-thinking of existing computational metaphors underlying our understanding of cognition and machine learning. The shift from traditional processing models to that of interconnected units suggests a new perspective of computing, in which collaboration of various pieces of hardware results in emergent intelligent behavior. Through conceptualizing a new means by which computational units can work together with each other, researchers make way for new structures centered on cooperation and adaptability rather than linear sequence of processing. This paradigm change works in harmony with the developments seen in machine learning hardware to provide environments where embedded intelligence can flourish. With the lines between hardware, software, and cognition blurring, machine learning for embedded systems is not a technological shift per se but a core paradigm shifts in the very essence of intelligence (Stein et al., 1999).

5.6.3. Ethical Considerations and Regulations

The integration of machine learning into embedded intelligence systems is of extreme ethical importance to be handled in a manner that enables responsible deployment. At the core of these matters lies the need for discrimination between various forms of artificial intelligence, e.g., Artificial General Intelligence and weak AI, as well as various methodologies, e.g., symbolic and connectionist AI. It is critical to the evolution of educated debate about ethical implications and social effects built into such technologies. Definitions of these disciplines properly and merely permit more rigorous academic discourse as well as for the interdisciplinarity of knowledge exchange needed to answer the advanced questions posed by embedded intelligence. These differences contribute to the development of a coherent framework of standards by which the ethical considerations around machine learning methods can be evaluated and guide researchers toward well-informed and responsible decision-making in their fields (Etzioni et al., 2019). In addition, with machine learning being used more firmly embedded in daily life, data privacy and security issues are the paramount ethical concerns. The implementation of such systems tends to entail the gathering and analyzing of tremendous amounts of personal information, which causes concerns regarding how this data is accessed, stored, and utilized. Ethical frameworks should place utmost priority on transparency and accountability in order to protect user data from potential abuse. Additionally, the developers need to ensure that their systems are framed within inclusivity and equity, ensuring that issues such as bias in training data and algorithmic decision-making are dealt with. This emphasis on ethical design serves not only individual rights but also the overall trust in machine learning technologies, so they serve a broad set of stakeholders without offending underrepresented groups (Apesoa-Varano et al., 2020). It is important that regulatory systems engage with negotiating the ethical landscape that comes with embedded intelligence in machine learning. Policy makers need to create standards that strike an optimal balance between protection and innovation, establishing norms that balance privacy, security, and ethical use. An effective regulatory framework involves convening technologists and social scientists to create a shared framework of ethical issues and users' rights. Because technologies are evolving rapidly, legislation has to be sufficiently agile so that it can respond to emerging problems while they occur and not erode public trust in AI systems. Engaging a wide range of stakeholders in the regulation process ensures that various perspectives are considered, which fosters the creation of policies that not only drive innovation but also defend ethical standards for machine learning for embedded intelligence (Apesoa-Varano et al., 2020).

SUMMARY

Machine learning for embedded intelligence, it is evident that this hybrid of technologies is revolutionary in various sectors. The advent of 6G wireless networks is an example, where machine learning algorithms are poised to revolutionize communication systems. Through the facilitation of unprecedented degrees of interconnectivity, these advancements have the potential to not only make device-to-device interactions more effortless but also allow for such new applications as autonomous vehicles and intelligent voice assistants. With the demand for innovative connectivity technologies continuing to soar, the application of machine learning in addressing both existing challenges in wireless communications as well as future ones cannot be overemphasized, as outlined by Abbas et al. (2020). The implication is towards smarter, more effective systems that learn to adapt to user needs and environmental conditions in a useful way. Above this, the combination of machine learning and cognitive science provides further understanding of how the potential of embedded intelligence can reshape our approach to technology's place in society. The cognitive approaches to technology place emphasis on interactive interactions, highlighting how users perceptions and decisions are influenced by the digital environment. Machine learning is able to improve these interactions by adapting experiences to preference and behavior. By being situated and ecological, as put forward by Clowes et al. (2017), we can see how these systems operate within collective and social contexts, resulting in more robust paradigms for investigating the influence of technology on cognition in human beings. This new vision opens the capability of embedded intelligence to facilitate not only task performance but also richer and deeper user experiences. In total, machine learning for embedded intelligence will revolutionize different dimensions of society by enhanced connectivity, maximum user experience, and new inventions. The ongoing advancement of these technologies places the necessity of research that is interdisciplinary spanning machine learning, cognitive science, and communication studies. As we persist in navigating the complexities of this integration, it is crucial that we remain attuned to the ethical considerations and social impacts that arise. Accordingly, we are able to realize the full potential of machine learning and build smart systems that are not only highly capable but also aligned with the higher aims of social health and solidarity. This forward-thinking vision will be crucial as we seek to create intelligent spaces that serve and enhance the human experience.

The addition of machine learning (ML) to embedded intelligence is a revolutionary move toward the automation and optimization of numerous systems. The initial focus should be on expanded use of artificial intelligence (AI) and ML to boost cybersecurity practices across connected devices. With more businesses going online, the heightened complexity and frequency of cyber-attacks highlight the importance of advanced security practices. The development of AI-powered cybersecurity tools has been found to have a remarkable rise in detecting and responding to easy and complex attacks in real time. This real-time feature is important given the fact that Nation-State-Affiliated Adversaries and Criminal Adversaries continuously update their approach, employing vulnerabilities as and when they arise, thus facilitating the need for cutting-edge ML applications in guarding digital infrastructure (Hussain, 2024). Apart from the issue of cybersecurity, there is the problem of machine learning being applied to improve network performance, particularly in IoT architectures. The intersection of AI and IoT technologies revolutionizes virtually the management of sensitive data within networks of wireless sensors (WSNs). Conceptual innovations, for example, such as a hybrid key management protocol, are characteristic of this convergence. Through the use of cloud services coupled with machine learning, specifically the XGBoost algorithm, these sites aim to optimize the security and efficiency of data communication. The combined higher-level capabilities enabled by this enable critical functions like smart network load balancing and dynamic key management in a way that data transmission becomes both efficient and secure in an era of time-sensitive communication being on the rise (Rao et al., 2024). Finally, the potential of machine learning goes way beyond mere data handling; it actually assists in producing predictive analytics and anomaly detection in embedded systems. With the incorporation of such concepts as User Behavior Analytics (UBA), systems are able to enhance security breach and operational anomaly prediction prior to occurrence. This anticipatory approach prevents risks and therefore enhances both security and performance results across numerous various applications. The coming together of predictive capability with real-time data analysis serves to underscore the applicability of ML in optimizing embedded intelligence. With increasing business reliance on data-driven insights, the requirement to refine ML algorithms and provide ethical application takes on ever-increasing significance. Therefore, summarizing all these aspects, we can realize that machine learning, intertwined with network management and cyber security, forms a pivotal basis to decide the fate of embedded intelligence (Hussain, 2024; Rao et al., 2024).

REFERENCES

1. Abbas, R., Ahmad, I., Ali, S., Samad, A., Amuru, S., et al. (2020). 6G white paper on machine learning in wireless communication networks. http://arxiv.org/abs/2004.13875 (accessed on 10th September 2025).
2. Abdou, M., Perot, E., Sallab, A. E., Yogamani, S., et al. (2017). Deep reinforcement learning framework for autonomous driving. http://arxiv.org/abs/1704.02532 (accessed on 10th September 2025).
3. Abtahi, A., Grzywaczewski, A., Agrawal, R., Al-Sultan, A., Asimov, R., Bernardo, A., Bezdek, J., et al. (2018). A novel big data analytics and intelligent technique to predict driver's intent. https://core.ac.uk/download/228154021.pdf (accessed on 10th September 2025).
4. Adil, M., Farouk, A., Abulkasim, H., Ali, A., Song, H., & Jin, Z. (2025). NG-ICPS: Next generation Industrial-CPS, security threats in the era of artificial intelligence, and open challenges with future research directions. *IEEE Internet of Things Journal, 12*, 1343–1367. https://www.semanticscholar.org/paper/6d5922275b70e57407d5565b61842db4c41e1f90 (accessed on 10th September 2025).
5. Aghi, D., Chiaberge, M., & Mazzia, V. (2020). Local motion planner for autonomous navigation in vineyards with a RGB-D camera-based algorithm and deep learning synergy. https://core.ac.uk/download/pdf/327177639.pdf (accessed on 10th September 2025).
6. Ali, M., Anjum, A., Devitt, J., Hill, R., et al. (2017). Towards in-transit analytics for Industry 4.0. http://arxiv.org/abs/1710.04121 (accessed on 10th September 2025).
7. Alienin, O., Gordienko, Y., Kochura, Y., Novotarskiy, A., et al. (2017). Performance analysis of open-source machine learning frameworks for various parameters in single-threaded and multi-threaded modes. http://arxiv.org/abs/1708.08670 (accessed on 10th September 2025).
8. Alsheikh, M. A., Lin, S., Niyato, D., Tan, H. P., et al. (2014). Machine learning in wireless sensor networks: Algorithms, strategies, and applications. https://core.ac.uk/download/35456291.pdf (accessed on 10th September 2025).
9. Anderson, B., Breughe, M., Charlebois, M., Cheng, J., et al. (2020). MLPerf inference benchmark. http://arxiv.org/abs/1911.02549 (accessed on 10th September 2025).
10. Apesoa-Varano, E. C., Gladstone, C., Kim, K. K., Lindeman, D., et al. (2020). Technology and caregiving: Emerging interventions and directions for research. https://core.ac.uk/download/323070528.pdf (accessed on 10th September 2025).

11. Ashouri, A. H., Cavazos, J., Killian, W., Palermo, G., et al. (2018). A survey on compiler autotuning using machine learning. http://arxiv.org/abs/1801.04405 (accessed on 10th September 2025).
12. Belpaeme, T., Cangelosi, A., Dautenhahn, K., Fadiga, L., et al. (2010). Integration of action and language knowledge: A roadmap for developmental robotics. https://core.ac.uk/download/1640754.pdf (accessed on 10th September 2025).
13. Bouganis, C. S., Kouris, A., & Venieris, S. I. (2018). Toolflows for mapping convolutional neural networks on FPGAs: A survey and future directions. https://core.ac.uk/download/189837587.pdf (accessed on 10th September 2025).
14. Chen, C., Del Bue, A., Murino, V., Zhang, J., et al. (2017). Manifold constrained low-rank decomposition. http://arxiv.org/abs/1708.01846 (accessed on 10th September 2025).
15. Civit Balcells, A., Domínguez Morales, M. J., & Luna Perejón, F. (2019). Wearable fall detector using recurrent neural networks. https://core.ac.uk/download/286563383.pdf (accessed on 10th September 2025).
16. Clowes, R. W., Heersmink, R., & Smart, P. (2017). Minds online: The interface between web science, cognitive science, and the philosophy of mind. https://core.ac.uk/download/131214931.pdf (accessed on 10th September 2025).
17. Delgado Prieto, M., Fernández Sobrino, Á., Fibla Biosca, P., Romeral Martínez, M., et al. (2019). Active learning-based laboratory towards engineering education 4.0. https://core.ac.uk/download/294830056.pdf (accessed on 10th September 2025).
18. Dewey, D., Russell, S., & Tegmark, M. (2015). Research priorities for robust and beneficial artificial intelligence. https://core.ac.uk/download/83234142.pdf (accessed on 10th September 2025).
19. Doctor, F., Iqbal, R., Mahmud, S., More, A., et al. (2020). Big data analytics: Computational intelligence techniques and application areas. https://core.ac.uk/download/228156656.pdf (accessed on 10th September 2025).
20. Dwivedi, Y. K., Kshetri, N., Hughes, L., Slade, E., Jeyaraj, A., Kar, A. K., Baabdullah, A. M., et al. (2023). Opinion paper: "So what if ChatGPT wrote it?" Multidisciplinary perspectives on opportunities, challenges and implications of generative conversational AI for research, practice and policy. *International Journal of Information Management, 71*, 102642–102642. https://doi.org/10.1016/j.ijinfomgt.2023.102642 (accessed on 10th September 2025).

21. Etzioni, A., Goodman, B., Coenen, C., Cath, C., Luxton, D. D., Waltz, D. L., Fisher, E., et al. (2019). Ethics of artificial intelligence demarcations. http://arxiv.org/abs/1904.10239 (accessed on 10th September 2025).
22. Evans, J. W., Knizhnik, J. R., McGowan, A. M. R., Moreland, J., et al. (2025). An EPIIC vision to evolve project integration, innovation, and collaboration with broad impact for how NASA executes complex projects. https://core.ac.uk/download/pdf/323104112.pdf (accessed on 10th September 2025).
23. Ferrer, E. C., Hardjono, T., Pentland, A., Rudovic, O., et al. (2018). RoboChain: A secure data-sharing framework for human-robot interaction. http://arxiv.org/abs/1802.04480 (accessed on 10th September 2025).
24. Garcia, N., Marques, G., Pires, I., Pombo, N., et al. (2020). Is the timed-up and go test feasible in mobile devices? A systematic review. https://core.ac.uk/download/288868644.pdf (accessed on 10th September 2025).
25. Hong, H. l. (1994). Machine learning and its applications in reliability analysis systems. https://core.ac.uk/download/9640661.pdf (accessed on 10th September 2025).
26. Hussain, M. J. (2024). A survey based on behavior analysis of artificial intelligence using machine learning process. *2024 4th International Conference on Sustainable Expert Systems (ICSES)*. https://www.semanticscholar.org/paper/a9754cd22c53d5c3735d47f1aad905b8733b0e91 (accessed on 10th September 2025).
27. Jalali, S., Laibacher, T., & Weyde, T. (2020). M2U-net: Effective and efficient retinal vessel segmentation for real-world applications. https://core.ac.uk/download/323057630.pdf (accessed on 10th September 2025).
28. Kepner, J., Kumar, M., Moreira, J., Pattnaik, P., et al. (2017). Enabling massive deep neural networks with the GraphBLAS. http://arxiv.org/abs/1708.02937 (accessed on 10th September 2025).
29. Mousavi, A., Ambrosetti, A., Askin, R., Banks, D., Banks, J., Beylkin, G., Blum, A., et al. (2013). Input variable selection in time-critical knowledge integration applications: A review, analysis, and recommendation paper. https://core.ac.uk/download/20572738.pdf (accessed on 10th September 2025).
30. Nguyen, A., Ngo, H. N., Hong, Y., Dang, B., & Nguyen, B. P. T. (2022). Ethical principles for artificial intelligence in education. *Education and Information Technologies, 28*, 4221–4241. https://doi.org/10.1007/s10639-022-11316-w (accessed on 10th September 2025).

31. Rao, S. M., Adinarayana, B., Vaddikasulu, K., Dontamsetti, S. P., & Ranganadh, P. S. V. (2024). Optimizing cloud IoT applications with machine learning driven hybrid key management protocols for WSN performance. *2024 2nd International Conference on Signal Processing, Communication, Power and Embedded System (SCOPES)*. https://www.semanticscholar.org/paper/732c3342756c6ffb368329e9e02b883334a62c71 (accessed on 10th September 2025).
32. Roffo, G. (2018). Feature Selection Library (MATLAB Toolbox). http://arxiv.org/abs/1607.01327 (accessed on 10th September 2025).
33. Sandford, R. (2009). Robots, drugs, reality and education: How the future will change how we think. https://core.ac.uk/download/4151800.pdf (accessed on 10th September 2025).
34. Shamsa, E. (2022). Self-aware resource management in embedded systems. https://core.ac.uk/download/521288193.pdf (accessed on 10th September 2025).
35. Shi, J. (2023). Analyses and optimizations of timing-constrained embedded systems considering resource synchronization and machine learning approaches. https://core.ac.uk/download/613822492.pdf (accessed on 10th September 2025).
36. Stein, L. A. (1999). Challenging the computational metaphor: Implications for how we think. https://core.ac.uk/download/pdf/84443.pdf (accessed on 10th September 2025).
37. Turney, P. (2000). Types of cost in inductive concept learning. https://core.ac.uk/download/pdf/84867.pdf (accessed on 10th September 2025).
38. Varaprasad, R., Chakkaravarthy, A. P., & Veeresha, M. (2024). A comprehensive analysis of intrusion detection system using machine learning and deep learning algorithms. *2024 International Conference on Intelligent Algorithms for Computational Intelligence Systems (IACIS)*. https://www.semanticscholar.org/paper/5458b2df4fda54be7d41a8d9d230d6a8ef394aac (accessed on 10th September 2025).

CHAPTER 6

NAVIGATION AND PATH PLANNING

Contents

Introduction 205
6.1. Historical Development of Navigation 210
6.2. Types of Navigation 215
6.3. Path Planning Algorithms 221
6.4. Applications of Navigation and Path Planning 226
6.5. Challenges in Navigation and Path Planning 231
6.6. Future Trends in Navigation Technology 236
Summary 241
References 243

INTRODUCTION

Navigation and path planning have become a key component of the robotics field, where navigating in environments with complexities is of extreme relevance. As robots are increasingly utilized in dynamic environments, optimal navigation systems are needed in order to ensure that they can operate safely and effectively and not run into any obstructions. The use of various computational methods such as fuzzy logic, genetic algorithms, and advanced search algorithms provides robust solutions capable of adapting to changing conditions. The systems maximize route selection while also preserving compliance with kinematic constraints imposed by the robotic vehicles. By their exploration of the intelligent integration of behavior for avoidance of obstacles and efficient optimization of paths, researchers aim to simulate navigation human-like strategies for improving the usability and reliability of robot systems in real-world systems, such as the developments achieved with fuzzy motion planning systems (El-Khatib et al., 2006).

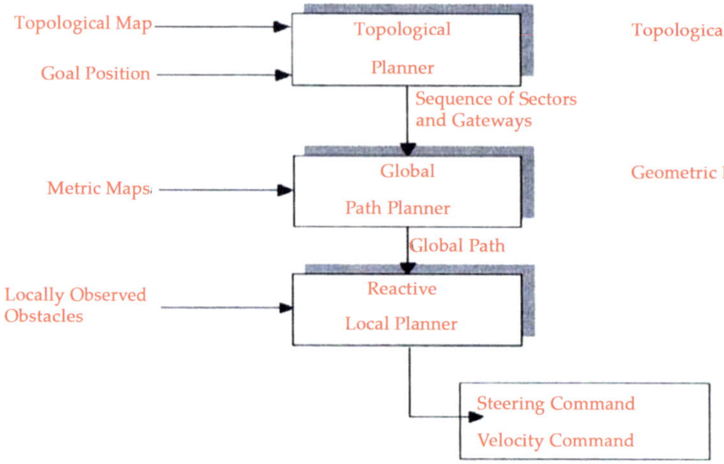

Figure 6.1. *Navigation and path planning.*

Source: https://www.researchgate.net/publication/221785982/figure/fig2/AS:305435950239747@1449833013462/Three-level-path-planning-and-navigation.png.

Breakthroughs in soft computing techniques have compelled improvements in navigation and path planning methods. Hybrid intelligent systems complexity has been described as possessing the capability of addressing complex problems that are recalcitrant to traditional approaches. These systems take advantage of the strength of different algorithms which, employed in combination, magnify performance beyond their constituent parts. For instance, integrating fuzzy logic with genetic algorithms and neural networks creates a more flexible and robust approach to task execution in robotics. All these collaborations are crucial in solving real-time navigation issues as they enable continuous reconfiguration to changing environments and local configurations of obstacles. This necessity has been addressed with innovative hybrid frameworks optimized for both movement and accuracy, therefore changing the way robots engage with their environment (Barczak et al., 2011). Overall, the ongoing enhancement of path planning and navigation technology is an illustration of the overall robotic demands for efficiency and flexibility. With real-world problems more complex in their challenges, the necessity of intelligent systems that can make decisions in real time is increasingly pressing. The interdisciplinary nature of research in the field highlights the importance of combining conventional and modern computational techniques. Not only are these integrated systems poised to enhance robot autonomy, but they also have the potential to expand the role of robotic technologies to a variety of uses in different industries, from manufacturing to search and rescue operations. The integration of several intelligent algorithms is breaking doors for future development and breakthrough in autonomous navigation, thereby underscoring the pivotal role that efficient path planning plays in dictating the path of robotics (Figure 6.1).

Definition of Navigation

In its definition, it is crucial to realize that navigation entails a series of complex steps required for successfully determining the course between two or more locations. Navigation, in effect, involves the movements of placing an object in space and determining the best routes to desired destinations. It is a mix of technological devices, such as Global Positioning Systems (GPS) and mapping technologies, and cognitive skills navigators use to make judgments in space. The blend of these factors makes it possible to recognize changes instantly in relation to changing environmental conditions or targets, making successful navigation dependent on both technology and human control. Furthermore, as environments become increasingly complex, such as city streets or interactive virtual spaces, navigation needs become more demanding, necessitating advanced path planning and decision-making strategies. Navigation in digital environments is used to refer to physical and virtual spaces. An example would be that the intelligibility of a virtual space has an enormous impact on how individuals move through virtual spaces. Research indicates that such areas as AlphaWorld are highly understandable at a local level, yet increasingly unintelligible at an expanding scale (Conroy-Dalton et al., 2002). This is an important factor in both the creation of urban spaces and in virtual models, requiring well-coherent structures in order to reduce navigation inefficiency. A navigable design must promote a clear understanding of spatial relationships and pathways, which can particularly cause issues in virtual environments where traditional navigational markers could be a waste of time. Thus, as virtual worlds become more aligned with physical realities, the fundamental principles of navigation must adapt to these dual realities. Essentially, the evolution of navigation systems relies on the adoption of mechanisms for assurance that further augment safety and effectiveness in navigation and path planning.

Upcoming innovations in cyber-physical systems (CPSs) underscore the need for frameworks like the Component-Based Simplex Architecture (CBSA), which assures the capacity of navigation systems to function well in real-time while continuing to meet provided safety standards (Clark et al., 2017). This approach supports not only adaptive navigation control but also contractually guarantees contracts that guarantee system-wide safety, e.g., collision avoidance or mission success with given parameters. These assurances are required since navigation systems are being embedded in progressively more advanced environments that need high safety and operational reliability. This convergence between navigation technology and assurance systems is a huge advancement that has the potential to enhance physical and virtual navigation processes equally, ultimately revolutionizing the way we conceptualize navigational accuracy and reliability.

Importance of Path Planning

Path planning is more than navigation; it is a key foundation upon which to make autonomous systems efficient and safe. With the best routes in any given environment, path planning addresses static and dynamic obstacles so that systems such as drones, robots, and autonomous vehicles can be made accommodating for real-time adjustment. An organized description of the environment, such as the hierarchical graph (H-graph) proposed for intelligent wheelchair navigation, organizes information in hierarchies of detail to facilitate high-level path searches. This kind of organized approach maximizes the systems information processing capacity and competency to make quick decisions to improve the experience and safety of the user in complex environments (Abascal et al., 2004). The effectiveness of path planning becomes increasingly obvious when there are multi-layered navigation cases, e.g., vertical building navigation path planning, in which navigations between floors can become less complex. The integration between human cognition and automatic path planning illuminates more about methods of enhancing navigational performance. Empirical evidence indicates that human drivers, such as UAV operators, deviate from optimum paths under theoretically ideal conditions due to many implicit factors, such as understanding of the situation and natural response to barriers. Inclusion of this implicit human knowledge (HHK) in path planning algorithms enables one to formulate better and more realistic navigational methods. This approach emphasizes that decision-making in the real world does not necessarily have to be in accordance with optimal algorithms but can be effective. Injecting human knowledge into machine learning for path generation exemplifies a hybrid strategy, wherein human experience shapes the planning mechanism to enhance overall navigational efficiency and safety (Alzugaray et al., 2016). In addition to this, owing to ongoing technology growth, the applicability of sophisticated path planning systems is illustrated through its applicability to important applications in areas where it is of key significance. Effective navigation in areas such as logistics, disaster relief, and healthcare is of paramount importance for maximum delay minimization and maximal outcomes. Path planning not only contributes to operational efficiency but also enhances the reliability of autonomous systems under unexpected circumstances. As seen from the ongoing research into path planning optimization and real-time integration based on learning, the field is poised to expand exponentially. This expansion does not only confirm the functionality of quality path planning systems but also shows the need for ongoing research towards creating and adapting these technologies in order to handle future navigation demands (Abascal et al., 2004; Alzugaray et al., 2016) (Figure 6.2).

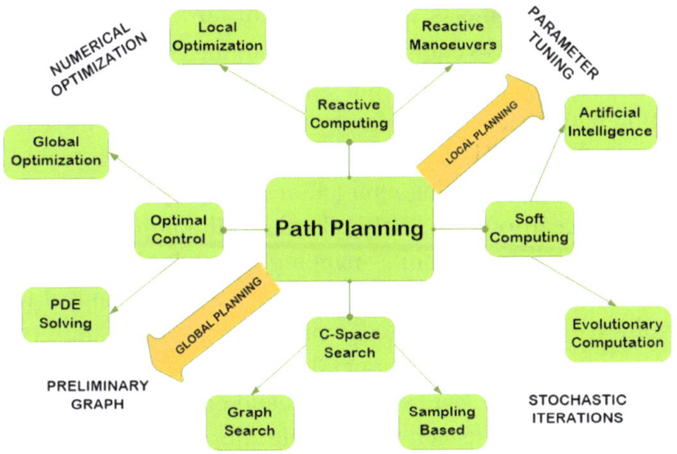

Figure 6.2. *Path planning methods.*

Source: https://www.researchgate.net/publication/363342911/figure/fig1/AS:1143 1281415052036@1746023950897/Schematic-showing-different-path-planning-methods.tif.

Overview of Technologies Used

The navigation and path planning environment is radically shaped by the intersection of emergent technologies enhancing situational awareness and operational performance. A key example of such a technological advancement is virtual reality (VR) application in air traffic control systems, as seen in the Retina Project. This project demonstrates the efficiency of VR in reducing the workload of air traffic controllers (ATCO) through simulating varying visibility conditions, such as optimal and reduced visibility caused by weather factors such as rain and fog. Results from the projects emphasize the need to provide transparent overlays and visually discriminable elements, such as color-coded flight strips, for improving readability and situational awareness of ATCOs under different operation conditions. Through the use of VR, the industry can tailor training and operational instruments, thereby optimize the process of navigation and enhance path planning functions for complex environments (Buyse et al., 2017). In addition to virtual reality, geospatial data capture and processing technologies play a critical role in path planning and navigation, particularly in disaster response and heritage documentation. ITHACA (Information Technology for Humanitarian Assistance, Cooperation and Action)–Politecnico di Torino is an example of how multidisciplinary solutions can improve geospatial mapping and data-synchronizing capabilities. The project is designed to prepare future architects and engineers to interact with terrestrial and aerial LiDAR, photogrammetry, and GNSS equipment. By enabling students to perform hands-on mapping and data collection in the Cinque Terre National Park, especially after environmental catastrophes, the project aims to build a robust battalion of "volunteer mappers." The resulting geospatial databases provide effective map updating, thereby enhancing navigation efficiency for rescue workers and municipalities (Boccardo et al., 2012). Besides, the application of unmanned aerial vehicles (UAVs) and mobile mapping technologies has revolutionized navigation data collection accuracy and velocity. The advanced technologies enhance traditional mapping methods with real-time data and greater spatial awareness. During disaster management, UAVs can map disaster-stricken areas in a short time to facilitate fast assessments and efficient response operations. The ultimate goal of all these technological developments is to create integrated platforms such as WebGIS for convenient sharing of geospatial data between interested stakeholders, e.g., emergency units and city authorities. Utilizing these technologies, the navigation and path-finding activities could be made dynamic and adaptive, developing smoother management of daily routines as well as crisis management (Boccardo et al., 2012).

6.1. HISTORICAL DEVELOPMENT OF NAVIGATION

The navigation route has traveled a far way since conventional means to sophisticated techniques used now. Initial sailors initially employed celestial navigation, using the stars, sun, and moon to navigate their paths through vast oceans. These early instruments, compressed into cultural traditions, sailed thousands of expeditions, as well as those ventured by the Polynesians in the Pacific. As civilizations began to develop, so did the need for the advanced navigation tools that eventually led to the creation of the magnetic compass in the 11th century. This made the sailors able to navigate even when it was cloudy, thereby expanding trade routes and enabling cultural exchange across greater distances. Each new advance in navigation was not only a gigantic technological step forward but also indicated the widening understanding of human interaction with the world, paving the way for modern navigation techniques in today's globally interconnected world. The coming together of navigation and technology has resulted in revolutionary breakthroughs, particularly with the invention of aerial surveillance methods. Aerial photography, which began to leave its imprint early in the 20th century, has contributed immensely to navigation accuracy and the reaping of geographical data. Early aerial surveys relied on oblique photographs that, by definition, pre-disposed bias. Modern advances in technology have provided a more integral solution. Developments in digital imaging, photogrammetry, and geographic information systems have enabled both navigators and archaeologists to cartographic large landscapes from the air. This advancement in data gathering serves not only the needs of navigation but is also a contribution to the overall objectives of understanding spatial environments through systematic mapping. The method of navigating on land and sea has therefore become increasingly reliant on sophisticated imaging and data examination methods, which highlight the demand for accuracy in journey planning.

Figure 6.3. *The evolution of navigation.*

Source: https://replogleglobes.com/app/uploads/2022/10/ancient-navigation-tools-1.jpg.

In contemporary navigation, integrated information systems are at the forefront with the intention of optimizing path planning and route optimization. Some of the recent projects, such as those involving multidisciplinary interdisciplinary research between institutions such as Politecnico di Torino and institutions such as ITHACA, emphasize the need for state-of-the-art geospatial data processing. By mixing traditional methods with modern tools such as UAVs and WebGIS systems, these attempts are aimed at providing accurate mappings and updating information critical for disaster management and heritage recording. All this is not limited to evidently demarcated navigational needs; they encompass a vast range of applications, proving how navigation remains a dynamic concept. As communities continue to experience the impacts of climate change and urbanization, learning from historical advances in navigation is key to the development of adaptive systems to guide sustainable path-making and effective disaster management approaches so that future generations can navigate in safety (Boccardo et al., 2012; Sevara et al., 2016) (Figure 6.3).

6.1.1. Ancient Navigation Techniques

Early navigation techniques are proof of human beings determination to explore and connect distant locations. Ancient civilizations charted their worlds with a series of techniques based on their discoveries about the occurrences of natural phenomena. Celestial navigation was the foundation of many ancient culture exploration endeavors, with the sun, moon, and stars serving as essential points of reference for travelers and sailors alike. The Polynesians, for instance, acquired the ability of wayfinding by utilizing their intricate understanding of tidal currents, wind directions, and constellations. This approach not only depicts cognitive skills of navigation but also highlights the importance of cultural diffusion because knowledge was transferred and modified over generations. These navigation techniques were the foundation for subsequent developments, demonstrating the vast impact of environmental perception on human navigation and path planning capability. Furthermore, incorporating cartography into navigation allowed ancient societies to expand their geographical knowledge. Maps provide an essential overview of regions and routes, enabling more effective planning of journeys and voyages. Early mapmakers in ancient Mesopotamia, for example, employed clay tablets to record their world, noting the locations of cities, rivers, and commercial routes. These maps not only allowed for useful navigation but also reflected the social priorities and values of the cultures that produced them. As cartography became more advanced, so did its method, incorporating details such as topography and landmarks to guide travelers to unknown destinations. The development of cartography, combined with mastered navigation methods, significantly increased the ability of early civilization to map uncharted areas, making the development of expanding trade routes and cultural exchange that created the basis of contemporary civilization possible. Additionally, antiquity's navigation methods became increasingly associated with technological innovation that raised efficiency and precision. Development of instruments such as the astrolabe and the compass were the pinnacles of navigational history. While the astrolabe allowed sailors to measure the altitude of heavenly bodies, improving their ability to determine latitude, the compass provided a reliable means of determining direction, taking the uncertainties of ocean travel. Furthermore, the application of path-planning methods, which evolved from basic navigation habits, helped early sailors navigate clear of danger, optimize route efficiency, and conserve energy on long voyages. Such advances are proof that ancient navigation habits were not merely rudimentary habits but rather complex systems that anticipated today's advanced navigation systems. The synthesis of celestial navigation, cartography, and instrument-making thus evidences the remarkable ability of ancient civilizations to adapt and refine their methods of navigation within the context of pathfinding.

6.1.2. Evolution of Navigation Tools

The development of navigation equipment has played a significant role in human exploration and transportation, ranging from rudimentary methods to advanced systems that involve advanced technology. Ancient seafaring societies used celestial bodies, as proved by the use of the sun and the stars to navigate. The development of the compass in the 11th century was a major milestone in navigation since it provided a sure point of reference no matter the visibility conditions. With the evolving requirements of society, so did the technology, with charts and maps enabling explorers to gain a general understanding of their world. With the invention of increasingly accurate clocks, such as the marine chronometer in the 18th century, the navigational precision improved further. This early history provides context for understanding how modern-day navigation technologies operate within the broader environmental and human advancement context. In modern times, the integration of technology has revolutionized navigation tools, with the first among these being the evolution of Global Positioning System (GPS) technology. Developed first for military applications, GPS has since become a ubiquitous technology pervasive in numerous applications, including vehicle navigation, aircraft operations, and marine activities. This technology involves the use of a network of satellites transmitting precise location information, to be decoded by handheld devices or in-vehicle systems. Additionally, artificial intelligence has begun to enhance navigation through adaptive algorithms that are sensitive to changing environments and user needs. The application of GPS and the likes is a significant departure from traditional practice, and thus, navigation becomes highly automated. This technological development has implications for efficient path planning for personal as well as business transport, in the sense that it can be adapted in real time with respect to current information. Secondly, the interplay between institutions and technological innovation has fostered novel approaches to the utilization of geospatial data in navigation. For instance, projects like those of Politecnico di Torino and ITHACA highlight the convergence of theoretical studies and practice in geospatial mapping. Such projects not only emphasize the importance of precision in data in navigation and planning routes, but also humanitarian purposes through the mapping of risk areas of disasters. With a combination of traditional surveying methods and the most recent technology like UAVs for aerial photogrammetry, data gathering is enhanced to enable faster and more precise updates to navigational charts. This combined strategy demonstrates the proficiency of conjoint navigation tools to respond actively in times of crises and thus marks a significant step in the evolution of navigation tools towards the time they will make optimal common commute as well as crisis management operations (Boccardo et al., 2012; Kopardekar et al., 2025) (Figure 6.4).

Figure 6.4. *Ultimate guide to survival navigation tools.*

Source: https://patriotprepped.com/wp-content/uploads/images/Survival-Navigation-Tools-Historical-Evolution-696x398.jpeg.

6.1.3. Impact of Technology on Navigation

The technological revolution has transformed navigation systems greatly through enhanced accuracy, efficiency, and safety in ground and air transportation. At the forefront of this revolution has been the advent of deep learning techniques, particularly in autonomous vehicles. Using advanced algorithms such as convolutional neural networks (CNN), these systems are capable of examining vast amounts of environmental data in real-time. This ability to analyze in real time greatly enhances path planning by allowing cars to better understand their environment and detect obstacles, which they can adjust their route for accordingly. As seen from a study demonstrating an accuracy rate of 98% in environmental perception and image recognition, deep learning plays an instrumental role in fine-tuning automated navigation. The practical applications of these developments lead to enhanced user safety and system dependability, demonstrating technology's marked impact on navigation techniques in today's transportation infrastructure (Cao et al., 2024). New technology like LIDAR is also at the forefront of advancing navigation and path planning in poor conditions. For applications such as autonomous airdrop systems, real-time atmospheric data acquisition using ground-based LIDAR significantly enhances the accuracy of landings. By measuring wind trends at multiple levels, such systems can transmit critical data that informs decision-making during the approach phase of an airdrop. Research has determined that the incorporation of LIDAR data offers a superior 40% improvement in accuracy compared with systems that are not provided with such feedback loops.

This shows that not only does technological integration improve autonomous operating capacities of systems but also addresses one of the primary sources of navigational failure—environmental uncertainty (Cacan et al., 2017). Thus, continuous integration of such advanced technologies is important in order to improve navigation performance and safety in dynamic conditions.

Moreover, technological innovation in navigation extended from commercial and private automobiles to other uses like ship and aviation navigation. In seagoing transportation, navigation by means of GPS and electronic navigation charts facilitates easier route planning, allows ships to avoid hazardous areas, and increases fuel economy. In addition, in flight, more sophisticated air traffic control and navigation systems in aviation help reduce the likelihood of midair crashes and make air travel more efficient overall.

As such technologies become increasingly sophisticated, they not only enhance safety and efficiency in travel enormously but also make it possible to integrate real-time information into navigation systems. The constant integration of navigation techniques and technological advancements showcases the transformative impact of these technologies in shaping the world for future mobility (Cacan et al., 2017; Cao et al., 2024).

6.2. TYPES OF NAVIGATION

The landscape of navigation technology has evolved to encompass various types that cater to particular operating requirements. At the forefront are traditional navigation systems that rely primarily on GPS and other space-based systems, providing geographical position data to allow one to navigate through varied environments. Such systems are susceptible to malfunctions where they get jammed by signals or experience interference in the air. On the contrary, inertial navigation with gyroscopes and accelerometers has a solution to provide constant monitoring of locations. Such navigation is particularly helpful in areas where satellite signals are not available, such as urban canyons or indoors. Overall, the efficiency of traditional and inertial navigation systems reflects the three-dimensional nature of navigation problems and the requirement for various solutions for various situational requirements. With advancements in automation and robotics, having sophisticated navigation systems embedded has also become critical.

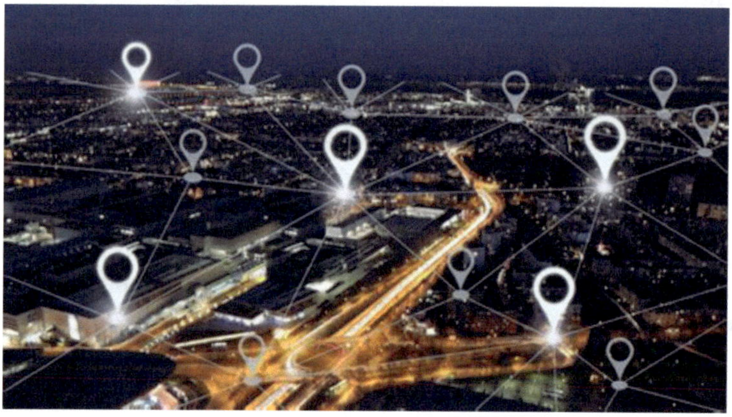

Figure 6.5. *Network GPS navigation in a modern city with future technology.*

Source: https://external-content.duckduckgo.com/iu/?u=https%3A%2F%2Ftse2.mm.bing.net%2Fth%2Fid%2FOIP.rYplaMiPjSQ-HYmieLmy_wHaED%3Fpid%3DApi&f=1&ipt=4954993441ad0216a37c376519e2ab0ecaf6f75b02274fb768f87d54141894e5.

Autonomous navigation, especially in mobile robots in dynamic spaces, has been a specific field that has been of interest. These robots utilize sophisticated approaches that are sensor-based, such as cameras and ultrasonic range finders, to move about and make decisions about locomotion. For example, in the case of partially structured worlds, neural networks help robots detect free space and move around obstacles effectively. The approach presented by Janglova et al. (2004) gives an idea of how the application of neural networks makes it possible for robots to adopt collision avoidance so that they can find a secure path in the presence of uncertain obstacles. Therefore, the applications of these higher-order navigation technologies highlighted the significance of inter-operability between diverse systems towards enhancing overall operational effectiveness. In the agricultural technology sector, the shift to intelligent navigation is best illustrated through precision farming practices. This navigation will often engage a combination of localization, mapping, and path planning, optimized for maximum field operation. Scientists are drawing focus towards integrating automation technology in farming, as stated by Aghi et al. (2020), where autonomous robots, guided by RGB-D camera data, are traversing vineyards with unprecedented saving of costs and resources. The twin-layer control system here is an indicator of an innovative solution of solving navigational problems, having backup algorithms that ensure robustness even in case of sensor failures. This adaptability complements the overall navigability of navigation in agriculture, emphasizing how modern autonomous systems are not only tasked with being able to read environmental cues but also with being capable enough to recover from potential disturbances. Developing on these adaptation strategies enables further understanding of path planning for complex agricultural settings and beyond (Figure 6.5).

6.2.1. Celestial Navigation

The historical significance of celestial navigation places it as one of mankind's oldest means of navigation, using astronomical observations to determine position on the surface of the Earth. The ancient mariners used the stars, the Sun, and the Moon as steady clues to navigate their way across vast and often level seas. By understanding the position of celestial objects, they could calculate their latitude and longitude with remarkable accuracy. This old technique proved particularly useful before the common application of modern navigational equipment, allowing explorers to cross unmapped seas and aiding commercial paths that contributed significantly to cultural exchanges. The understanding of the motion of celestial bodies not only found practical use in navigation but also stimulated advancement in keeping time and the science of astronomy, laying foundational blocks for the navigation systems of today. In contemporary use, celestial navigation has also seen resurgence popularity as an essential component of many astronomical missions, with specific emphasis being in space exploration. Spacecraft today not only rely on the traditional celestial bodies as references but also on sophisticated algorithms that integrate dynamic data on lunar and planetary orbits. Lunar missions have the potential offered by integrating celestial navigation with emerging technologies demonstrate satisfactory effectiveness in mission planning and correction of trajectory. Specifically, research on lunar navigation like this illustrates how the power of passive radio frequency technologies supplemented by celestial markers can significantly enhance the reliability of navigation systems for missions in which GPS is inadequate. While technology advances, therefore, the principles of celestial navigation remain viable, mapping the course to possible future exploration off Earth and enabling mission planners to create complex route plans for difficult journeys. One such application of celestial navigation principles for modern robotic and autonomous platforms is an example of how this ancient method continues to improve with time. Combining celestial navigation methods with inertial navigation, researchers are developing new techniques that facilitate precise position location in environments hostile or inaccessible to standard GPS satellites. Moon surface navigation research, highlighted in recent studies, infers that transit through passive and inertial processes can enable the enhancement of exploration missions. Furthermore, such advancements place more focus on the utilization of celestial navigation in achieving mission success and safety in uncertain environments. When we are going towards Mars and beyond missions in the future, studying and adopting such ancient methods of navigation will be the key to well-planned routes and success in space missions (Abbeduto et al., 2025) (Figure 6.6).

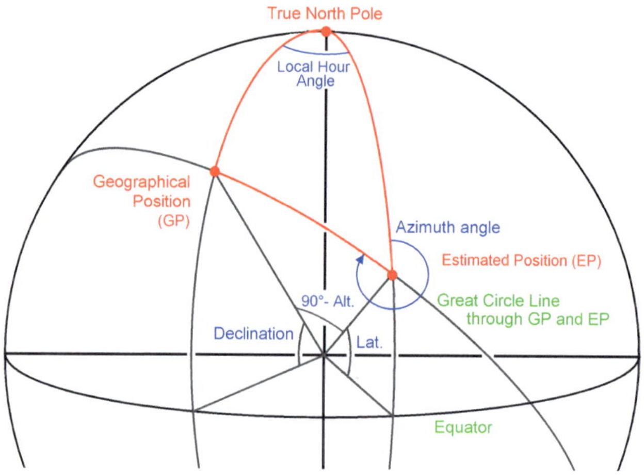

Figure 6.6. *Celestial navigation.*

Source: http://geoffrey-kolbe.com/C-Nav/images/Navigational%20Triangle.jpg.

6.2.2. Terrestrial Navigation

The field of terrestrial navigation encompasses a variety of techniques which are important for positional computation and movement on the ground surface of the Earth. Natural features, celestial, and primitive tools have long been at the core of guiding navigators. With the advancement of technology, modern-day terrestrial navigation relies on global positioning system (GPS), Mapcarta maps, and electronic navaids. These innovations enable greater accuracy and efficiency in plotting route courses and finding one's way, both on and off road, and in outdoor activities. And the integration of GPS with computer mapping programs has transformed the way people perceive their environment, enabling them to lay down routes in real-time and react to changing circumstances. This change underscores the significance of understanding the past and current applications of navigation as key components of effective path planning in a variety of fields, from city planning to recreational navigation. Scientists working towards accurate positioning have overcome some challenges, particularly the interference brought about by atmospheric effects, for instance, the ionospheres effect on satellite signals.

Figure 6.7. *Terrestrial and coastal navigation.*

Source: https://northeastmaritimeonline.com/wp-content/uploads/2022/11/Terrestrial-Navigation-3m.jpg.

The ionosphere, a component of the upper atmosphere, can result in time delays of navigation signals emitted by satellites that create errors during calculations of exact locations. By researching total ionospheric electron content (TEC) with methods such as Faraday polarization measurements, scientists have devised techniques to reverse such errors. Besides enhancing terrestrial navigation precision, such research also illustrates the imperative connection between navigation technology and atmospheric science. It is essential to know how these interactions influence navigation ability in a bid to create more sound systems with the capacity for both general and special navigational needs, thereby enhancing the path-planning research area. The future of terrestrial navigation is ready for radical transformation as emerging technologies continue to advance navigational techniques and approaches. Projects like NASAs LunaNet highlight the importance of building scalable navigation and communication structures, not just for the facilitation of lunar exploration but also on terrestrial ground. Understanding how structures like these can be used on Earth, researchers can innovate to further navigation capabilities on different settings. The search for lunar resources and the development of a human base on the Moon will be a proving ground for these technologies, allowing for necessary improvements in navigation. As earth navigation improves, the lessons learned through projects such as LunaNet will drive the development of even more sophisticated path planning systems that can respond to a vast array of terrestrial and extraterrestrial barriers in order to improve our capacity to traverse difficult terrain (Figure 6.7).

6.2.3. Electronic Navigation

The advent of electronic navigation has transformed the science of navigation and path planning, particularly for autonomous vehicles. Traditional navigation has relied on manual data entry and basic mapping, which were cumbersome and error-prone in dynamic conditions. Electronic navigation employs advanced algorithms and real-time data processing, enhancing the ability to navigate difficult terrain. As an example, a two-level navigation architecture consisting of a Sugeno-type fuzzy motion planner and a fuzzy controller based on a modified proportional navigation illustrates the direction of more clever navigation algorithms. The system not only allows vehicles to plan the best trajectories but also navigates obstacles cleverly according to real-time sensor data and kinematic limitations. The simulation data validates the performance of the systems under varying environments, referencing its capacity to make decisions in a matter of seconds that might be out of a human navigator's reach. In maritime usage, electronic navigation is reiterated by the development of Unmanned Surface Vehicles (USVs). These intelligent boats depend profoundly on sophisticated global path planning technologies that leverage electronic charts for optimized navigation. New advancements include an enhanced A* algorithm that is tailored for optimizing sailing costs based on various factors, such as sailing safety and the environment. Through the utilization of S-57 electronic charts to construct an octree grid model, researchers have enhanced the navigation efficacy without compromising safety through weighted sailing routes. The algorithm minimizes risk exposure to potential hazards and reduces path planning complexity, marking a significant development in maritime navigations safety and reliability. The promising results from simulation indicate that the new methods can produce routes not only efficient but tailored to the complexities of crossing open waters. Furthermore, electronic navigation consequences extend beyond convenience; they represent a fundamental shift in the manner transport interfaces with its environment. The sophisticated combination of motion planning and obstacle avoidance is crucial for autonomous navigation-based applications such as robotics and unmanned vehicles. Just as the application of fuzzy logic into navigation applications, these technologies replicate human decision-making and therefore ensure increased levels of operational accuracy in complex environments. Such adaptability is crucial for real-time applications since it allows vehicles to adapt dynamically their courses following newly emerging issues. Collectively, these electronic navigation advancements play an important role in path planning methodologies in numerous disciplines so autonomous systems may operate safely and efficiently under a broad range of operating limitations.

6.3. PATH PLANNING ALGORITHMS

In robotics, path planning algorithms play a key role in optimal navigation across various environments. Such algorithms are supposed to identify optimal paths considering some constraints such as dynamic obstacles and the physical characteristics of the robot. One of the significant ways is the application of pseudospectral optimization methods. These methods have gained popularity due to the computational efficiency and convergence rate they offer, which allows computation of trajectories taking into account a robot's dynamic properties. With the aid of these approaches, researchers can modify existing methods to support sensing and real-time avoidance of obstacles within real-world applications, thereby allowing the expansion of the use of robots in unstructured environments (Corke et al., 2010). Still another creative approach combines fuzzy logic with path planning to enhance navigation in dynamic scenes. The system internally organizes itself into two functional layers: a Sugeno-type fuzzy motion planner and a fuzzy controller which is a modified version of proportional navigation-based fuzzy controller. Employing human-like routing, the fuzzy motion planner utilizes visual perception to navigate around obstacles in a clever manner. Besides, the adaptive proportional navigation controller automatically adjusts the robot's path in real-time to facilitate effective obstacle avoidance and satisfy kinematic constraints. The two-layer structure is demonstrated to be versatile in navigating through intricate environments, with the research being an apex of path planning method formulation. The findings in simulation demonstrate its capability to process several scenarios effectively, thereby validating its reliability in static as well as dynamic obstacle situations (El-Khatib et al., 2006). The use of advanced path planning algorithms, along with enhancing the effectiveness of robotic motions, opens the door for their use in many fields. Within autonomous vehicle systems, for example, they are essential to ensuring safety and efficiency while moving through cities full of unknown variables. With continued improvement in algorithms that address the synthesis of path planning and real-time obstacle recognition, researchers strive towards developing systems that simulate human decision-making. Such advancement would lead to more intelligent, dependable robot systems for handling dynamic and complicated environments. The future of path planning and navigation solutions will be influenced by the interplay between theoretical development and practical implementation as the area continues to advance (Figure 6.8).

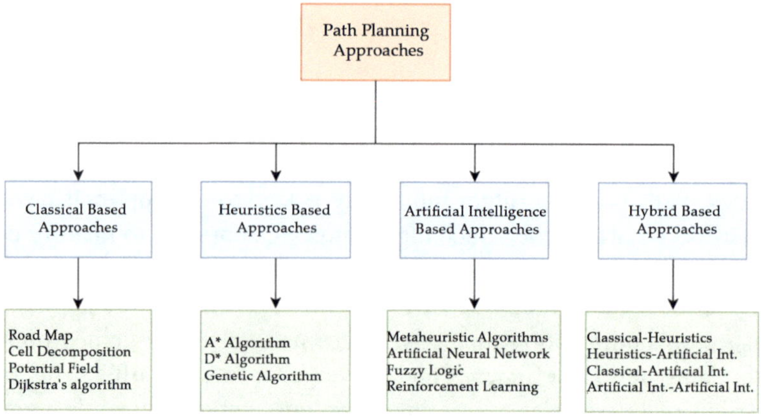

Figure 6.8. *Path planning algorithms for mobile robots.*

Source: https://api.intechopen.com/media/chapter/a043Y00000yJC4SQAW/a093Y00001fjnDaQAI/media/F2.png.

6.3.1. A* Algorithm

The A* algorithm is a cornerstone in path planning and navigation due to its efficiency and universality in the resolution of complex routing problems. A best-first search is utilized, combining the strengths of Dijkstra's algorithm and heuristics so that the shortest path may be found. By employing a cost function that encompasses both the traversed actual distance and an estimate of distance to the destination, A* balances optimally between exploring and exploiting the search space. The two-level focus allows the algorithm to discard fewer promising routes at an initial stage, thereby being particularly well-suited in situations where there is extensive branching as well as potential obstacles. Therefore, the A* algorithm has found application in many areas, including robotics, AI pathfinding, and city simulation, where fast and stable route-finding is crucial. In virtual environment research nowadays, the relevance of the A* algorithm increases with the richness of 3D worlds. With rich graphical descriptions and virtual agents, effective pathfinding is necessary. As illustrated in recent research, the increasing concreteness of virtual worlds and cities with intricate designs necessitate automatic navigable path extraction and their traversability (Belo et al., 2010). The challenge is not merely to discover the shortest path but to adapt to real-time variation in dynamic environments, e.g., unexpected obstacles or dynamic ground. The ability of A* algorithm to utilize elastic heuristics enables real-time adaptation, making it invaluable to guarantee efficient navigation in such an adaptive environment. Moreover, the usage of the algorithm in autonomous navigation of vehicles underlines its ability to operate under various traffic conditions while respecting environmental limitations. A new approach blends trajectory optimization with A* principles to handle dynamic maneuvers in autonomous situations such that vehicles are able to drive through obstacles without violating traffic conventions (Barber et al., 2017). With the integration of not only the geometry of the environment but also behavioral aspects of other road users, A* enhancement realizes improved safety and optimality of motion. According to this, as urban populations continue to grow denser, the evolution of the A* algorithm remains at the forefront in advocating for reliable path planning methods that effectively address a range of issues facing autonomous systems (Figure 6.9).

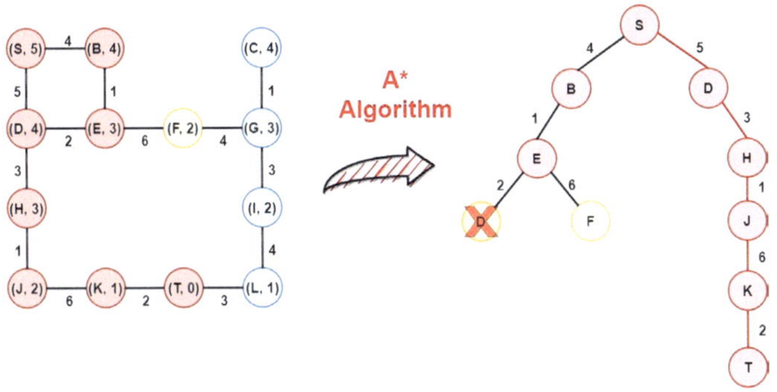

Figure 6.9. *A* Algorithm.*

*Source: https://miro.medium.com/v2/resize:fit:1113/1*3bzre-TGcFlxc0j9aeWM-mQ.jpeg.*

6.3.2. Dijkstra's Algorithm

Dijkstra's algorithm is the all-encompassing paradigm for navigation and pathfinding, most famously effective in its application for finding the shortest distance between nodes in a weighted graph. At its very core, its functionality relies on the exploration of each path in a sequential, step-by-step manner, incrementally calculating the shortest-known distance to adjacent nodes until all nodes are examined. This algorithm works on a greedy principle, always expanding the most likely to be best node in each step, effectively reducing the sum of the travel cost. The success of Dijkstra has made it established across different domains, ranging from routing in networks to robotics, where accurate navigation is important.

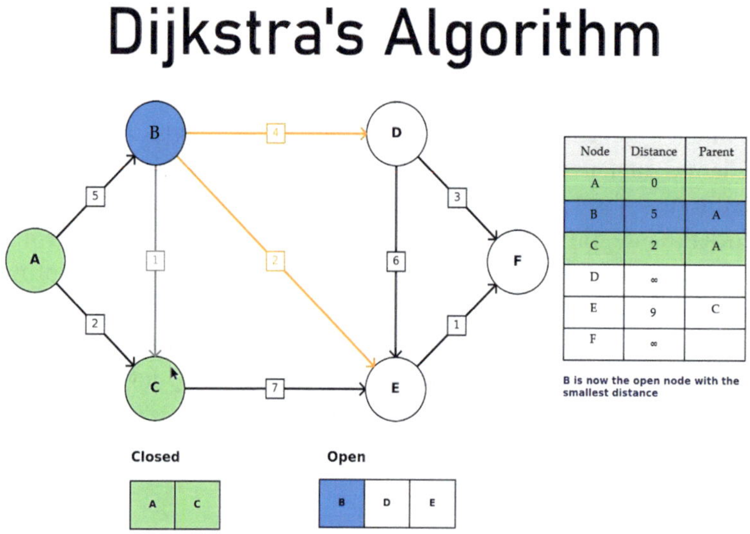

Figure 6.10. *Dijkstra's algorithm.*

Source: https://i.ytimg.com/vi/71Z-Jpnm3D4/maxresdefault.jpg.

The computational efficiency of the algorithms makes it possible to handle bigger datasets, making it ideal for real-time navigation applications where quick responses are necessary. But the effectiveness of Dijkstra's Algorithm may be called into doubt in complex, dynamic worlds, particularly in emergency evacuations. For instance, while Dijkstra's Algorithm is a good technique for discovering paths, real-time decision-making applications and occupancies changing pose additional variables. Literature suggests that evacuation routing algorithms are significantly influenced by the initial distribution of evacuees, the type and location of disasters, and the intrinsic nature of the environment (Bi et al., 2015). In such scenarios, the traditional approach of Dijkstra may not be the best, and modifications or other algorithms capable of responding dynamically to changing conditions or constraints faced in evacuation would be required. The limitations of Dijkstra's Algorithm in navigating risk zones further illustrate the need for more advanced adjustments in path planning methodologies. In applications where the robot or evacuee will be required to move through risk zones of varying intensities, a simple implementation of Dijkstra's might not be sufficient since the optimal path cannot be guaranteed to possess optimal substructure—sub-paths in an optimal path might not be optimal individually (Hou et al., 2017). To counter this, researchers have developed improved implementations of Dijkstra's Algorithm that include risk factors within their cost functions, trading path length against exposure to risk. These modifications prove the continuous development of navigation algorithms and illustrate the multi-disciplinary push for enhanced path planning, proving the applicability and utility of Dijkstra's Algorithm across both traditional and modern fields of use (Figure 6.10).

6.3.3. Rapidly Exploring Random Tree (RRT)

The invention of the rapidly exploring random tree (RRT) has significantly advanced the boundaries of navigation and path planning in robotics. By utilizing a sample-based algorithm, RRT effectively constructs paths in high-dimensional, complex spaces. The exceptional ability of RRT is to efficiently search the search space by gradually constructing a tree that branches out to new regions. This ability allows RRT to navigate dynamic environments with obstacles efficiently, critical in mobile robotic applications. With this, the algorithm demonstrates an efficient solution to planning trajectories after collision avoidance constraints and plays a significant role in applications where traditional planning fails in their reliance on global information. In practical use, RRT effectiveness has been improved by modifications, such as the RRT* variant, which assures optimal planning of paths. The RRT* algorithm, aside from searching the space, enhances the paths to yield suboptimal solutions in the long term, considerably improving the efficiency of navigation procedures. These enhancements have been a key component in the development of systems responding to complex operational environments. For instance, in a reported 3D path planning system, RRT* is used concurrently with point cloud data acquired with sensors like the Kinect V2. Through this approach, robots can be controlled to move optimally and safely within complicated environments through the proper combination of path planning and obstacle detection to generate practicable results for real-time applications (Castillo et al., 2016). Furthermore, the integration of RRT with other methods, such as the path integral control framework, reveals how RRT can solve challenges brought about by uncertainty in robot motion. The path integral control methodology reconfigures traditional optimal control problems into a statistical inference landscape, allowing RRT to leverage broader sampling strategy. This combination enhances RRT's ability to search through dense regions by generating several, potentially optimal paths and decreasing the likelihood of converging to local minima. The incorporation of homology-embedded sampling strategies further enhances this approach by providing the ability to detect varied motion plans through different homology classes. With such advancements, RRT becomes an integral component of modern robotics with dynamically achievable and collision-free motion planning to assist in dealing with complex navigation issues (Choi et al., 2018).

6.4. APPLICATIONS OF NAVIGATION AND PATH PLANNING

The integration of path planning and navigation technologies is transforming industries, including agriculture. Although farming is evolving to newer paradigms such as 3.0 and 4.0 agriculture, autonomous machinery has emerged as a method of choice. The technologies not only streamline farming processes but also improve the use of resources and reduce human labor. Autonomous agricultural field machines greatly rely on sophisticated navigation capabilities, which include tasks like localization, mapping, and path planning. For instance, a paper recently presented shows a low-cost local motion planner implemented in vineyards, using an RGB-D camera along with a dual-layer control algorithm. This model illustrates how farm robots could efficiently navigate semi-structured environments with continuous learning from real-time data, thereby making the farming operations more efficient and sustainable (Aghi et al., 2020). Navigation and path planning have wider implications in robotic systems beyond agriculture, particularly in dynamic environments. A nice example is illustrated in the design of a Sugeno-type fuzzy motion planner for car-like nonholonomic robots, where the importance of real-time navigation in a cluttered environment comes into play. The system architecture allows the robot to process visual information in the same way as human decision-making while navigating around obstacles. Utilizing a combination of fuzzy logic-based decision-making and proportional navigation-based real-time behavior optimization, the robot can move around stationary and dynamic obstacles while respecting its kinematic constraints. Results of several simulations demonstrate the effectiveness of such combined approach, presenting promising applications in diversified environments with the overall enhancement of robotic mobility and task performance (El-Khatib et al., 2006). Further, the uses of path planning and navigation extend into autonomous cars, where efficiency and safety are paramount concern. With the complexities of driving through city streets that are crowded with pedestrians, bicyclists, and other motor vehicles, effective path planning systems are paramount. Artificial intelligence and machine learning-based techniques play a central role, enabling autonomous vehicles to easily sense and react to environmental stimuli. By leveraging advanced algorithms for processing big sensor data, these vehicles can make informed routing decisions while optimizing against factors like travel time and collision avoidance. Not only does this enhance operating safety but also towards the bigger picture for smart city infrastructures, where navigation and path planning solutions are part of the means to improve urban mobility and reduce traffic congestion (Figure 6.11).

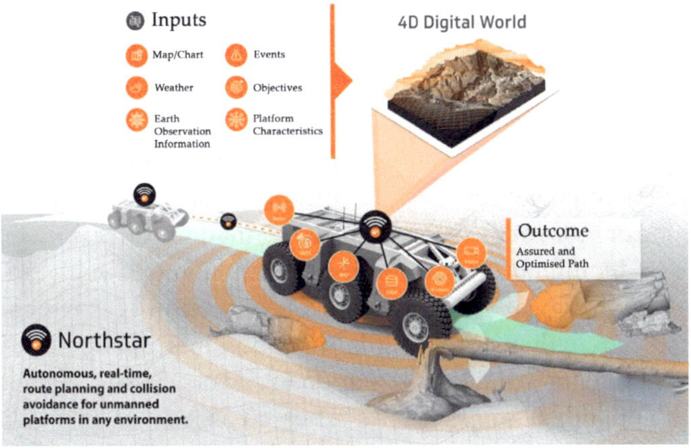

Figure 6.11. *Path planning for autonomous vehicles.*

Source: https://www.unmannedsystemstechnology.com/wp-content/uploads/2021/06/Autonomous-route-planning-for-unmanned-platforms-1024x655.jpg.

6.4.1. Autonomous Vehicles

The autonomous vehicle technology has greatly altered the navigation and path planning dynamics, especially in heavy traffic environments. Advanced algorithms with Model Predictive Control (MPC) embedded within them help significantly to make the vehicles drive efficiently and be stable in their formation. A case in point is a strategy that has been proposed to allow a fleet of unmanned aerial and ground vehicles to move together without subjecting them to highly accurate external localization processes. Instead, the approach uses top-view visual relative localization of air vehicles, and this makes the vehicles more capable of orienting themselves within their surroundings. In addition to facilitating easier navigation, the method adapts to dynamic obstacles as well as sustaining clear visibility among teammates, an essential condition for stability in different operation environments (Abdessameud et al., 2013).

In addition to more effective navigation methods, the integration of a leader-follower formation driving algorithm is a significant improvement in path planning for autonomous vehicles. The new algorithm comprises a convex hull representation of the overall 3D formation that maps the vehicle group onto a given path. With this representation, the vehicles can follow non-collision paths even in static and dynamic cluttered environments. The algorithm design ensures constant visibility, as required for effective top-view localization, enhancing the dependability of the navigation across various scenarios. This leads to the achievement of navigation robustness, which has immense advantages towards the application of large numbers of micro-scale robots in reconnaissance and surveillance missions (Krajnik et al., 2013). Moreover, fault tolerance for these navigation structures further improves the fault tolerance of autonomous vehicles in unstructured environments. Through the inclusion of a fault detection and recovery mechanism, the introduced solutions can handle cases when one or more of the vehicles in the formation suffer from breakdowns. This resiliency highlights the capability for autonomous vehicle fleets to be utilized in mission-critical applications in which reliability is paramount. Their versatility and strength point towards their possible application in the real world, the steppingstone to their integration in sectors such as logistics, military applications, and urban transport. As research is continued to thrive, these advancements do not just reveal to us the enhanced capability of autonomous navigation systems but also predict a future where such technologies are a normal phenomenon.

6.4.2. Robotics and Drones

Integration of robotics and drones with applications significantly enhances navigation and path planning functionality. These developments are largely the result of the use of sophisticated algorithms that make it possible to have optimal route selection for mobile robots in unstructured environments. For instance, aerial images obtained by using drones may be employed to create large cost maps that outline the difficulty of traversing different kinds of landscapes. By assigning certain costs to various terrains, navigation algorithms are in a position to accurately identify the optimal route for a robot and, ultimately, improve its operational efficiency and safety. The A* algorithm, in particular, operates by employing the cost map to arrive at the shortest and safest route, demonstrating the importance of accurate data in enabling valid navigation. Overall, the complementarity between robots and drones in path planning significantly eases operations in challenging terrain to increase smooth flow of robots where human intervention may not be feasible (Shrivastava, 2025). The choice of path planning algorithms is a significant aspect of the operation and efficiency of robot systems. There are algorithms with different strengths based on the specific operational scenario, ranging from a static scenario to a dynamic one. Deterministic algorithms like A* and Dijkstra's algorithms tend to be applied in structured environments where there is a predictable outcome and optimal path routing. Probabilistic algorithms like Rapidly exploring Random Trees Star (RRT*) and Probabilistic Roadmap (PRM) perform optimally under unstructured or unexpected environments.

Figure 6.12. *Researchers developing drones that can build structures.*

Source: https://regmedia.co.uk/2022/09/23/nature_scan_build_drone.jpg.

This adaptability allows drones and mobile robots to access and navigate difficult terrain that may have obstacles or dynamically varying conditions. Through investigating various measurements such as execution time and resource usage, scientists can inform a proper selection of algorithms to facilitate improved decision-making when developing and deploying robotic systems (Aksoy et al., 2024). The ongoing work in navigation and path planning techniques presents the rapidly evolving aspect of drone and robotics technology. With the rise in the need for autonomous systems in every industry—spanning from the environment to logistics—the path planning advancements become the focal point of these technologies. To demonstrate this, incorporating simultaneous localization and mapping (SLAM) capabilities allows robots and drones to navigate independently in new locations without being equipped with a map of the local area. This is not only more effective but allows real-time analysis of data and obstacle avoidance. Therefore, the intersection of robotics and drones in navigation is not merely a tech play; it is an embodiment of the capacity to drive transformative change in other sectors. Through continuous optimization of algorithms and integration of newer technologies, the future of robotics path planning and navigation is promising (Shrivastava, 2025) (Figure 6.12).

6.4.3. Maritime Navigation

Maritime navigation involves a host of sophisticated techniques and resources for the purpose of attaining safe and effective movement over water. One of the most important aspects of the process is the capability to identify best routes that minimize the risk of collisions, particularly in tight waterways where various ships may concurrently exist. Sophisticated algorithms have been developed for further optimizing trajectory planning with the help of artificial potential fields and speed vector analysis. In such a context, the proposed algorithms not only adhere to current collision avoidance guidelines and global maritime navigational rules but are also effective within advanced maritime settings. Such versatility renders them necessary to enhance navigational safety in addition to efficiency, demonstrating the delicate margin between human abilities and computerized systems in modern maritime navigation (Han et al., 2008). While the maritime industry upgrades navigation equipment with added intelligence, the trend of Unmanned Surface Vehicles (USVs) is in focus. Developing USVs heavily relies on cutting-edge global path planning technologies that reflect their level of autonomy and strategic acumen. Of particular note is the enhanced A* algorithm, which is highly optimized for sailing expenses through electronic charts. By using the creation of an octree grid model of the environment from S-57 electronic charts, this algorithm takes into account many factors such as sailing safety weight and pilot number in order to minimize risk when sailing. Its ability to provide safer routes and facilitate smoother paths signifies a certain improvement in maritime navigation technology with regards to both efficiency and safety in an ever-differentiating marine environment (Li et al., 2017). Moreover, the use of these high-tech navigation solutions is imperative in order to counteract the problems that plague today's maritime operations. As traffic picks up, particularly in busy ports and straits, the necessity for precise navigation and route planning has never been more urgent. By utilizing cutting-edge methods, has also drawn from algorithmic learning for automation, the maritime authorities are able to efficaciously reduce the incidence of maritime accidents and enhance the operational efficiency of commercial vessels. By implementing such new-generation practices, not just are seafarers' lives saved, but marine ecosystems are also preserved, showing a large-scale responsibility in the context of maritime navigation. In this regard, it is evident that ongoing use and research in regard to the latest navigation technologies are fundamental to the future of safe and sustainable shipping (Figure 6.13).

Figure 6.13. *Future of maritime navigation.*

Source: https://images.marinelink.com/images/maritime/w800/photo-northrop-grumman-sperry-marine-108240.jpg.

6.5. CHALLENGES IN NAVIGATION AND PATH PLANNING

The navigation and path planning environment is filled with a number of challenges that significantly affect autonomous system effectiveness. Environmental variability is one of the largest challenges, particularly in dynamic environments where obstacles are prone to shifting unexpectedly. For instance, controlling small autonomous robots in racing situations calls for not only quick decision-making but also real-time adaptation to changing conditions, such as moving gates and light changes. Sophisticated perception and planning for such a need are delivered by integrating convolutional neural networks (CNNs) into the control systems. This is attainable through proper navigation by creating models that have the ability to adapt in a flexible manner without the need for extensive real-world fine-tuning, ensuring robustness under uncertain conditions. This flexibility is crucial when it comes to maintaining performance, which also requires powerful systems that can function on their own as there is uncertainty in their environment (Dosovitskiy et al., 2019). Another substantial challenge that affects navigation and path planning is the need for good state estimation. In urban environments, for example, mobile wheeled-legged robots must be constantly surveying their surroundings in order to navigate effectively. The richness of such environments, with heterogeneous terrain and moving hazards, demands an acute strategy that synthesizes rich sensor feedback with high-level decision algorithms. By applying model-free reinforcement learning algorithms, navigation systems can learn and adapt their paths even under diverse and unforeseen obstacles. This convergence of locomotion and navigation has been useful in enhancing operational performance. The ability to transition smoothly between one form of movement to the next is imperative for handling the complexities of urban navigation, which demonstrates the importance of advancing solid state estimation methods in modern autonomous systems (Bjelonic et al., 2024). Furthermore, the convergence of hierarchical control schemes represents a dominant problem in effectively navigating dynamic environments. In advanced environments, the utilization of a single control method might lead to suboptimal path design and performance under stress. Through hierarchical reinforcement learning, researchers have been able to develop systems that are capable of changing their strategies based on real-time feedback from the operating environment. Through adaptive integration, the system can improve its navigation strategies optimally. However, this complexity also comes with its own challenges, including the smooth transfer of control between control layers and computational balances. Ultimately, developing ways to reconcile such complex components is important to enable autonomous systems, such as wheeled-legged robots, to successfully perform kilometer-scale navigation missions. These advancements have the potential for improved logistics and operational versatility within diverse urban landscapes, underscoring the need to address these intricate navigation challenges (Bjelonic et al., 2024) (Figure 6.14).

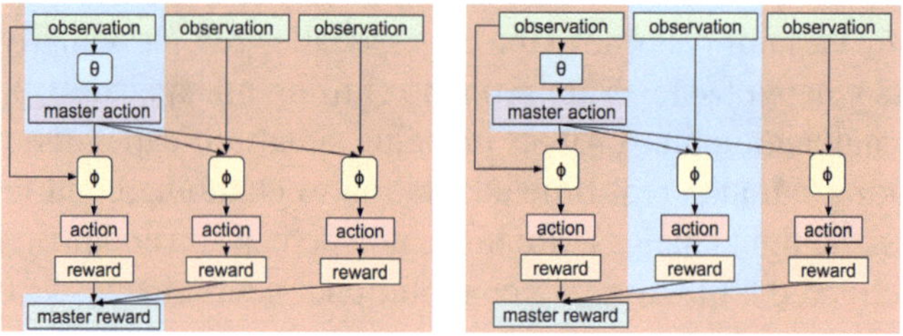

Figure 6.14. *Hierarchical reinforcement learning.*

*Source: https://miro.medium.com/v2/resize:fit:1104/1*BKMPDpNe0gBQTRUFRYkNwA.png.*

6.5.1. Environmental Factors

Environmental factors play a significant role in path planning and navigation, particularly in the case of autonomous systems operating in uncertain environments. Terrain heterogeneity, weather, and visibility are just a few of the factors that may play critical roles in influencing autonomous system operation decisions. For instance, an appropriately formulated mission planning framework must be able to account for spatio-temporal heterogeneities in the operating environment so that autonomous vehicles can properly react to unforeseen circumstances. Here, application of high-level reactive mission planners can direct the vehicles to desired destinations as well as schedule tasks based on environmental updates. Such adaptability enhances the ability of the system to realize its goals within time limits, as highlighted by the hierarchical dynamic framework in the literature (Atyabi et al., 2017). Additionally, visibility conditions have been shown to greatly impact navigation strategies, particularly in aviation and underwater contexts. The research in the Retina Project puts considerable emphasis on situational awareness under varied atmospheric conditions, to underscore how good-visibility situations are fundamentally distinct from those that are degraded by rain or fog. Effective path planning calls for real-time adjustments to these visibility constraints since the performance of air traffic controllers (ATCs) in actual operation may rest on their ability to interpret environmental parameters accurately. By maintaining visual presentation consistent in form and emphasizing priority information for urgency, ATCs are able to enhance situational awareness, thereby improving general safety and decision-making procedures under unfavorable conditions (Buyse et al., 2017). Integrating environmental factors into models of navigation necessitates a dimensional methodology that factors in both the static and dynamic nature of the operational environment. This involves not only real-time path optimization but predictive modeling of likely environmental changes. Optimized algorithms such as biogeography-based optimization can facilitate the generation of optimal routes that can respond to changing environments and thereby enhance mission success rates. The combination of situational awareness, terrain adaptability, and real-time data analysis, which is successful, speaks to the complexity of work involved in the development of robust navigation systems capable of performing in a range of environmental forces. By continued focus on these various environmental factors, researchers and engineers can advance path planning methods so that they not only remain robust but also efficient in autonomous functioning across diverse domains.

6.5.2. Computational Complexity

Computational complexity is the determining aspect in navigation and path planning to be able to identify the efficiency and viability of algorithms employed by autonomous agents. With environments to be navigated by robots becoming ever more dynamic and complicated, the computational cost of traditional path planning rises significantly. Algorithms such as Rapidly exploring Random Tree (RRT) and Dijkstra's algorithm are plagued by the burden of dense obstacle fields, taking enormous amounts of computational power to determine viable paths. Such inefficiency has the ability to significantly limit a robots functionality within real-time applications because heavy amounts of computational delay can result in suboptimal path decisions that do not account for changing elements in the environment. Thus, computational complexity solving not only optimizes performance but also enhances the reliability of navigation systems, which is very important for efficient robotic applications in real tasks. Recent innovations in the hopes of overcoming these issues have introduced frameworks specifically designed to decrease path planning complexity without compromising efficiency. For example, the Topomap framework is a new method that integrates spatial data from vision SLAM systems into a three-dimensional topological map for path planning. Through point cloud to occupancy transformation of noisy sparse point clouds, the framework mitigates computation-inefficient requirements of conventional planning algorithms. Specifically, the used representation in Topomap not only streamlines the global planning process but also reflects comparable efficiency without significantly increasing computation time or storage needs. The bearing of such innovation underlines the necessity for reducing computational complexity so as to improve real-world applicability of robotic navigation systems under difficult conditions, as seen in encouraging test results on mobile robot platforms. Though promising, the trend towards reducing computational complexity in path planning and navigation continues to challenge existing paradigms in robotics to their limits. The development of hybrid intelligent systems that integrate fuzzy logic, genetic algorithms, and other computational methods shows an ongoing effort to satisfy more challenging task specifications. Existing methods, however, must address constraints related to optimality and real-time performance, which enhances the issue of robustness of existing solutions. The development of hybrid solutions with advanced capabilities is a sign of the ability to overcome issues linked with traditional hybridization. For instance, integrating algorithms such as A* and Voronoi diagram calculation with fuzzy systems and genetic algorithms is able to offer more capable and adaptable navigation methods. From further innovation within these bounds, researchers can ensure progress that not only reduces computational complexity but also enhances the effectiveness of robotic path planning in unfamiliar environments.

6.5.3. Safety and Reliability Issues

For the field of navigation and path planning, safety and reliability are of central concern especially considering how the systems being used are being altered by increased technology. Effective path planning not only facilitates effective transport but also aims to avert danger posed by accidents and system failure. In air traffic management, for instance, focus has historically been placed on mitigating the risk of accidents through aggressive safety policies and programs. The transition to more sophisticated frameworks from traditional methods is imperative to address the challenges of today, especially with the added challenge of air traffic systems. As stated by Brooker et al. (2007), safety standards have also become more stringent, requiring ongoing innovations in operational methods as well as safety procedures to ensure that the ultimate goal of minimizing accident risk is always attained. With the introduction of new paradigms such as NextGen in the US and the SESAR in Europe, the likelihood of safety and reliability issues increases.

Figure 6.15. *High-speed civil transport.*

Source: https://upload.wikimedia.org/wikipedia/commons/c/c0/High_Speed_Civil_Transport_%28HSCT%29.jpg.

The NextGen and SESAR are initiatives towards enhancing air traffic management through the incorporation of cutting-edge technologies and procedures. This advancement comes with enormous challenges, for instance, system resilience and managing human-automation interaction. The complexities ushered in by advanced systems necessitate upgraded risk assessment processes able to be adaptive to future demands as traditional methods can no longer suffice. Measures of confidence, such as the Human-in-the-Loop simulations with seeded errors identified by Brooker et al. (2007), must be put in place to maintain the credibility of these new systems and ensure they are able to operate safely under a congested airspace. Also, ensuring safety and reliability in navigation and path planning is beyond the technology innovation; it requires an integrated design approach and operational procedures. As noted in, effective integration of flight and propulsion control technologies is crucial to fulfilling operational safety, especially for systems such as complex transport systems of high-speed civil transports (HSCTs). Control systems not only need to provide safety but also consider special operating problems such as high-temperature application and the need for fail-safe operation. Having technology development and certification assisted by regulatory bodies is essential since problems left unaddressed compromise not just system performance but safety as well. A full appreciation for safety and reliability issues is, therefore, directly dependent upon navigation and path planning performance, and hence emphasizes the necessity for continuous improvement and surveillance in system design (Figure 6.15).

6.6. FUTURE TRENDS IN NAVIGATION TECHNOLOGY

With navigation technology improvements, some of the key trends that will change the face of autonomous systems are some of the major developments. One of the key developments is the integration of artificial intelligence (AI) and machine learning algorithms into navigation systems. The two technologies enhance path planning functions by allowing systems to learn from experience while navigating through changing environments. For instance, the application of deep learning algorithms is able to improve obstacle detection and evasion significantly, resulting in more efficient navigation for mobile robots. With the ability to process huge volumes of data in real time, the future navigation systems will not just generate routes but also constantly optimize them based on changing conditions. This evolution towards intelligent navigation systems is part of the broader trend in robotics towards greater autonomy and responsiveness in path planning, pushing the boundaries for unmanned vehicle operation. In addition, the advent of unmanned vessels is proof of machine learning and high-level path planning algorithms' deployment in marine applications. These ships, today a revolutionary force in oceanic exploration and resource management, rely extensively on sophisticated navigation technologies. Autonomous ship path planning requires careful planning in the modeling of the environment, algorithm construction, and collision avoidance to ensure safety when operating at sea. Future development will most probably be focused on integrating multi-robot collaborative path planning, which will enhance efficiency and effectiveness in managing complex maritime scenarios. By integrating the strengths of various path planning algorithms, unmanned ships are able to maximize their operating capacities, whether conducting environmental monitoring or assisting maritime search-and-rescue operations. The interplay between unmanned systems and navigation technology is thereby setting the stage for thrilling innovations in autonomous navigation. Aside from AI-based innovation, the incorporation of sensors and real-time data streams into navigation systems is another critical trend that will shape the navigation technology of the future. Sustained advances in sensor technology are enabling vehicles to gather more accurate environmental data, which has a positive effect on the decision-making functions of navigation. This is augmented by increasing IoT device deployments that provide insightful data about ambient situations, hence making better decisions about paths to follow. As noted, future navigation technology lies in the co-use of different path planning algorithms with real-time data integration in order to build robust systems that effectively navigate complex environments. Collectively, these advancements create a new era in navigation technology that has the capability to not just optimize path planning but also enhance the safety and reliability of autonomous systems across all their uses, ultimately transforming how we approach and implement navigation in the modern world.

6.6.1. Integration of AI and Machine Learning

The integration of machine learning (ML) and artificial intelligence (AI) into path planning and navigation methods has shown phenomenal progress, making it possible to develop adaptive and autonomous technologies. An example is the use of biologically inspired navigation systems, such as the robot *Psikharpax*, employing a reinforcement learning algorithm to modify its navigation methods in real time. By using AI, the robot can acquire and change its approach based on previous experiences in various conditions, mimicking the navigation capacity of natural living things. The ability to store acquired strategies for use in maximizing its overall performance in unfamiliar settings is a reflection of the significant role of the integration of AI and ML into robotics and navigation processes. As these technologies advance, they allow machines to be more autonomous and efficient and therefore resolve previously unsolvable complex navigation challenges. In environments with significant obstacles and dynamic change, the integration of AI and ML allows real-time assessment and adjustment of navigation strategies. The EVA Retriever project is one instance of this ability through intelligent voice-controlled systems capable of precise autonomous navigation. This free-flying autonomous robot is designed to locate and retrieve objects, such as tools or astronauts, which may get lost in space. The primary functions of the system are target searching, rendezvous maneuver planning, and grappling operations with continuous tracking and avoidance of obstacles. Using machine learning algorithms, the EVA Retriever can improve its performance as it gains experience, which enables it to navigate complex routes and interactively adjust its behavior to respond to novel obstacles. This hierarchy of system intelligence is significant to effective motion in uncertain worlds, validating the critical role of AI and ML in modern path planning solutions. Beyond mere robotic movement, the implications of AI and ML in navigation are significant opportunities for constructing human-machine collaboration. As can be seen in the advanced models of navigation robots such as *Psikharpax* and the EVA Retriever, these technologies enable more friendly interactions between human beings and machines. This integration enforces the creation of systems not just that are capable of carrying out correct navigational actions but ones that also learn from human input and feedback, creating a symbiotic synergy. The effectiveness of such systems lies in their ability to mimic cognitive processes, such as decision-making and adaptability, like human behavior. The expansion of AI and ML in navigation is ultimately revolutionizing the interaction between autonomous systems and humans, driving towards a future where autonomous agents can augment human capabilities in hybrid environments. Such a combination of approaches holds out the promise of improved safety, efficiency, and performance of navigation systems in a wide range of fields and applications and is thus central to current research and development.

6.6.2. Advancements in Sensor Technology

Sensor technology developments have, in particular, impacted areas of navigation and path planning to enhance the autonomy and efficiency of robotic systems. High-level sensors like RGB-D cameras and LIDAR are now more than ever integral to autonomous navigation tasks. These sensors provide valuable information about the environment, allowing machines to successfully map their surroundings and localize themselves within changing environments. It is this same localization and mapping capability that enables robots to perform advanced path planning tasks in both structured and unstructured environments. For instance, as demonstrated in current studies, the use of low-cost RGB-D cameras enables effective navigation within farms to be accomplished, reflecting the promise of significant cost savings in human labor and resource consumption (Aghi et al., 2020). Furthermore, the integration of advanced algorithms alongside sensor technology is critical in enhancing path planning and navigation systems. An example of this includes the use of the dual-layer control algorithm in autonomous farm vehicles, which serves to illustrate how the usage of such algorithms enhances the reliability and performance of the navigation system.

Figure 6.16. *Advances in sensor technologies.*

Source: https://www.mdpi.com/sensors/sensors-20-06783/article_deploy/html/images/sensors-20-06783-g001.png.

While one algorithm constructs movement control from depth and disparity maps, another stable secondary algorithm ensures continuity of operation through real-time adaptation to potential failures. Far beyond just enhancing movement in dangerous terrain, this hardware-software collaboration optimizes movement as well as ensuring continuous learning and adaptation through incoming data from working environments. These technologies reflect the intense interaction between sensor technological advancements and sophisticated algorithms designed to take advantage of this data (Balcells et al., 2016). As sensor technology continues to expand, its reach on path planning and navigation is both widespread and profound. Advancements in sensor fusion, where multiple sources of information are combined into an even richer sense of the world around us, illustrate this trend. By combining the strengths of various sensors, robots can achieve more integrated situational awareness and more informed decisions regarding their paths. Such a function is essential in dynamic or unknown environments where quick adaptation has to be undertaken. Growing sensor and clever algorithm integration, particularly in mobile robots and autonomous vehicles, represents a watershed to more intelligent and capable navigation systems. With scientists exploring new ways and pushing existing technologies, the future is brighter for path planning and navigation (Aghi et al., 2020) (Figure 6.16).

6.6.3. Development of Smart Navigation Systems

The development of smart navigation systems has created a shift in many industries, particularly agriculture and autonomous driving. The integration of robotics and automation has revolutionized operations, particularly in precision agriculture, where unmanned equipment is being utilized extensively to maximize efficiency and reduce the costs of operating operations. The smart systems utilize a combination of mapping, localization, and path planning techniques to navigate through semi-structured and unstructured environments. For instance, one new development was aimed at a cost-effective local motion planner for autonomously navigating vineyards, based on RGB-D camera technology and new control algorithms. This approach not only enhances operation capacity but also addresses the necessity of adaptive algorithms with error recovery capabilities, illustrating the precarious balance between machine learning and operations in real-world farm settings (Aghi et al., 2020). Parallel to the developments in farming, intelligent navigation systems have advanced significantly in mobile robots. Wireless sensors serving as active landmarks used for the localization of vehicles form a system methodology that simplifies unmanned navigation operations. By employing multiple Bloom-filter routing information compression, scientists have come up with a system that maximizes space usage and still achieves efficient routing mechanisms. This technology therefore significantly reduces the resource demand typically inherent in traditional navigation systems, allowing more sustainable use of mobile robots in various environments, e.g., wheelchair navigation. This new approach focuses on the importance of the balance between resource constraint and functional performance in intelligent navigation systems with a clear trend towards maximizing navigational intelligence (Cheng et al., 2013). With the advancement in smart navigation systems, their application also increases and diversifies. Efficient implementation of these systems not only optimizes navigation processes but also encourages increased machine autonomy across domains. Through the integration of advanced algorithms with real-time decision-making facilities, such systems learn adaptively from the environment and consequently reinforce their path planning capabilities. Therefore, smart navigation systems are a requirement in environments where operation efficiency, safety, and productivity are paramount. The effort in research and experimentation being put into this field has the potential to revolutionize industries like agriculture, transportation, and logistics toward increasingly automated futures. This conjunction of application and technology highlights the imperative path of smart navigation system design toward changing machines' perception and interaction with their environment.

SUMMARY

Path planning and navigation have evolved swiftly, especially in urban areas where situational complexity is a clear challenge. Autonomous systems, as depicted by the DARPA Urban Challenge, illustrate the need for capable situational reasoning in order to navigate through intricate urban configurations. The finite state model presented in the context of Team Caltech's Alice robot highlights the manner in which well-structured logic helps improve the decision-making process within real-world situations. These types of methodologies not only help promote following traffic laws, but they help to make autonomous navigation systems safer and more efficient overall, working against potential failures that were noted in early implementations (Burdick et al., 2008). All these improvements point to the importance of embedding robust algorithms in handling diverse urban environments ultimately resulting in improved navigation technologies. Further, the application of neural networks for path planning marks a paradigm shifts in how autonomous robots interact with their environment. Employing sensors to gather data about their environment and using two particular neural networks, robots can construct collision-free paths under diverse obstacles. This new approach greatly enhances robots' navigation abilities within partially structured environments. The ability of the first neural network to recognize "free" space using ultrasound data and a second that is able to spot safe direction of movement is a case of heightened adaptability that is critical to use within real-world environments (Janglova et al., 2004). As simulations show, such approaches not only optimize path planning but ensure operational safety and therefore validate the effectiveness of intelligent control systems in robotic navigation. The future of navigation and path planning is to continue optimizing such models and therefore advocate for setting up systems that can effectively cope with dynamic environments. With ongoing research and development in technology, increasingly powerful algorithms and machine learning techniques can be integrated to further enhance navigation system reliability and robustness. It will be essential to employ a multidisciplinary strategy with elements of robotics, artificial intelligence, and cognitive science to address the challenges that exist in different environments. Looking ahead, lessons learned from successful installs and documented failures (Burdick et al., 2008; Janglova et al., 2004) will play an integral part in the future development of autonomous navigation, ushering in a new era of smart systems that are able to navigate and plan courses in progressively complex environments. In path planning and navigation, multi-agent system complexity demands advanced control algorithms that can ensure safe coordination. Particularly, development of the combined linear quadratic regulation and reciprocal velocity obstacle (LQR/RVO) satellite formation control algorithm stands out as a major advancement in this field. The new algorithm allows each satellite to maneuver independently without directly communicating with other satellites while avoiding potential collisions. The RVO component is employed to actively prevent collisions, and the LQR component is tasked with propelling the spacecraft to their respective targets. Several test cases have validated the efficacy of this method, such that the method not only guarantees safe maneuvers but also optimizes fuel consumption while in operation, as observable in the harmony that is evident in performance and resource efficacy (Ceriotti et al., 2017). Another significant consideration in navigation throughout virtual spaces is the user's cognitive load. With increasing sophistication of 3D interfaces, knowledge of how resources of information are arranged becomes essential in effective wayfinding. Such an interface's layout must be responsive to the user's interaction with information, as different arrangements can have a huge impact on their navigation. A distributed resources model has been proposed by research aiming to decrease cognitive load and hence expand user capacity to access necessary navigational support in virtual spaces. Through the experimentation of its efficacy in assisting users of complex spaces by applying and evaluating many wayfinding tools, this model enables a direct correspondence between navigational tool efficiency and cognitive requirements (Hart et al., 2006). The ongoing exploration of virtual and real-world navigation systems emphasizes the intricate balance between user interaction and technology. Advances in control algorithms, such as the LQR/RVO for space travel, emphasize the need to balance safety and efficiency in navigation. Simultaneously, advances in 3D interfaces emphasize how the user experience may be optimized using intelligently designed means to reduce the cognitive load. Together, they reflect a larger phenomenon in route planning and navigation: that more integrated approaches need to be created that reconcile technology with human needs. As these fields of study continue to evolve, so too do their implications for a wide range of industries—anything from aviation to entertainment—offer promising potential for future research and creation.

REFERENCES

1. Abascal, J., & Cagigas Muñiz, D. (2004). Hierarchical path search with partial materialization of costs for a smart wheelchair. https://core.ac.uk/download/288003098.pdf (accessed on 10th September 2025).
2. Abbeduto, L. J., Amdursky, M. E., Breseke, D. K., Broadbent, M., et al. (2025). Lunar navigation study, summary volume: Final report, Jun. 1964–May 1965. https://core.ac.uk/download/pdf/85253632.pdf (accessed on 10th September 2025).
3. Abdessameud, A., Christensen, A. L., Liu, C., Mayne, D. Q., Xiao, F., Heredia, G., Heredia, G., et al. (2013). Fault-tolerant formation driving mechanism designed for heterogeneous MAVs-UGVs groups. https://core.ac.uk/download/18533283.pdf (accessed on 10th September 2025).
4. Aghi, D., Chiaberge, M., & Mazzia, V. (2020). Local motion planner for autonomous navigation in vineyards with a RGB-D camera-based algorithm and deep learning synergy. https://core.ac.uk/download/pdf/327177639.pdf (accessed on 10th September 2025).
5. Aguilar Castillo, W. G., & Morales, S. G. (2016). 3D environment mapping using the Kinect V2 and path planning based on RRT algorithms. https://core.ac.uk/download/87654089.pdf (accessed on 10th September 2025).
6. Aksoy, N., Çakıl, F., & Tekdemir, I. G. (2024). Performance analysis of deterministic and probabilistic path planning algorithms in complex environments. *2024 Innovations in Intelligent Systems and Applications Conference (ASYU)*. https://www.semanticscholar.org/paper/73087c9b258d35346319f7d261003b82eb170195 (accessed on 10th September 2025).
7. Alzugaray López, I., & Sanfeliu Cortés, A. (2016). Learning the hidden human knowledge of UAV pilots when navigating in a cluttered environment for improving path planning. https://core.ac.uk/download/81581485.pdf (accessed on 10th September 2025).
8. Atyabi, A., Powers, D. M. W., Sammut, K., Yazdani, A., et al. (2017). A hierarchal planning framework for AUV mission management in a spatio-temporal varying ocean. http://arxiv.org/abs/1604.07898 (accessed on 10th September 2025).
9. Barber, D., Best, A., Manocha, D., Narang, R., et al. (2017). AutonoVi: Autonomous vehicle planning with dynamic maneuvers and traffic constraints. http://arxiv.org/abs/1703.08561 (accessed on 10th September 2025).
10. Barczak, A. L. C., Fatahillah, A., Reyes, N. H., & Susnjak, T. (2011). A reconfigurable hybrid intelligent system for robot navigation. https://core.ac.uk/download/148639235.pdf (accessed on 10th September 2025).

11. Belo, O., Deusdado, L. D., & Fernandes, A. R. (2010). Path planning for complex 3D multilevel environments. https://core.ac.uk/download/55631610.pdf (accessed on 10th September 2025).
12. Bi, H., & Gelenbe, E. (2015). Cloud enabled emergency navigation using faster-than-real-time simulation. http://arxiv.org/abs/1412.7059 (accessed on 10th September 2025).
13. Bjelonic, M., Hutter, M., Lee, J., Miki, T., et al. (2024). Learning robust autonomous navigation and locomotion for wheeled-legged robots. http://arxiv.org/abs/2405.01792 (accessed on 10th September 2025).
14. Blöchliger, F., Dymczyk, M., Fehr, M., Schneider, T., et al. (2018). Topomap: Topological mapping and navigation based on visual SLAM maps. http://arxiv.org/abs/1709.05533 (accessed on 10th September 2025).
15. Boccardo, P., Chiabrando, F., Facello, A., Gnavi, A., et al. (2012). Training of crisis mappers and map production from multi-sensor data: Vernazza case study (Cinque Terre National Park, Italy). https://core.ac.uk/download/pdf/11431904.pdf (accessed on 10th September 2025).
16. Brooker, P. (2007). Air traffic safety: Continued evolution or a new paradigm. https://core.ac.uk/download/138278.pdf (accessed on 10th September 2025).
17. Burdick, J. W., Du Toit, N. E., Murray, R. M., Wongpiromsarn, T., et al. (2008). Situational reasoning for road driving in an urban environment. https://core.ac.uk/download/4885309.pdf (accessed on 10th September 2025).
18. Buyse Sánchez, M., & Ramos Isidoro, D. (2017). Symbolic representation of scenarios in Bologna airport on virtual reality concept. https://core.ac.uk/download/87650625.pdf (accessed on 10th September 2025).
19. Cacan, M., Costello, M., Scheuermann, E., Slegers, N., et al. (2017). Utilizing ground-based LIDAR measurements to aid autonomous airdrop systems. https://digitalcommons.georgefox.edu/cgi/viewcontent.cgi?article=1074&context=mece_fac (accessed on 10th September 2025).
20. Caluwaerts, K., Dollé, L., Favre-Félix, A., Girard, B., et al. (2012). A biologically inspired meta-control navigation system for the Psikharpax rat robot. https://core.ac.uk/download/55802668.pdf (accessed on 10th September 2025).
21. Cao, J., Chang, J., Li, X., Li, H., et al. (2024). Research on the application of computer vision based on deep learning in autonomous driving technology. http://arxiv.org/abs/2406.00490 (accessed on 10th September 2025).

22. Carlin, C. M., Lambregts, A. A., & Ray, J. K. (2025). High-speed civil transport flight- and propulsion-control technological issues. https://core.ac.uk/download/pdf/42813729.pdf (accessed on 10th September 2025).

23. Ceriotti, M., Channumsin, S., & Radice, G. (2017). Distributed spacecraft path planning and collision avoidance via reciprocal velocity obstacle approach. https://core.ac.uk/download/96884020.pdf (accessed on 10th September 2025).

24. Cheng, Y., Ji, Y., Jiang, P., Wang, Z., et al. (2013). Design of a multiple bloom filter for distributed navigation routing. https://core.ac.uk/download/151156669.pdf (accessed on 10th September 2025).

25. Choi, H. L., Ha, J. S., & Park, S. S. (2018). Topology-guided path integral approach for stochastic optimal control in cluttered environment. http://arxiv.org/abs/1603.05099 (accessed on 10th September 2025).

26. Christe, S. D., Cooper, L. V. D., Gramling, C. J., Israel, D. J., et al. (2025). LunaNet: A flexible and extensible lunar exploration communications and navigation infrastructure. https://core.ac.uk/download/pdf/288485269.pdf (accessed on 10th September 2025).

27. Civit Balcells, A. (2016). A short curriculum of the robotics and technology of computer lab. https://core.ac.uk/download/322845099.pdf (accessed on 10th September 2025).

28. Clark, M., Grosu, R., Phan, D., Schierman, J., et al. (2017). A component-based simplex architecture for high-assurance cyber-physical systems. http://arxiv.org/abs/1704.04759 (accessed on 10th September 2025).

29. Conroy-Dalton, R. (2002). Is spatial intelligibility critical to the design of largescale virtual environments? https://core.ac.uk/download/1669766.pdf (accessed on 10th September 2025).

30. Corke, P., Hillier, N., & Martin, S. (2010). Practical application of pseudospectra optimization to robot path planning. https://core.ac.uk/download/10903310.pdf (accessed on 10th September 2025).

31. Dosovitskiy, A., Kaufmann, E., Koltun, V., Loquercio, A., et al. (2019). Deep drone racing: From simulation to reality with domain randomization. http://arxiv.org/abs/1905.09727 (accessed on 10th September 2025).

32. El-Khatib, M. M., & Hamilton, D. J. (2006). A layered fuzzy logic controller for nonholonomic car-like robot. https://core.ac.uk/download/9037328.pdf (accessed on 10th September 2025).

33. Erickson, J. D., Healey, K. J., Hess, C. W., McFadin, L., et al. (2025). An intelligent, free-flying robot. https://core.ac.uk/download/pdf/42821678.pdf (accessed on 10th September 2025).

34. Gorman, F. J., Jr., & Soicher, H. (2025). Plasma spheric effects on one way satellite timing signals. https://core.ac.uk/download/pdf/10315982.pdf (accessed on 10th September 2025).

35. Han, D., Lee, B. S., & Xue, Y. (2008). Automatic collision avoidance of ships. https://core.ac.uk/download/9023332.pdf (accessed on 10th September 2025).

36. Hart, J., & Smith, S. P. (2006). Evaluating distributed cognitive resources for wayfinding in a desktop virtual environment. https://core.ac.uk/download/66632.pdf (accessed on 10th September 2025).

37. Hou, B., Salzman, O., & Srinivasa, S. (2017). Efficient motion planning for problems lacking optimal substructure. http://arxiv.org/abs/1703.02582 (accessed on 10th September 2025).

38. Janglova, D. (2004). Neural networks in mobile robot motion. https://core.ac.uk/download/322385954.pdf (accessed on 10th September 2025).

39. Jian, H. (2024). Path planning algorithm for mobile robots. *Transactions on Computer Science and Intelligent Systems Research*. https://www.semanticscholar.org/paper/31cd24ddcc35ae7d7cb46939c2d36ce4f4a12ea6 (accessed on 10th September 2025).

40. Kopardekar, P. H., & Sridhar, B. (2025). Towards autonomous aviation operations: What can we learn from other areas of automation? https://core.ac.uk/download/pdf/286393131.pdf (accessed on 10th September 2025).

41. Krajnik, T., Preucil, L., Saska, M., Vanek, P., et al. (2013). Navigation, localization and stabilization of formations of unmanned aerial and ground vehicles. https://core.ac.uk/download/20321114.pdf (accessed on 10th September 2025).

42. Lerner, A., Nareyek, A., Nareyek, A., Watt, A., Lintermann, B., Cruz-Neira, C., Jones, C., et al. (2010). Developing serious games for cultural heritage: A state-of-the-art review. https://core.ac.uk/download/9599306.pdf (accessed on 10th September 2025).

43. Li, B., Liang, X., Wang, Y., Yu, H., et al. (2017). Research and implementation of global path planning for unmanned surface vehicle based on electronic chart. http://arxiv.org/abs/1802.03221 (accessed on 10th September 2025).

44. Lu, L., & Zhang, Z. (2024). Research and application of path planning for unmanned ships. *Scientific Journal of Technology*. https://www.semanticscholar.org/paper/2e1d9d5610260ed461d513216a024bccbf3f8d7c (accessed on 10th September 2025).

45. Sevara, C., & Verhoeven, G. (2016). Trying to break new ground in aerial archaeology. https://core.ac.uk/download/55850378.pdf (accessed on 10th September 2025).
46. Shrivastava, A. (2025). Path planning for robot navigation. *International Journal for Research in Applied Science and Engineering Technology*. https://www.semanticscholar.org/paper/2537bf816b998ad5faca0f96e9ba6cc10833a373 (accessed on 10th September 2025).
47. Thalmann, D. (2014). Populating 3D cities: A true challenge. https://media.neliti.com/media/publications/176357-EN-populating-3d-cities-a-true-challenge.pdf (accessed on 10th September 2025).

CHAPTER 7

COMMUNICATION AND NETWORKING IN AUTONOMOUS SYSTEMS

Contents

Introduction 249
7.1. Wireless Communication Protocols 252
7.2. Vehicle-to-Vehicle (V2V) Communication 257
7.3. Machine-to-Machine (M2M) Communication 262
7.4. Internet of Things (IoT) Integration 267
7.5. Challenges in Communication and Networking 272
7.6. Future Trends in Communication Technologies 277
Summary 283
References 286

INTRODUCTION

With an era of rapid technological development, convergence in communication and networking has been at the forefront of autonomous system design. Advances in wireless communication protocols have also introduced groundbreaking transformations in the manner devices interact with one another, resulting in a significant boost in the efficiency and trustworthiness of the systems. Technologies such as Wi-Fi, 5G, ZigBee, and LoRa play a vital role in providing seamless connectivity among autonomous system entities. 5G's high bandwidth and low latency, for instance, support the real-time data exchange required for autonomous vehicles in dynamic environments. ZigBee and LoRa are also invaluable as low-power, long-range communications that are required in cases where conventional networks cannot work. Such flexibility of this communication technologies architecture opens doors to more sophisticated applications across various fields, further enriching the autonomous systems ecosystem.

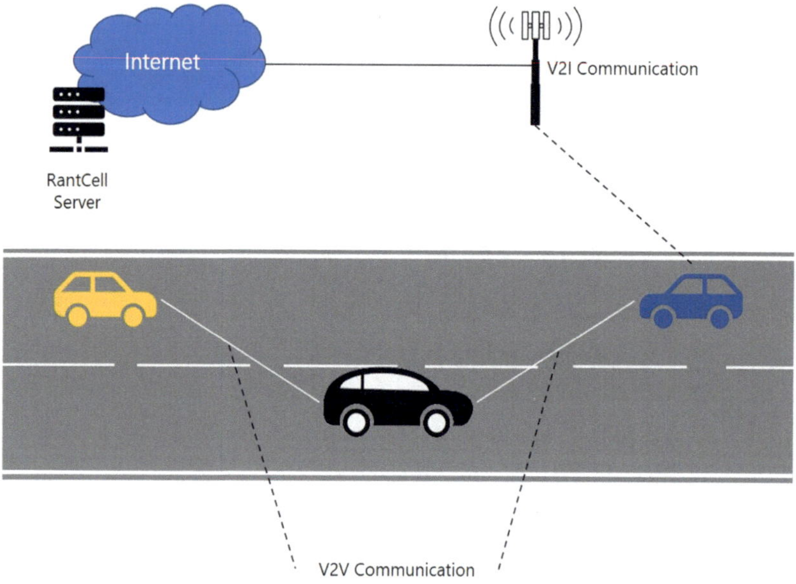

Figure 7.1. *Communication and networking in autonomous systems.*

Source: https://rantcell.com/images/V2I-and-V2V-Communications.jpg.

Networking automobiles and equipment with advanced communication networks enables the innovative applications of Vehicle-to-Vehicle (V2V) and Machine-to-Machine (M2M) communications. Communication patterns in these are fundamental to enhancing safety and operational effectiveness of autonomous systems. V2V communication enables vehicles to share information on their speed, direction, and intended movements, thereby reducing the chances of accidents and enhancing traffic capacity. On the other hand, M2M communication establishes device-to-device interaction without human interference, where responses and decision-making processes are computerized in order to optimize processes. As the Internet of Things (IoT) continues to integrate within such systems, the synergy between V2V and M2M is maximized even further. This consolidation not only maximizes the strength of data sharing but also facilitates end-to-end monitoring and management of autonomous systems, developing a robust ecosystem that lays the groundwork for future innovations. The progress of communication technologies in autonomous systems reflects the broader directions of digital transformation, which are reshaping industries. While corporations are embracing these innovations, the implications of business models and performance metrics need to be carefully evaluated. As organizations transition from digitization to digital transformation, organizations are being asked to form dedicated organizational structures that can accommodate the advanced requirements of such emerging technologies (Peter et al., 2019). Furthermore, the DT technology depicts that the combination of cyber and physical spaces can be used to facilitate real-time simulation and prediction, which can enhance the operational efficiency (Tao et al., 2019). This convergence of networking and communication technologies in autonomous systems not only hastens development but also requires the reshaping of strategic frameworks across industries. The resultant challenge lies in leveraging the technologies while dampening potential risks, ensuring a sustainable future for autonomous systems in the digital realm (Figure 7.1).

Importance of Communication in Autonomous Systems

The sophisticated autonomous system environment needs a robust communication infrastructure to function and integrate in varied environments. Wireless communication protocols such as Wi-Fi, 5G, ZigBee, and LoRa are essential in facilitating both V2V and M2M communications to address various operational needs. Wi-Fi and 5G provide high-speed data, sufficient for data-heavy applications in cities, while ZigBee and LoRa provide low-power counterparts, hence supporting efficient communication in rural regions. The underlying issue remains achieving a balance among transmission range, power consumption, and bandwidth to support vehicles' ability to communicate with each other and with nearby infrastructure. Proper use and adaptation of these protocols have a direct impact on the reliability and responsiveness of autonomous systems, hence ensuring safer and more efficient transportation networks, boosting operational efficiency in scenarios where real-time data sharing is critical. With the growing number of Internet of Things (IoT) devices being deployed, communication among autonomous systems is even more pertinent. The IoT ecosystem enables enormous quantities of sensors and devices to share, collect, and transmit useful data that informs decision-making systems. In autonomous vehicles, this information might carry road conditions, traffic patterns, or danger data. The effectiveness of this integration relies on V2V and M2M communication protocol performance, which must enable instant and accurate data exchange to deliver enhanced situational awareness and decision-making. For instance, the deployment of robust V2X communication infrastructures that consider vehicular and infrastructure data can yield significant traffic management and accident prevention benefits. Beyond safety, however, these advancements have enormous promise for a revolutionary enhancement in overall traffic efficiency, and the case for ongoing development of communication infrastructures to facilitate autonomous systems is an appealing one. Through the exploration of applications within real-life settings of these wireless communication technologies, testbeds are a useful innovation and assessment platform. These infrastructures allow researchers to conduct field operation trials that quantify the value and efficiency of wireless vehicular cooperative systems. The City of Bristol V2X communication testbed is an example; it supports the collection of sensor data from various urban environments that can be utilized to assess the effectiveness of communication protocols in real-world environments. Tests of these mass trials show high packet delivery rates, allowing evidence for the construction of models that support effective communication between vehicles and infrastructure. However, problems still remain, such as instability of transmission rates under some conditions. Addressing these issues in the context of autonomous vehicles is necessary for the achievement of stable and resilient operation, underpinning the need for rigorous testing and iterative optimization toward the achievement of sustainable, intelligent transportation systems (Cartolano et al., 2008; Tassi et al., 2018).

7.1. WIRELESS COMMUNICATION PROTOCOLS

The evolution of wireless communication protocols has acted as a driving force in the evolution of the communication infrastructures necessary for autonomous systems to operate optimally. Technologies like Wi-Fi, 5G, ZigBee, and LoRa have emerged as building blocks to manage the massive amounts of data generated by the systems. For instance, Wi-Fi facilitated fast internet in congested areas, and 5G transformed low-latency communication, which supports real-time communications crucial to ensure safety in self-driving cars. ZigBee and LoRa have also claimed space in low-power, long-distance communications crucial for Internet of Things (IoT) deployment in smart cities. These protocols not only enable enhanced connectivity but also the necessary bandwidth and reliability to support frictionless interaction between devices embedded in autonomous ecosystems. In the autonomous systems ecosystem, V2V and M2M communication offer paradigm shifts wrecked by the integration of advanced wireless protocols. V2V technology encourages direct vehicle-to-vehicle interaction, allowing for the exchange of information regarding road conditions, traffic patterns, and possible hazards to improve traffic efficiency and general safety. This wireless exchange of information becomes particularly important in the deployment of autonomous vehicles, where there is a need for instantaneous decision-making based on shared data. M2M communication facilitates different devices and machines to communicate on their own, using protocols specific to a specific purpose. This feature has vast applications in sectors such as industrial automation and intelligent agriculture, where sensors can transmit valuable information and respond to changing conditions in real-time. As the convergence of robotics and wireless sensor networks (WSN) matures into a new and flourishing field known as Robotic Wireless Sensor Networks (RWSN), the role of advanced communication protocols becomes increasingly apparent. In this paper, a networked multi-robot system is collaborating to accomplish some sensing tasks under stringent communication performance constraints. An example of the potential of advanced algorithms in utilizing mobility to enhance packet routing performance using robotic routers is one of them. Furthermore, as noted in earlier research, existential issues concerning connectivity, localization, and robust information flow remain dominant in RWSN studies. Solving these issues necessitates further research and innovation in wireless communication technology, affecting the performance and reliability of autonomous systems in several important ways (Balasubramanian et al., 2018; Cartolano et al., 2008) (Figure 7.2).

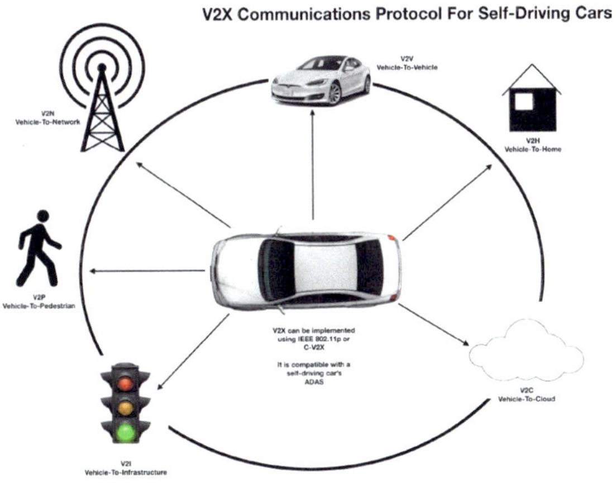

Figure 7.2. *Wireless communication protocols playing a pivotal role in advancing communication frameworks necessary for the effective operation of autonomous systems.*

*Source: https://miro.medium.com/v2/resize:fit:4060/1*fFwV0zFTAHAlasJ1evB7iw.png.*

7.1.1. Overview of Wi-Fi Technology

The expanding communications environment emphasizes wireless protocols' significance, topped by Wi-Fi. Since its initial iterations, with their limited speed and distance, Wi-Fi technology has evolved to the sophisticated standards of today such as Wi-Fi 6 and onwards. These assist in facilitating massive data transmissibility and holding more devices in a single network environment, which autonomous systems require. Interestingly, Wi-Fi operates on several frequency bands, so it is compatible with other wireless protocols like 5G, ZigBee, and LoRa. All these protocols have a niche for themselves; for instance, while 5G has the best place in high-speed data transfer for mobile phones, ZigBee and LoRa are employed in low-power applications, ideally suited for the application in smart sensors in the Internet of Things (IoT).

Thus, Wi-Fis integration with these technologies allows for a more flexible and interoperable wireless communication framework essential for developing sophisticated autonomous systems. As autonomous systems increasingly depend on robust communication channels, the role of Vehicle-to-Vehicle (V2V) and Machine-to-Machine (M2M) communication becomes paramount. Wi-Fi technology provides a foundational layer for these interactions, enabling vehicles to share real-time information about their surroundings, thereby enhancing road safety and efficiency. This responsiveness is especially crucial in cases that require timely distribution of information, such as collision avoidance. Additionally, M2M communication also enhances the capabilities of Wi-Fi by providing interaction between various devices and machines, making it possible for smart cities, industrial automation, and smart manufacturing. The seamless interaction between the devices maximizes use of all available bandwidth and ensures optimal performance. With the growing amount of autonomous vehicles and connected devices, Wi-Fi technology remains at the forefront as a facilitating technology, establishing a dynamic and responsive communication platform required for the nurturing of innovation and development in autonomous systems. Notably, the integration of Wi-Fi technology under the broader umbrella of the Internet of Things (IoT) indicates the need for secure communication protocols. With increasingly more devices getting connected, issues of communication security and integrity are born. Study of Secure Device Pairing (SDP) schemes suggests a requirement to counteract these shortcomings to maintain trust and reliability in autonomous systems (Fomichev et al., 2017). Further, as discussed in the latest literature on programmable wireless networks, there exists a significant shift towards the development of network architectures that facilitate increased flexibility and security with a view to counter the alleged ossification of static systems (Ahad et al., 2014). Therefore, with Wi-Fi technology constantly advancing, its use in the IoT and autonomous systems must prioritize security protocols to an extent that with connectivity on the rise, confidence and security are not compromised in this ever-changing technological landscape.

7.1.2. The Role of 5G in Enhancing Connectivity

With an increasingly integrated world becoming a reality, the demand for contemporary wireless communication protocols cannot be overstated. Among them, 5G is particularly notable due to the unprecedented advancements it has brought compared to its predecessors, particularly in speed, latency, and capacity. Unlike traditional protocols such as Wi-Fi and ZigBee with limited range and bandwidth, 5G enables seamless connection with more support for a wider array of applications, notably autonomous systems. This is crucial as vehicles and machines increasingly depend on the sharing of real-time data for optimal performance. With autonomous technology gaining traction, the cutting-edge feature provided by 5G enables Vehicle-to-Vehicle (V2V) and Machine-to-Machine (M2M) communications to become faster, more reliable, and more efficient. All these improvements play a vital role in making autonomous technologies safer and more efficient, thereby making roads safer and industries smarter. The presence of the Internet of Things (IoT) similarly showcases the paradigmatic existence of 5G in autonomous systems. With 5Gs ultra-reliable low-latency communication (URLLC), the devices can transmit data in near real-time, supporting sophisticated applications such as smart city infrastructure and high-level traffic management systems. 5G's wider bandwidth means that enormous volumes of data can be handled simultaneously without jamming, which is particularly beneficial in networked environments where multiple diverse devices talk simultaneously. This paradigm shift holds out the possibility for improved efficacy in energy consumption, public safety, and management of resources. Along with emerging technologies such as machine learning that enhances data use and network efficiency, 5G networks guarantee not just enhanced connectivity but also the platform for new innovations to do with autonomous systems (Abbas et al., 2020). As the need for stronger and more resilient networks keeps rising, 5G finds itself more at the heart of addressing the shortcomings existing within traditional cellular architectures. The legacy models, which prefer to stringently interleave network and data access points, are not able to accommodate the growing demands of data-centric services and diverse deployment scenarios (Imran et al., 2016). The shift to a more dynamic control-data separation architecture (CDSA) allows for greater efficiency and resource utilization, which are essential for autonomous system scalability in the future. By providing a specialized coverage layer for data services, 5G networks can support the dynamic nature of services that are requested in connected circumstances. This development shows how with increased connectivity, 5G not only provides enhanced existing applications but also enables future-proof development of intelligent autonomous systems, empowering them with a strong foundation for integration and use (Figure 7.3).

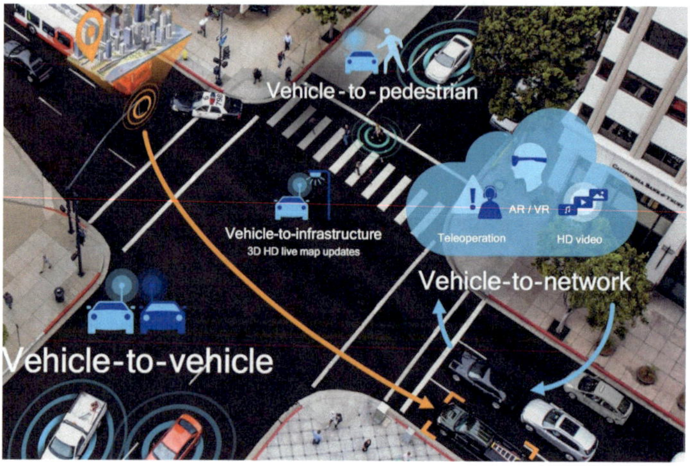

Figure 7.3. *Wireless communication protocols in autonomous systems.*

Source: https://etn-sas.eu/wp-content/uploads/2020/03/V2X_QualComm-1536x1000.jpg.

7.1.3. Comparison of ZigBee and LoRa for Low-Power Applications

ZigBee and LoRa are two distinct solutions in the area of low-power wireless communication that cater to the specific needs of the IoT application, most notably in autonomous systems. ZigBee, operating under IEEE 802.15.4, is specifically designed for short-range, low-data-rate communication and is optimal for applications with mesh networking and low latency. Its architecture enables devices to effectively communicate within a limited cluster, so secure connections between a high concentration of intelligent devices are achievable. Although ZigBee excels in the support of local networks and applications such as automation in homes, it suffers from a limitation of range, which, in most cases, is up to 100 meters indoors and slightly higher outdoors, thereby restricting its application where high coverage is necessary. Hence, knowing its limitation is crucial while planning its incorporation into large IoT networks that are based on real-time data transfer and responsiveness. In contrast, LoRa employs a spread-spectrum technology that provides a significantly increased range as opposed to ZigBee, which tends to be several kilometers in the open terrain. This aspect makes LoRa especially suitable for applications that demand sparse sensor networks spread over wide geographical regions, including environmental monitoring and smart agriculture. Its power-hungry character is necessary because devices can last years on a single battery, a good characteristic for maintenance-free remote systems. LoRas absence of optimal value when transmitting data at a higher rate is perfectly suited to applications where transmissions are not frequent, ideal for the needs of machine-to-machine (M2M) communication applications designed to conserve energy. This overall efficacy explains why LoRa becomes an option of strategy for many IoT applications where range and longevity become primary considerations. The higher level of communication requirements in self-governing systems necessitates having to think about the involvement of protocols like ZigBee and LoRa in the broader context of wireless technology innovation, including Wi-Fi and 5G. Although ZigBee and LoRa are focusing on niche segments based on varying metrics of functionality, their adoption within smart campuses and smart cities highlights their essential role as enablers of interconnected systems. The IT infrastructure used in these contexts is a developing pattern similar to that of smart cities, requiring secure connectivity between numerous IoT nodes and gateways. As academic discourse has noted, the interplay of technologies such as blockchain would further evolve these frameworks by providing security and transparency, both vital to university planners and IoT engineers alike. Thus, a comprehensive comparison not only describes the strengths and weaknesses of ZigBee and LoRa but situates them within the larger context of communications and networking for self-driving systems to offer a symbiotic operation of advanced modern infrastructure.

7.2. VEHICLE-TO-VEHICLE (V2V) COMMUNICATION

With advancements in autonomous vehicles revolutionizing the vehicular communication domain on an ongoing basis, the convergence of various wireless communication protocols is proving increasingly important. Wi-Fi, 5G, ZigBee, and LoRa represent a few among the most popular technologies supporting Vehicle-to-Vehicle (V2V) communication. Each protocol possesses unique characteristics proving appropriate for varying dimensions of vehicular networking. For example, high-throughput and ultra-low-latency 5G is particularly promising for real-time vehicular communication that can accelerate decision-making and improve road safety. Technologies like ZigBee and LoRa, however, focus on low power consumption and high range and are designed for particular applications where bandwidth requirements are not high, but reliability is paramount. The extensive range of protocols puts tremendous emphasis on the value of a synergistic approach that leverages the strengths of each to provide the combined effect of V2V communication in autonomous systems. The integration of V2V communication with machine-to-machine (M2M) technologies further enhances the value of potential applications in autonomous systems. This harmony provides a ground where vehicles not only communicate with one another but also seamlessly interoperate with a broad range of IoT devices. For instance, the integration of V2V communication protocols allows vehicles to share critical information, such as road conditions, road obstructions, and weather forecasts, thereby stimulating situational awareness. This interconnectedness can lead to safer roads and smoother traffic flows.

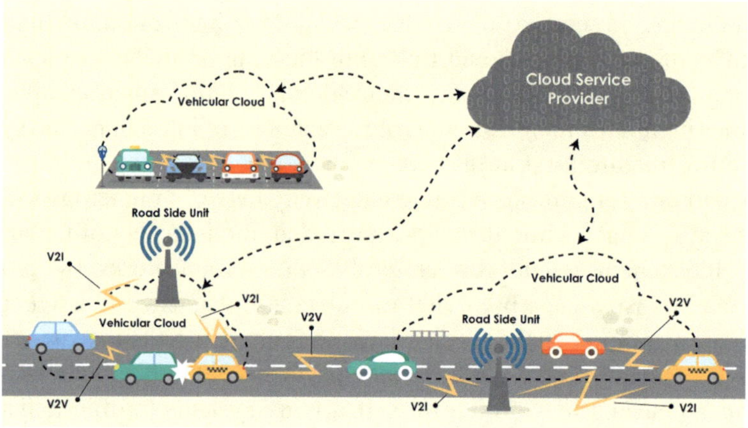

Figure 7.4. *Vehicle-to-vehicle (V2V) communication.*

Source: https://www.benthamopen.com/contents/figures/TOTJ/TOTJ-14-86_F6.jpg.

By facilitating a robust data exchange among vehicles and IoT devices, enhanced V2V systems are central to the creation of smart road networks with the objective of achieving streamlined traffic flow as well as mitigating hazards. Especially, the contribution of evolutionary technologies like Cellular Vehicle-to-Everything (C-V2X) cannot be overstated when it comes to V2V communication. C-V2X aims to surpass the limitations of legacy V2V and V2I networking with less complex communication channels that reduce latency and enhance data security via cellular-based technology. The transition to LTE-V2X and subsequently to 5G-V2X marks a great leap, addressing core challenges such as ultra-high reliability and increasing demand for high throughput in automotive communication. With its security-centric new architecture, 5G-V2X is well on its way to becoming the future-proof backbone for V2V communication, enabling safe and efficient information sharing between the infrastructure and vehicles (Guizani et al., 2019). Evolving V2V reality shows a dynamic interplay of technology and infrastructure moving toward safer and more efficient autonomous systems (Figure 7.4).

7.2.1. Definition and Significance of V2V Communication

In the evolving situation of autonomous vehicles, the significance of communication without interruption cannot be overemphasized. V2V (Vehicle-to-Vehicle) communication comes to the forefront as an essential component of this ecosystem through wireless protocols such as DSRC and 5G. These technologies enable real-time data transfer between vehicles so that vehicles can share information about traffic updates, hazardous situations, and path optimization. As cars communicate with each other in real time, they not only improve their own safety but also the safety of the traffic as a whole. By integrating V2V communication and protocols such as Wi-Fi, ZigBee, and LoRa, the collective initiative to create a sturdy and connected ecosystem is clearly demonstrated. Aside from that, V2V communication also provides an underlying foundation for machine-to-machine (M2M) communication, a part of the entire Internet of Things (IoT) system wherein cars act as nodes communicating with other infrastructures in order to create efficient operations. The far-reaching impacts of V2V communication are beyond the aspect of vehicular communication; these come within the context of the overall smart city and intelligent transportation system plan. The promise of cars that are outfitted with next-generation communication technologies is to engage in a dynamic dialogue with the world around them, such as traffic signals and roadside devices. Through enhanced situational awareness, V2V communications significantly decrease threats to avoid crashes and make traffic flow more efficient. Also, as technology advances in leveraging high-definition video and data collection in urban environments, the transmission of information in real-time comes into focus. This communication involves not only traffic movement but also safety in public as well. By providing abilities that are able to prevent crime and give a sense of security to drivers, V2V communication plays a multi-functional role, highlighted by applications such as cooperative home illumination systems, addressing car-linked crime and general security (Kargl et al., 2014). As requirements for scalable and reliable communications keep on growing, persistent challenges in V2V communication necessitate collaborative research endeavors. At the top of the agenda of burning issues is ensuring quality of service in transmitting data through varying network conditions, especially in the case of vehicular IoT, where availability of resources and heterogeneity of capability among devices generate uncertainty. To mitigate these issues, different protocols and technologies are tested for their suitability in particular settings. As an instance, video streaming within vehicular networks as a part of V2V dynamics has been observed to receive considerable examination, responding to bandwidth usage concerns and quality monitoring (Abdullah et al., 2017). This ongoing discourse of V2V communication not only enriches scholarship on decentralized networks but also on real-world implementation in autonomous systems, making the pivotal importance of fault-tolerant vehicular communication structures in smart mobility an indubitable fact.

7.2.2. Safety Applications and Benefits of V2V Systems

Advancements in wireless communication technologies are paramount to enhancing the safety features and benefits of vehicle-to-vehicle (V2V) systems. Wi-Fi, 5G, ZigBee, and LoRa are some of the major technologies that act as root protocols for rapid data exchange between vehicles. For instance, 5G offers low latency and high-speed communications, which are essential for real-time exchange within dynamic road traffic. V2V systems employ such protocols to facilitate the sharing of critical information among vehicles, such as speed and direction, which can improve situational awareness significantly. Such real-time sharing not only avoids accidents but also aids in traffic management to reduce congestion and pollution. As pointed out by Cartolano et al. (2008), the improvement of traffic management on roads through such state-of-the-art communication technologies emphasizes the necessity of an end-to-end approach to vehicular communication systems in order to enable secure and efficient traffic flow. Apart from V2V communications, machine-to-machine (M2M) communication also enhances the connectivity ecosystem of autonomous systems. By closing the loop between various IoT devices and vehicles, M2M creates an ecosystem where vehicles can receive information about road conditions, weather warnings, and future dangers. Such an end-to-end communication system enhances the overall safety level by enabling predictive analytics, which can notify drivers in advance of potential hazards such as sudden braking or road obstructions. The incorporation of IoT in V2V systems is a precursor to a new paradigm of automobile design, where cars are not only a mode of transportation but a part of an interconnected system. From Denis et al. (2014), it is apparent that the convergence of these technologies will help create an intelligent, secure driving experience while addressing road safety and environmental sustainability. While the safety benefits of V2V systems are considered to surpass the need for traditional redundancy, the advantages also go beyond safety. As the transport landscape changes in the modern era, the capabilities of V2V systems to increase fuel efficiency, reduce traffic jamming, and lower emissions are considered a fundamental benefit. As vehicles exchange information and coordinate movements, they can avoid wasteful motion, thereby consume less fuel and lower their carbon footprint. Furthermore, continuous advancements in V2V communication can allow for coordinated maneuvers such as platooning, which can provide further traffic congestion reductions. Interestingly, the inclusion of V2V systems in cities is aligned with the larger goals for sustainable infrastructure development. Based on the strategy laid out by Cartolano et al. (2008), proper deployment of these types of technologies opens the door for cleaner, safer traffic control solutions and emphasizes the social importance of vehicle-to-vehicle communication system development in making the road safer (Figure 7.5).

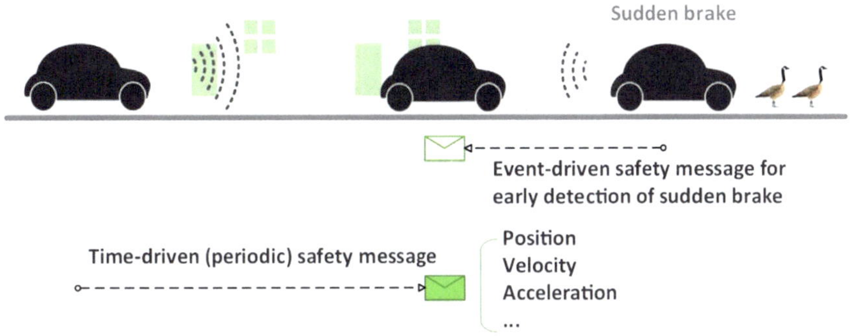

Figure 7.5. *Example of safety applications based on V2V communications.*

Source: https://external-content.duckduckgo.com/iu/?u=https%3A%2F%2Fwww.researchgate.net%2Fprofile%2FNing-Lu-8%2Fpublication%2F264564008%2Ffigure%2Ffig2%2FAS%3A613935090786305%401523384940427%2FExample-of-safety-applications-based-on-V2V-communications.png&f=1&nofb=1&ipt=d272bf05f95064b19dff4f274061c36f6e6055d5cb86870fe65e2393a6da9dee.

7.2.3. Challenges and Limitations in V2V Communication

The technological context for autonomous systems has a significant influence on communications strategies, particularly with respect to vehicle-to-vehicle (V2V) communication. Wireless communication protocols such as Wi-Fi, 5G, ZigBee, and LoRa are significant in facilitating data communication between vehicles to enhance safety and efficiency. These protocols offer some benefits—such as high data rates and wide coverage—while each protocol also has some particular challenges. For example, 5G can provide higher bandwidth, but its deployment is not geographically uniform, limiting its availability for V2V systems. Similarly, LoRa has improved performance in far-end communication but lacks bandwidth, inhibiting the volume of data exchange. The heterogeneous protocols thus need customized integration in the V2V system to ensure a reliable and stable network, especially as autonomous technologies develop. Incorporating the Internet of Things (IoT) into the V2V environment enhances these difficulties with multiple layers of complexity regarding device interoperation and information handling. Vehicle-to-vehicle and infrastructure-to-vehicle communication is dependent on solid machine-to-machine (M2M) interactions. Inconsistencies in communication protocols may interfere with this synchronization, leading to delays or useless data transmission. Spearheaded by the urgency of real-time decision-making ability, such limitations are bound to affect the overall effectiveness of V2V systems. Besides, as the number of devices connected increases, the network is prone to congestion and interference, also rendering communication an issue. Lacking supportive solutions to these issues, the potential of a robust, networked transportation system continues to be a mirage, and it brings into perspective the necessity of ongoing research and development. Field operational testing reveals important findings about the real-world challenges of deploying V2V communication systems. While initial tests indicate improved control of road traffic and increased vehicle awareness, sustained system performance over the long term is a cause of significant concern. For instance, as reflected in the results of various trials, packet transmission rate stability is crucially important, particularly when employing single-core equipment in urban environments. These limitations may result in information loss or delayed communication, which are unsafe in high-stake, high-response conditions. Furthermore, as identified in recent studies, the efficiency of wireless vehicular cooperative systems needs to be evaluated longitudinally and wide-ranging in nature to realize their maximum potential for traffic management (Cartolano et al., 2008). These operational assessments are essential to guide the design and tuning of communication protocols so that V2V communications can become more dependable in different environments (Tassi et al., 2018).

7.3. MACHINE-TO-MACHINE (M2M) COMMUNICATION

With the rapidly shifting realm of autonomous systems, integration of diverse communication protocols has been essential to ensured operation. Enabling effective data exchange between devices, wireless communication technologies like Wi-Fi, 5G, ZigBee, and LoRa enhance machine-to-machine (M2M) communication, essential for real-time operation and decision-making. Wi-Fi and 5G offer quick connectivity sufficient for urban environments where there is heavy interactivity of devices. Conversely, ZigBee and LoRa are best suited for low-power applications, with the ability to effectively manage sensor networks requiring long-distance communication without large amounts of energy consumption.

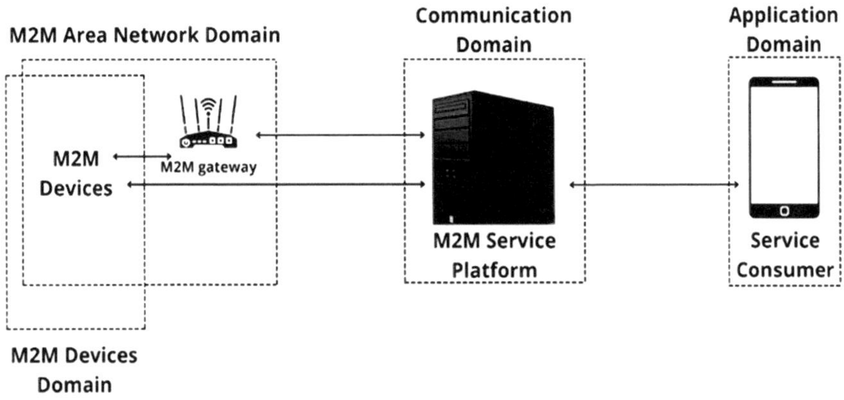

Figure 7.6. *Importance of machine-to-machine communication for manufacturing.*

Source: https://www.machinemetrics.com/hs-fs/hubfs/machine-to-machine-communication-m2m.png?width=750&name=machine-to-machine-communication-m2m.png.

The unique capabilities of these protocols allow for various applications such as smart cities and remote monitoring systems, thereby proving their significant contributions to the improvement of the functionality of autonomous systems. While M2M communication is gaining steam, its role in vehicle-to-vehicle (V2V) communication speaks volumes about its application in transportation and traffic management. Enabling cars to communicate with each other and the surrounding infrastructure, V2V communication encourages secure and efficient driving. M2M-based systems can share information about road conditions, traffic signals, and upcoming threats, thereby allowing vehicles to take correct decisions in real-time. Therefore, not only does this technology improve the performance of individual vehicles but also that of transportation networks as a whole. With autonomous vehicles becoming increasingly common, the use of resilient M2M communication infrastructures serves to emphasize the need for robust and secure networks that can handle the requirements of new transportation systems. The integration of M2M communication within the broader Internet of Things (IoT) framework is reflective of the transformative power of such technologies across industries like manufacturing and energy management. Within the smart manufacturing environment of Industry 4.0, M2M communication lies at the heart of enabling real-time analytics of data that enable efficiency and innovation. As such, visualization of manufacturing information allows In-Transit Analytics, a procedure which may be able to uncover information important to maintaining competitive advantage (Ali et al., 2017). Similarly, M2M communication in sectors such as energy management facilitates grid reliability and efficiency by means of near-real-time communications among grid elements that ultimately steer interoperability in smart grids (Chin et al., 2011). The convergence of these technologies serves as the basis for a more integrated and responsive setting, driven by the need for better performance across sectors (Figure 7.6).

7.3.1. Explanation of M2M Communication in Autonomous Systems

In each contemporary autonomous system, coexistence of a large number of wireless communication protocols is imperative towards enabling seamless interaction among machines. Among the most prevalent of such protocols are Wi-Fi, 5G, ZigBee, and LoRa. Each protocol has unique roles to play in the framework; for instance, Wi-Fi and 5G are concerned with high-speed data communication and low latency needed for real-time car navigation in an autonomous vehicle. Correspondingly, ZigBee and LoRa are appropriate for low-power, long-range communication for IoT devices requiring permanent connectivity without depleting power supplies. The synergistic integration of the two communication protocols supports an effective network for vehicles and machines to exchange critical operating information with ease, generating enhanced situational awareness and harmonized behaviors in variable conditions. The concept of machine-to-machine (M2M) communication is common in the area of autonomous systems, particularly vehicle-to-vehicle (V2V) communication. Not only can cars utilize such communication to share information regarding their movement speed, direction, and potential hazards, but they can also make roads safer and more efficient. Autonomous cars can become networked nodes that communicate with and support one another through M2M communication. Further, the robust data transmission facilitated by M2M systems is used to fuel an overarching database that guides predictive analytics and real-time responsiveness in transportation systems. This shift towards networked machines is a fundamental development within autonomous vehicle technology and serves as a foundational component within the broader context of smart city programs and intelligent transportation systems. Besides, integrating the Internet of Things (IoT) into autonomous systems is symbolic of the significant advancements brought about by M2M communication.

IoT sensors embedded in infrastructure and vehicles enhance the general performance of autonomous systems by providing real-time data utilized to make decisions. One such instance is the way intelligent traffic lights can adapt based on real-time traffic information offered by vehicles, maximizing traffic flow and minimizing congestion. In addition, such integration brings to the fore challenges that arise from security and privacy concerns, particularly with more devices on the network. Addressing these challenges is crucial towards realizing the full potential of autonomous systems and ensuring compliance with communications standards and protocols. With further research in the sector, integrating M2M communication strategies into IoT technology will forever redefine efficiency of operations in transport and otherwise (Asadi et al., 2014; Chin et al., 2011).

7.3.2. Applications of M2M in Industrial Automation

The advent of wireless communication protocols has significantly touched the face of industrial automation through Machine-to-Machine (M2M) applications. Wi-Fi, 5G, ZigBee, and LoRa are some technologies that enable devices to connect with ease, enhance operational efficiency and flexibility in manufacturing environments. The capability of 5G technology, for instance, enables the transfer of high-speed data and low latency to support real-time communication of machines. On the other hand, ZigBee and LoRa possess low power with long range, which makes them suitable for industrial sensor networks. These wireless technologies allow M2M communication systems to merge, facilitating various machines, sensors, and devices to exchange information with one another and, therefore, enhance production activities and minimize downtime. Improved communication naturally brings improved decision-making, predictive maintenance, and resource usage, which align with the goals of optimal automation in the industrial economy. The integration of Internet of Things (IoT) technologies further expands the capability of M2M communications in industrial automation. IoT devices like sensors, actuators, and smart meters gather and exchange data in real time, and it allows the manufacturers to monitor system performance and respond to variances in real time. The use of IoT with M2M platforms makes it possible for machines not just to exchange data but also to send important information to data analysis systems. This synchronization enhances industrial procedures with insights on the delicacy of manufacturing procedures, enables intelligent automation, and enables real-time adjustments based on data analysis. In addition, the application of IoT in the industrial environment is supplementary to the trend towards Smart Grids, where smart energy management solutions form the foundation of efficient consumption and distribution. The transition towards

Smart Grids legitimizes the collaborative environment of modern industrial settings, where M2M and IoT are imperative for operational efficiency (Chin et al., 2011; Copeland et al., 2019). Vehicle-to-Vehicle (V2V) communication is also crucial in industrial automation through M2M protocols. The technology provides direct communication between vehicles, which proves useful in manufacturing and logistics sectors where the coordination of the fleet is crucial. M2M solutions in V2V communication help organizations to better coordinate routes, save fuel, and enhance safety through elimination of the risk of accidents through early warning and sharing of knowledge. Moreover, the interconnection of M2M communications with the existing transport management systems leads to enhanced asset tracking as well as enhanced inventory control. As a result, industrial automators can achieve higher operational efficiency and cost savings. These advances highlight the paradigmatic impact that M2M applications have on industrial automation, thus leading to a more integrated and reactive industrial landscape. The potential of this technology is far-reaching, paving the way to smarter and better production systems in today's competitive environment (Figure 7.7).

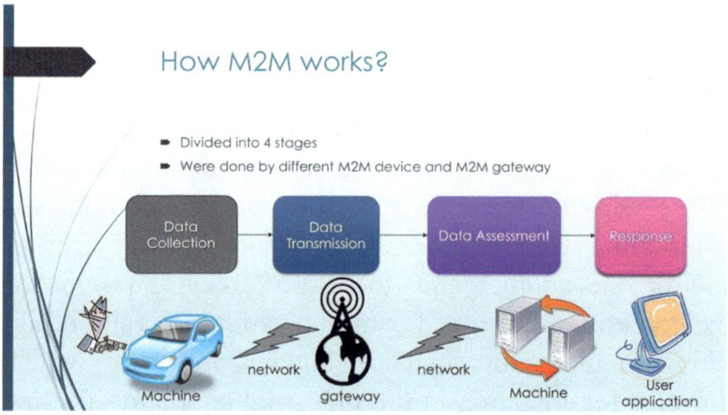

Figure 7.7. *Applications of M2M in industrial automation.*

Source: https://image1.slideserve.com/2382634/how-m2m-works1-l.jpg.

7.3.3. Security Concerns Associated with M2M Communication

Interconnectedness provided by M2M communication brings with it a realm of possibility across numerous industries, particularly in the context of autonomous systems and the Internet of Things (IoT). The expanded network surface does include the propensity to amplify threats to security. Wireless communication protocols such as Wi-Fi, 5G, ZigBee, and LoRa bring with them serious vulnerabilities. These protocols are open to interception, unauthorized access, and data tampering, which could heavily undermine the operation of autonomous systems. For example, in smart grid scenarios, where M2M communication plays a key role in energy management, violations may have an impact on operational integrity and consumer privacy. As outlined by Chin et al. (2011), these challenges need to be addressed for the successful deployment of secure and interoperable smart grid networks, pointing to some protection necessities that are paramount for robust security policies attuned to the nature of M2M environments. M2M communication and vehicle-to-vehicle (V2V) communication form the central part in the functional process of autonomous vehicles. This process is, however, accompanied by very stern security vulnerabilities that cannot be overlooked. Interception of sensitive information exchanged between vehicles can pose potential risks of accident or illegitimate tracking of movements. Additionally, to make the vehicle-to-vehicle communication real-time necessitates quick data transfer mechanisms, which can be exploited to compromise the system by malicious entities. The emergent security threats of V2V communications identified in the existing literature necessitate novel encryption and authentication protocols to enhance such systems against exploitation. As implied in the findings on emerging technologies (Asadi et al., 2014), handling the security concerns of such communication technologies will be essential toward realizing maximum benefits from autonomous transportation systems. M2M communication being part of the broader IoT ecosystem focuses on the utmost importance of an integrative cybersecurity approach. With numerous devices communicating on their own, data integrity and potential vulnerability to weaknesses are of utmost priority. Interdependencies built in IoT platforms often render security systems complicated, where any violation in one layer creates cascading impacts on the integrated systems. Hence, securing M2M communication needs an integrated system encompassing not only technical measures but also standardized protocols to secure data. Effective cybersecurity is paramount for effective implementation of smart cities and industrial automation, where M2M communication cannot be avoided. Highlighting the necessity for coordinated standardization procedures, as discussed by Chin et al. (2011), offers a beginning point for improving the security framework governing these advanced communication networks and increasing the resilience of autonomous systems against emerging cybersecurity threats.

7.4. INTERNET OF THINGS (IOT) INTEGRATION

The coordination of advanced communication protocols has played a pivotal role in the success of Internet of Things (IoT) integration in autonomous systems. Wi-Fi, 5G, ZigBee, and LoRa are just a few technologies that provide the required frameworks for efficient wireless communication to assist devices with data exchange at high speeds and with minimal delay. For instance, 5G technology boosts bandwidth dramatically and reduces latency, and so is ideally suited to real-time application in autonomous vehicles, where millisecond response can mean safety or danger. Similarly, ZigBee and LoRa are crucial in low-power applications, allowing smart sensors and devices to communicate where energy efficiency is the priority.

Figure 7.8. *Internet of things (IoT) integration.*

Source: https://www.checkboxtechnology.com/wp-content/uploads/2020/09/1_Wt-dyFZNDahpaoaxMGSCXIw.jpeg.

This diverse collection of protocols makes it possible for the adaptable ecosystem to support various IoT applications towards the final goal of highly autonomous systems where connectivity plays a central role. In autonomous technology, other communication modalities such as Vehicle-to-Vehicle (V2V) and Machine-to-Machine (M2M) also justify the significance of incorporating IoT. V2V communication provides vehicles the ability to share essential information of speed, direction, and road conditions, hence providing a better situational awareness and contributing to safer traffic and efficiency. M2M communication provides autonomous communication between devices and machines in industries, bringing together a platform where efficiencies in operations can be enhanced significantly. For instance, in logistics and supply chain, M2M can automate inventory levels by sending real-time data on items in stock. Therefore, the integration of these modes of communication creates processes more efficient, decreases downtime, and ultimately leads to a more powerful and adaptable operation setup. As autonomous system architecture continues to evolve, IoT integration usage receives valuable data that can be utilized to make strategic decisions. Supply chain digitization, as can be seen from existing trends, highlights the need for having an entirely networked foundation that incorporates all stakeholders, from manufacturers to consumers. Successful implementation of IoT technology, as explained by recent studies, allows for real-time monitoring and tracking, encouraging data insights into logistics movement and inventory management (Tiwari et al., 2024). In addition, the arrival of Industry 5.0 also emphasizes the co-operation of human beings and machines to the extent that innovation in IoT supports greater productivity without displacing human resources (Nahavandi, 2019). Not only does this co-operation strengthen a dynamic manufacturing culture but it also creates problems to be addressed cautiously through visionary policies and technological advances further supporting co-operation between human beings and machines (Figure 7.8).

7.4.1. Role of IoT in Enhancing Autonomous Systems

Development of autonomous systems increasingly relies on sophisticated communication technologies that facilitate smooth vehicle-environment interactions. A crucial step in this process involves the use of wireless communication protocols like Wi-Fi, 5G, ZigBee, and LoRa, each offering distinct advantages to specific applications. Wi-Fi and 5G offer low latency and high data rate and are consequently very well-suited for real-time applications such as decision-making and navigation in autonomous vehicles. ZigBee and LoRa are based on low-power solutions suitable for sensor networks utilized for monitoring the vehicle status and environment, however. These diverse channels of communication enhance the use of the Internet of Things (IoT) within autonomous systems, significantly enhancing their ability to exchange important information and operation feedback, making vehicle operations safer and more effective. To drive the performance of autonomous vehicles in real-time operations at a faster rate, two critical frameworks—vehicle-to-vehicle (V2V) communication and machine-to-machine (M2M) communication—are employed. V2V communication provides for direct vehicle-to-vehicle exchange in which vehicles can exchange information regarding speed, direction, and potential hazards. This vehicle network provides for enhanced situational awareness and decision-making capabilities by offering a collaborative environment of road users. M2M communication comes to the aid of this by connecting various devices and sensors in a vehicle to collect information and process it from various sources. In combination, these paradigms of communications make the ecosystem more interconnected so that autonomous vehicles are able to dynamically adapt to changes in their surroundings, hence lowering the risks of collisions and improving road safety as a whole. The combination of IoT and sophisticated communication protocols greatly improves the potential of autonomous systems such that they have greater perception, trustworthiness, and operating efficiency. Notably, IoT integration allows data from myriad sources to be aggregated in considerable volumes, which is of paramount importance for successful decision-making in autonomous operations. In addition, utilization of privacy-preserving techniques, explored in recent studies, enhances security of data while maintaining operational performance. Application of advanced learning models facilitates successful processing of data without divulging sensitive information. As autonomous systems continue to advance, the integration of secure and reliable communication channels and IoT will be key to overcoming the current challenges and realizing the full potential of autonomous technology in making the horizon for smarter and safer transport systems (Bolaji-Adetoro et al., 2025; Sabeer et al., 2025).

7.4.2. Data Exchange and Interoperability in IoT Networks

In the rapidly evolving communication technology landscape, wireless protocol convergence is the most important to enable effortless data exchange in Internet of Things (IoT) networks. Wi-Fi, 5G, ZigBee, and LoRa are some of the protocols available that offer varying advantages and are utilized for distinct applications. For instance, Wi-Fi allows bandwidth for applications requiring large data transfers, while ZigBee is applied for low-power, short-distance transmission. Conversely, 5G networks deliver more mobile connectivity with the support of a wide range of devices at one time and enabling applications that involve real-time data transmission. Similarly, LoRa technology is appropriate for long-distance communication in rural environments and IoT applications involving sensors. All such diverse protocols necessitate a responsive and networked topology for the functioning of autonomous systems employed in various settings. Operation of IoT networks relies greatly on the interoperability of communication frameworks, particularly in applications like Vehicle-to-Vehicle (V2V) and Machine-to-Machine (M2M) communications. V2V communication enhances road safety and efficiency as it allows vehicles to share real-time information about obstacles, upcoming maneuvers, and traffic situations. Information sharing not only aids safer driving but facilitates innovation in autonomous vehicle technology. Similarly, M2M communication facilitates direct conversation between devices independently of human intervention, significantly streamlining operations across industries such as manufacturing, healthcare, and energy management. By providing robust interoperability among disparate devices and communication protocols, these technologies facilitate seamless communication between autonomous systems, providing their capabilities and efficiency with a significant enhancement in real-world applications. The integration of IoT into cloud computing systems also enhances the potential for data interoperability and sharing, as witnessed through the Cloud of Things (CoT) idea. The CoT framework represents a scalable platform that addresses some of the key issues posed by the massive volume and high complexity of data generated in IoT systems. Middleware technologies are a central interstitial in this architecture that facilitates intercommunication between multiple different heterogeneous devices with different interfaces and operating systems. By standardizing data interchange and greater interoperability between devices, middleware encourages a homogeneous operating environment for IoT applications, as covered in the growing body of literature on these frameworks (Bacciu et al., 2017; Farahbakhsh et al., 2017). Accordingly, resolving these interoperability issues becomes critical to the achievement of effective, responsive autonomous systems that can fully capitalize on the potential of IoT integration and connectivity (Figure 7.9).

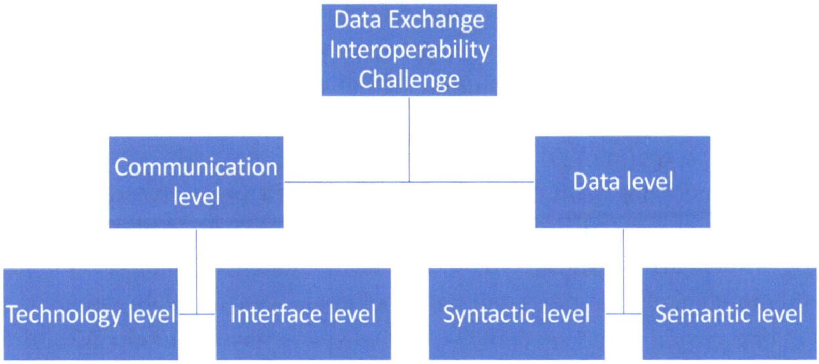

Figure 7.9. *Data exchange interoperability challenge.*

Source: https://www.mdpi.com/sensors/sensors-18-04404/article_deploy/html/images/sensors-18-04404-g001.png.

7.4.3. Impact of IoT on Real-Time Decision-Making

The integration of advanced wireless communication protocols has tremendously enhanced the functions of the Internet of Things (IoT) to enable real-time decision-making in various industries. Wi-Fi and 5G features provide high data transmission speeds, enabling devices to easily communicate with each other and provide the bandwidth needed for bandwidth-hungry applications. ZigBee and LoRa, nonetheless, have specific uses by offering low power and long-range communications capabilities, appropriate for applications requiring long battery life and over large geographic regions. The convergence of the diverse communication protocols creates an ecosystem where devices can communicate real-time data, thereby allowing automated systems to process data and respond to changes in conditions at velocity, significantly impacting decision-making processes in autonomous systems. Also, Vehicle-to-vehicle (V2V) and machine-to-machine (M2M) communication increases real-time decision making by directly sharing data between connected devices. V2V communication, which is also very important in automotive applications, enables vehicles to share information about traffic conditions, road conditions, and vehicle conditions in real time. This pre-emptive sharing of information improves situational awareness, making it possible to automatically adjust driving behavior to improve safety and efficiency. Similarly, M2M communication plays a critical role in an industrial setting because it facilitates the transmission of data from machines to central systems. All these exchanges are equal to efficient processes and reduced operational delay, pointing out that more connectivity and data transfer have a direct result on timely and knowledge-driven decisions. The integration challenges of IoT into autonomous systems are also complex and worth noting. As IoT devices grow in numbers, so does the demand for effective data management and privacy solutions. Challenges such as cybersecurity and data privacy are important threats that must be overcome if confidence in IoT systems is to be maintained. Energy efficiency and scalability remain major challenges, especially with the adoption of IoT in big networks. To realize the full potential of real-time decision-making made possible by IoT, these challenges have to be addressed. It has been evidenced that the integration of forthcoming technologies such as Blockchain can mitigate some of these problems by providing an assurance-enabled space for data exchanges, thereby increasing the credibility of IoT systems and their contribution to autonomous functional frameworks (Abdolee et al., 2019; Buyya et al., 2017).

7.5. CHALLENGES IN COMMUNICATION AND NETWORKING

In the rapid-moving wave of autonomous systems, communication and networking are confronted with serious challenges to performance and reliability. Wireless protocols such as Wi-Fi, 5G, ZigBee, and LoRa all introduce differing strengths and weaknesses to data transmission in dynamic environments. For instance, while 5G provides high-speed connectivity with support for real-time applications, its coverage can be unpredictable in rural or secluded areas. Meanwhile, LoRa has its strengths in long-range communication with minimal power consumption but lacking in data rate and latency. This is a scenario where one has to consider the intricacies of protocol compatibility and varied requirements of autonomous systems, particularly as they increasingly depend on resilient and robust networking solutions to operate unproblematically within a range of environments.

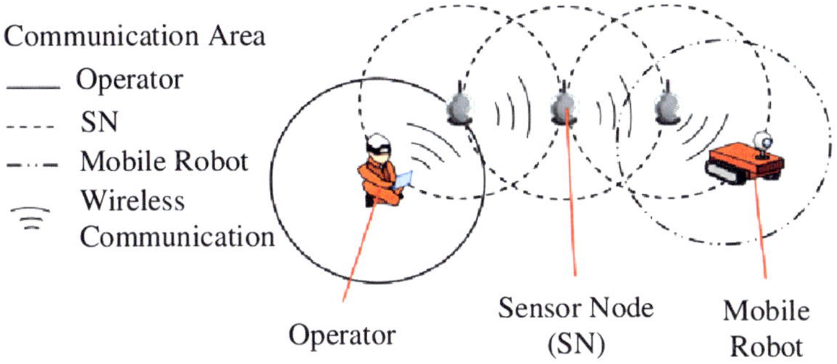

Figure 7.10. *Robot wireless sensor network (RWSN).*

Source: https://www.researchgate.net/profile/K-Koyanagi/publication/334943063/figure/fig1/AS:787789293563904@1564835012425/Robot-Wireless-Sensor-Network-RWSN.png.

Vehicle-to-Vehicle (V2V) and Machine-to-Machine (M2M) communication integration provides a specifically appealing mechanism of enhancing the safety and efficiency of self-driving networks. Both models support the exchange of critical information between automobiles and devices to guarantee that activity such as traffic control and collision prevention is feasible. Application of the systems comes with its setbacks notwithstanding its potential, such as issues of interoperability and latency being among the top priority items. Successful V2V communication relies on real-time information exchange to minimize the risks, but with high latencies, response times are delayed. Additionally, the successful development of M2M communication systems calls for establishing standardized protocols so that all manufacturers and devices support one another. This situation underscores the need for comprehensive research into enhancing the performance as well as the reliability of such communication networks. The lofty idea of unifying these communication systems under the broad umbrella of the Internet of Things (IoT) is an ambitious undertaking that holds a lot of promise. With a growing number of IoT devices, scalable communication solutions are the greatest requirement of the time. Autonomous systems can leverage technologies such as Robotic Wireless Sensor Networks (RWSN) in order to optimize communication channels and data capture, particularly in environments where conditions are likely to fluctuate. However, the marriage of wireless sensor networks and robotics also has accompanying challenges, notably connectivity, localization, and routing. Meeting these areas is key to achieving a robust and vibrant communication network that meets standards.

Further, long-term trials and collaborations towards these new technologies, like those highlighted by Cartolano et al. (2008) and Balasubramanian et al. (2018), will be important in defining their complete potential as well as resolving existing issues in the area of communication and networking for autonomous systems (Figure 7.10).

7.5.1. Latency Issues in Autonomous Systems

There are numerous communication protocols that serve to overcome latency challenges in autonomous systems, especially as the technologies go mainstream. Wi-Fi, 5G, ZigBee, and LoRa all possess different strengths and weaknesses when it comes to range, speed, and power consumption. 5G, for instance, is most important when it comes to reducing latency significantly, supporting ultra-reliable low-latency communications (URLLC) required in autonomous vehicles. On the contrary, while Wi-Fi networks have high throughput, they are subject to congestion and interference, leading to fluctuating latency. ZigBee and LoRa, intended for low-power and wide-area communication, are efficient in evading latency over vast geographical regions and thus better suited for IoT devices. The design optimization of such protocols is a fundamental aspect of the overall network strategy within autonomous systems, directly affecting operational effectiveness and safety, where every microsecond could be crucial. The nature of vehicle-to-vehicle (V2V) and machine-to-machine (M2M) communication also raises the stakes for overcoming latency issues in autonomous systems. These modes of communication are vital in enabling the real-time exchange of information among devices and automobiles in a position to make timely decisions in dynamic settings. For example, V2V communication can alert vehicles to dangers or traffic situations ahead to enable anticipatory adjustment of driving plans. The performance of V2V, however, will rely on being able to transmit messages with zero latency to ensure timely delivery of information to drivers. Similarly, M2M communication is the foundation for the operation of millions of IoT devices within autonomous systems, where low-latency links are essential to facilitate seamless interoperability. Thus, continuous development of communication protocols, particularly with the advancement of 5G technology, is inevitable to counteract these latency problems and enhance overall system responsiveness. In addition, the integration of Internet of Things (IoT) into autonomous systems introduces additional layers of complexity, which exacerbate latency issues. IoT sensors generate enormous volumes of data, and these must be processed and transmitted in an efficient manner to enable timely operations of autonomous systems. The greater reliance on machine learning algorithms as outlined in the literature needs reliable wireless networks with the capability of supporting real-time exchange of data across various layers of the network from the physical layer through to the application layer. Successful wireless communications are included in making the low-latency requirements called out for applications like autonomous vehicles and industrial automation. Since 5G technology will emerge as one of the prominent solutions, studies have concluded its potential for stable improvement and latency minimization in V2X communications, a critical component of the environment in which autonomous systems operate. Future work in optimizing such communication systems remains of paramount importance in addressing the inherent latency issues of today (Abbas et al., 2020; Guizani et al., 2019) (Figure 7.11).

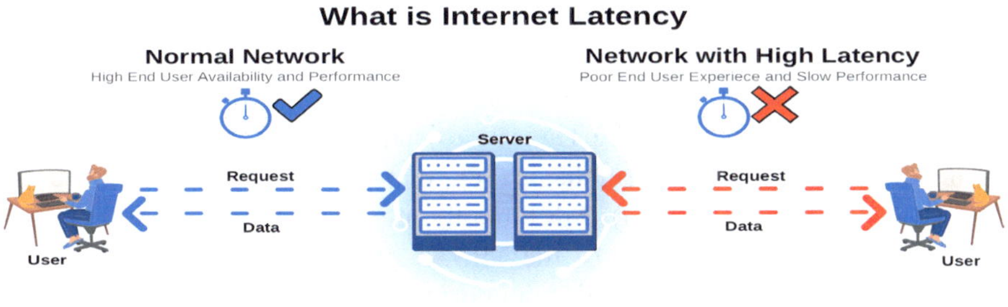

Figure 7.11. *Latency issues in autonomous systems.*

Source: https://external-content.duckduckgo.com/iu/?u=https%3A%2F%2Fcdn.broadbandsearch.net%2F2023%2F1%2F1673515133298-1_What_is_Internet_Latency.png&f=1&nofb=1&ipt=08cb122527d9cd3491a590fea3b69d8bdd7cdca7edc2702fcc649527be8304e8.

7.5.2. Reliability and Robustness of Communication Networks

The exponential expansion of autonomous systems demands strong communication networks that can manage a variety of operational problems. Among all the wireless communication protocols, Wi-Fi, 5G, ZigBee, and LoRa are crucial in providing undisturbed connectivity in various environments. They are the backbone of smooth data exchange and real-time communication in autonomous systems. For instance, while Wi-Fi and 5G offer high data rates and low latency essential for vehicle-to-vehicle (V2V) communications, ZigBee and LoRa support low-power and long-range applications suited for sensor networks in remote areas. Their successful integration into autonomous networks underlines the importance of having robust communication frameworks capable of adapting to varying demands and obstacles that these systems encounter in dynamic conditions. With the importance of V2V and machine-to-machine (M2M) communication in mind, these systems are highly reliant on the reliability of the underlying communication network. V2V communication enables the exchange of safety-related data, such as traffic or hazard warnings, to enhance situational awareness and reduce the chances of accidents. M2M communication enables smooth device-to-device interaction within the Internet of Things (IoT). This is particularly important in scenarios involving autonomous systems where sensor networks continuously monitor environmental conditions to facilitate decision-making. As network architectures and protocols change, ensuring reliable connections are made strong is important for maintaining system performance and operational integrity, leading in the end to enhancing safety and efficiency in autonomous operations. The integration of reliable communications infrastructure into autonomous systems is not just data transport but also addresses failure resilience and safety. As emphasized in the Technical Topic Notes of the Tigar's project, network resilience assurance remains a challenge with the emphasis on demands of safety and dynamic test plans (Bloomfield et al., 2020). Additionally, next-generation technologies in space environments, like those of the MISAT program, outline opportunities for wireless sensor network-based, miniaturized and distributed systems. Such technologies facilitate robust communication structures, which are required to maintain continuous operation for terrestrial and extraterrestrial autonomous systems (Gill et al., 2007). Ultimately, these innovations create credibility that ensures autonomous devices and cars are capable of operating effectively even under unforeseen circumstances, therefore enabling technologies to be integrated seamlessly for the future.

7.5.3. Regulatory and Standardization Challenges

The integration of autonomous systems with available communication infrastructures raises severe regulatory and standardization challenges, particularly in the wake of increased adoption of wireless communication protocols such as Wi-Fi, 5G, ZigBee, and LoRa. These technologies provide necessary features in autonomous systems, including vehicle-to-vehicle (V2V) communication and machine-to-machine (M2M) transactions, which are essential for real-time sharing of information and business efficiency. However, differences in operational norms across regions complicate interoperability and require complex regulatory coordination to ensure seamless integration. With several manufacturers employing different protocols, variable communication practices could undermine the effectiveness of autonomous systems. This situation requires concerted effort by players in the industry and regulators to adopt uniform standards that enable consistency and reliability of communication so autonomous systems can function effectively across environments and jurisdictions. In addition to this, the rapid pace of technologies such as the

Internet of Things (IoT) also calls for a need to undertake a review of regulatory regimes to address the unique challenges such emerging technologies pose. Integration of IoT in autonomous systems enhances the level of complexity of data and demands robust security measures to protect against the risk of vulnerabilities in networked devices. This is particularly true in deployments like smart grids, where optimization of energy consumption is based on real-time communication among system components. As the study on smart grid communication indicates, addressing interoperability problems while ensuring standards compliance is necessary to develop trust in autonomous technologies. Regulatory bodies must weigh these new threats, such as data privacy and cybersecurity, to spur innovation in communication practice, effectively recasting autonomous systems' regulation within the general technological context (Chin et al., 2011). Moreover, with telecommunication technologies advancing with new technologies such as cognitive radio and software-defined radio, existing rules may fail to reflect the specifics of these innovations. The policy implications of these developments, as discussed in a number of research projects, focus on the need for adaptive regulatory frameworks that can respond to greater competition in markets and technology change. The challenges posed by standardization are heightened by the accelerated rate of innovation and conflicting interests of different players in the industry. For this reason, effective coordinated standardization grows more imperative. It is only by collective efforts that regulatory bodies and technology developers are able to offer a harmonized working environment for autonomous systems and hence make proper use of their potential to improve communication networks' connectivity, efficiency, and safety (Pujol et al., 2025).

7.6. FUTURE TRENDS IN COMMUNICATION TECHNOLOGIES

The technology of communications is evolving rapidly, fueled by the growth in the demand for connectivity and integrating autonomous systems. Wi-Fi, 5G, ZigBee, and LoRa are dominant protocols that facilitate enhanced data transmitting capabilities within diverse surroundings. Wi-Fi keeps providing high connectivity for urban areas, whereas low latency and high bandwidth of 5Gs are necessary for cases in which data has to be processed in real-time, which is increasingly relevant in autonomous vehicles. Moreover, ZigBee and LoRa are best in cases where low power consumption and long-range communication are desired and hence are most appropriate for industrial IoT applications. As these technologies converge, they facilitate efficient architecture for autonomous systems to communicate better in various contexts, thereby occupying a pivotal position in shaping the patterns for the future in the field of communication technologies. Vehicle-to-Vehicle (V2V) and Machine-to-Machine (M2M) communication technologies are a revolutionary innovation in networked systems.

Figure 7.12. *Future communications fast technology.*

Source: https://thumbs.dreamstime.com/b/future-communications-fast-technology-g-network-connection-concept-high-speed-mobile-wireless-142338090.jpg.

These structures allow for direct communication between devices and cars without the presence of an intermediary infrastructure, enhancing traffic control and safety. For instance, V2V communication enables cars to share vital information such as speed, direction, and possible hazards in real-time, thus minimizing the possibility of congestion and accidents. Similarly, M2M communication allows for devices to autonomously exchange information, which is essential in smart city remote monitoring and control and industrial remote monitoring and control. The convergence of these methods into autonomous systems is an indicator toward an integrated and efficient environment, making way for a boost in efficiency and functionality within smart ecosystems. The potential of integrating the Internet of Things (IoT) with existing communication networks holds the promise of a new era of connectivity and automation. Convergence in this sense not only enables devices to talk but also to gather, process, and react to data autonomously, revolutionizing operational paradigms in industries across the board. As conceptualized by current research, the future development in programmable wireless networks can assist in overcoming the complexities of bringing together disparate technologies in a synergistic manner. Efforts made nowadays are also looking at architectures such as software-defined networking and cognitive radio networks to bring improved flexibility and ease of management in wireless communications (Ahad et al., 2014). Furthermore, integration of IoT functionalities with advanced communication protocols can also enable more efficient traffic management solutions, such as the ability of wireless vehicular cooperative systems to revolutionize road traffic management in a sustainable way (Cartolano et al., 2008). As they continue to unfold, they promise to bring about thrilling advances in the area of autonomous systems (Figure 7.12).

7.6.1. Advancements in Wireless Communication Protocols

The evolution of wireless communication standards has significantly changed the character of autonomous systems. With the advent of standards such as Wi-Fi, 5G, ZigBee, and LoRa, there are now more robust connectivity solutions that are individually tailored for specific applications in autonomous systems. One such example is 5G technology, which offers higher bandwidth and reduced latency, characteristics needed in real-time communications for autonomous vehicles. Conversely, ZigBee and LoRa are low-power wide-area networks, thereby ideal for utilization in long-range communication at low energy consumption. This relationship between the wireless protocols has brought the potential of straightforward communication among autonomous entities that has enabled advanced applications such as Coordinated Multi-Robot Systems, where communication efficacy and performance operation are vital. Thus, as such technologies grow, they stand to significantly enhance the reliability and performance of autonomous systems by providing an even more integrated environment for operation. In the vehicle-to-vehicle (V2V) and machine-to-machine (M2M) communication, advancement in wireless protocol further enhances the efficiency of autonomous systems. V2V communication allows vehicles to exchange crucial information such as speed, direction, and road conditions, providing a safer and more efficient use of the road. On the other hand, M2M communication facilitates intercommunication between devices and machines without human intervention, allowing remote observation and collaborative reaction to environmental changes. Based on definitions in the context of Robotic Wireless Sensor Networks (RWSN), autonomous networks' convergence with wireless protocols indicates colossal untapped potential for operational efficiency in cooperative control and learning. Further, the creation of algorithms tailored to optimize such communication processes is another layer of intricacy in how information is passed on between autonomous devices and vehicles and ultimately results in smarter transport systems and a sustainable city. The union of the Internet of Things (IoT) and robust wireless communication protocols is a promise of a future of autonomous systems that is yet to be. IoT devices, due to their array of sensors and communication capabilities, can produce tremendous amounts of data that can be accessed, manipulated, and analyzed in real-time to provide information on the performance of the system as well as the environment. Sensing device miniaturization, thanks to Micro Electro-Mechanical Systems (MEMS) and Very Large-Scale Integration (VLSI) technology advances, has made it possible to deploy efficient wireless sensor networks. As revealed by recent studies, the marriage of RWSN with IoT offers opportunities for innovative solutions leveraging the performance of communications in pursuing dynamic sensing goals in autonomous systems (Balasubramanian et al., 2018). Thus, as IoT continues to evolve with such wireless communication advancements, the prospect of intelligent environments adjusting reasonably to the needs of human beings becomes more attainable (Hong et al., 2006).

7.6.2. The Potential of Edge Computing in Autonomous Systems

The functionality of autonomous systems greatly depends on stable and effective communication platforms. Such a scenario can be simply understood when examining how wireless communication protocols like Wi-Fi, 5G, ZigBee, and LoRa enhance the operations of such systems.

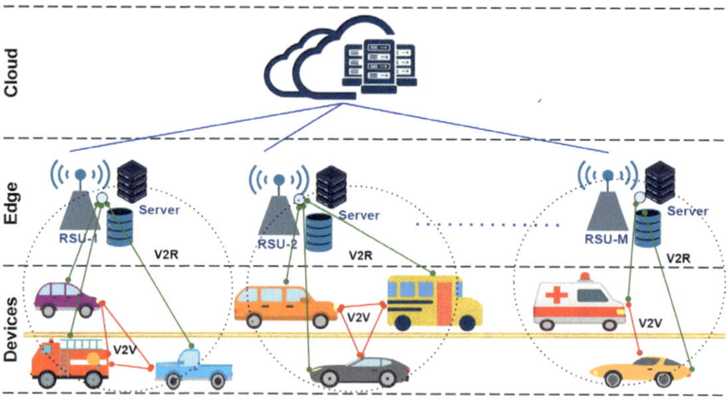

Figure 7.13. *A simplified tier topology of edge computing-based connected autonomous vehicles.*

Source: https://www.researchgate.net/publication/363000983/figure/fig1/AS:11431281127450401@1679035868847/A-simplified-tier-topology-of-Edge-computing-based-connected-autonomous-Vehicles-7.png.

Each protocol has its own advantages; for instance, 5G possesses low latency combined with high bandwidth, which is essential for real-time data transfer in autonomous vehicles, while ZigBee and LoRa enjoy energy efficiency and range, respectively. The application of Internet of Things (IoT) technology has also sped up communication further by facilitating smooth data exchange between devices. This convergence of advanced communication protocols is in line with the edge computing philosophy, enabling faster data processing and response times while reducing reliance on centralized cloud infrastructure. These technologies thus underscore their important roles in shaping the autonomous future of autonomous systems. Developing continuously from traditional practice, autonomous systems now incorporate cutting-edge technology to optimize communication channels. Vehicle-to-vehicle (V2V) and machine-to-machine (M2M) communication are a quintessential example of this technology, as they enable peer-to-peer data transfer between devices and vehicles without the need for intermediary systems.

Peer-to-peer interactions allow for increased situational awareness, which allows for real-time adaptation to changing circumstances, a critical capability for autonomy. \By taking advantage of edge computing, V2V and M2M communications can tap into localized data processing, which reduces latency and increases responsiveness. Also, because there are so many different communication protocols available, the systems can be optimized for their specific applications, building robust networks that cater to various usage demands. This ground-breaking potential of edge computing with regard to V2V and M2M communication reflects the significance of leveraging these technologies in ways that will ensure maximum autonomous systems to enhance security and efficiency. Exploring the synergies of merging edge computing with autonomous systems is revealed to hold a plethora of benefits, particularly in performance in operations and resources. Fog computing architecture as a middle layer for the integration of edge devices and the cloud optimizes the efficacy of autonomous systems by computing data in different layers. This multi-layered architecture addresses the vulnerability of conventional cloud infrastructure such as bandwidth constraint and latency in data, thereby opening up real-time decision-making possibilities for autonomous vehicles. Moreover, architectures such as Robotic Wireless Sensor Networks (RWSN) show how these innovations can lead to improved connectivity and localization, essential in addressing more complicated communication performance requirements in autonomous systems (Balasubramanian et al., 2018). Therefore, the integration of communication and edge computing is at the core of building the capabilities of autonomous systems, showcasing an essential line of development toward making transport and other such critical domains smarter and more responsive (Jukan et al., 2016) (Figure 7.13).

7.6.3. Emerging Technologies and Their Impact on Networking

Development of communication technologies has significantly changed the manner in which independent systems interact and establish effective networks. With mounting pressure for effective and high-speed connectivity, wireless communication standards such as Wi-Fi, 5G, ZigBee, and LoRa become the focus to accelerate data exchange and operational efficiency. These protocols are utilized for particular purposes; for instance, 5G offers higher data rates and lower latency, which is essential in real-time applications for autonomous vehicles and industrial control. Furthermore, ZigBee and LoRa offer long-range, low-power communication and hence are ideal for IoT devices where energy consumption is essential. In application for autonomous systems, these protocols offer seamless operations with devices capable of communicating effectively, hence enhancing the overall networks reliability and robustness. The convergence of various communication technologies eventually leads to more synchronized and responsive autonomous movements. At the same time, the establishment of vehicle-to-vehicle (V2V) and machine-to-machine (M2M) communication is a cornerstone in the advancement of autonomous systems. V2V communication enables vehicles to communicate with each other to share critical information regarding the surroundings, thereby improving security and traffic efficiency. By minimizing the delay in the exchange of information, these kinds of technologies minimize accident risk as well as enhance smoother logistics in transportation. M2M communication also enables interconnectedness between various devices and machines within industrial facilities and fosters real-time monitoring and control. With an interconnected strategy, utilization of resources becomes maximized much better, downtime and operational costs decrease, and predictive maintenance as well as automated decision-making mechanisms are facilitated amidst complex networking environments. The synergy of V2V and M2M communication enhances the capability of autonomous systems to operate successfully and smartly. With each other, the integration of forthcoming technologies such as optical wireless communication (OWC) is equally impressive under the future network scenario of autonomous systems. As noted, the insufficiency of the existing radio frequency (RF) spectrum addresses the growing requirement for high-data-rate applications, particularly under the introduction of 5G services. To respond, OWC offers a feasible solution by accessing much wider bandwidth, essentially solving the RF spectrum crisis. Visible light communication and free space optical communication open the way towards much wider bandwidth availability and much greater data rates, eventually allowing seamless interactions among devices connected in autonomous networks. This technological development not only enhances communication efficiency but also puts forward the crucial imperative for continued research in areas of overlap between robotics and wireless sensor networks, as the demand for high-performance networks becomes increasingly pressing in our rapidly digitizing world (Balasubramanian et al., 2018; Chowdhury et al., 2018).

SUMMARY

The mixed topology of networking and communication in autonomous systems emphasizes the ubiquitous contribution of multiple wireless communication protocols. Wi-Fi, 5G, ZigBee, and LoRa all have distinct advantages, facilitating data exchange between devices in the rapidly evolving environment of autonomous vehicles and smart infrastructure. For instance, 5G networks enable very low latency and high data rates that are essential in real-time decision-making for vehicle-to-vehicle (V2V) and vehicle-to-infrastructure (V2I) communications. Nevertheless, devices like ZigBee and LoRa are tailored for application in a specific use case with low power consumption and long distance, essential for integration in Internet of Things (IoT). Hence, an in-depth understanding of these protocols is necessary as autonomous systems increasingly rely on advanced networks of devices harmoniously and efficiently interacting with one another for enhanced operational safety and efficiency in transport networks. Moreover, the integration of machine-to-machine (M2M) communication methodologies into autonomous systems only enhances the potential for innovation in transport networks even further. This paradigm shifts the focus from individual device performance to a cohesive interaction among multiple interconnected devices that can potentially exchange data, learn from one another, and support different scenarios. Notably, the integration of V2V and M2M communications allows vehicles to exchange critical data involving road conditions, accident alerts, and route optimization in real-time. By allowing cars to exchange information with other cars and infrastructure components, these systems play a major role in enhanced traffic management and lower congestion. For this purpose, strong standards and protocols must be developed to facilitate maximum interoperability among devices, allowing seamless communication and effective information distribution through multiple platforms while avoiding the potential for information overload. In conclusion, with autonomous systems making their way more and more into our daily lives, effective networking and communication solutions cannot be overemphasized. The development and optimization of wireless vehicular cooperative solutions have the potential to improve traffic management, which subsequently leads to safer and more sustainable transport environments. As indicated by ongoing research projects, innovations in architecture will address the challenges posed by dynamic and insecure operating environments, crucial for advancing personal mobile computing. This new paradigm demonstrates how wireless communication technologies can reshape our interaction with interfaces and environments, emphasizing the need for long-term evaluation of the benefits and challenges inherent in these systems. Continuing research and on-field experiments testing, as highlighted by Cartolano et al. (2008), will be instrumental in unleashing the transformative potential of these technologies, helping usher a globalized future where efficiency and safety take center stage in transport systems.

Autonomous development necessitates robust communication frameworks, and hence the use of wireless communication protocols such as Wi-Fi, 5G, ZigBee, and LoRa becomes more and more critical. These protocols provide a foundation for efficient data transmission to enable devices to interconnect in sophisticated environments. Wi-Fi and 5G facilitate high-speed connectivity necessary for real-time engagement, particularly in applications such as vehicle-to-vehicle (V2V) communication where the shared information may be real-time enough to enhance safety and operational efficiency. LoRa and ZigBee, respectively, are, however, aimed for low-power, extended-range communications and are therefore suitable for sensor network and Internet of Things (IoT) integration applications. The combined use of these protocols underscores the potential for creating more integrated and responsive autonomous systems, which can leverage real-time data to enhance decision-making processes, thus facilitating the flow of interactions between vehicles and infrastructure without hiccups. The integration of V2V and machine-to-machine (M2M) communication in autonomous systems basically transforms the transportation and automated process scenario. This kind of communication allows vehicles to share important information such as speed, direction, and weather conditions, thus enhancing integrated driving. M2M communication plays a very active role in machine-to-machine data exchange, which can enhance operation efficiency across industries from manufacturing to logistics. By leveraging such communication characteristics, autonomous vehicles are in a better position to respond to dynamic conditions, respond to potential threats, and minimize their travel paths. Such an overall approach not only improves efficiency and safety but also facilitates the development of smart transportation systems, wherein the infrastructure and the vehicles operate in harmony to deliver improved traffic management and reduced congestion. The integration of Internet of Things (IoT) improves the performance of autonomous systems by facilitating the possibility of communication among a range of interconnected devices. The higher the number of devices that are interconnected, the higher the ability of real-time data collection and analysis, with the potential for more intelligent decision-making processes. Remote monitoring and control of autonomous vehicles is one of the primary areas of use where IoT devices deliver vital information to enhance safety and functionality. The possible integration of communication-assisted sensing (as envisioned by next-generation 6G specifications) has the potential to optimize the synergy between data exchange and surrounding world perception and make the autonomous systems more capable. In combining these advances, autonomous systems take center stage in technology development, potentially transforming various industries while adapting to new issues related to privacy, trust, and system dependability in an increasingly networked world (Liu et al., 2022; Yogesh et al., 2022).

REFERENCES

1. Abbas, R., Ahmad, I., Ali, S., Samad, A., Amuru, S., et al. (2020). 6G white paper on machine learning in wireless communication networks. http://arxiv.org/abs/2004.13875 (accessed on 10th September 2025).
2. Abdolee, R., Maroufi, M., & Tazekand, B. M. (2019). On the convergence of blockchain and Internet of Things (IoT) technologies. http://arxiv.org/abs/1904.01936 (accessed on 10th September 2025).
3. Abdullah, A. H., Aliyu, A., Aslam, N., Cao, Y., et al. (2017). Towards video streaming in IoT environments: Vehicular communication perspective. https://core.ac.uk/download/160606310.pdf (accessed on 10th September 2025).
4. Ahad, N., Ahmed, N., & Qadir, J. (2014). Building programmable wireless networks: An architectural survey. http://arxiv.org/abs/1310.0251 (accessed on 10th September 2025).
5. Ali, M., Anjum, A., Devitt, J., Hill, R., et al. (2017). Towards in-transit analytics for Industry 4.0. http://arxiv.org/abs/1710.04121 (accessed on 10th September 2025).
6. Asadi, A., Mancuso, V., & Wang, Q. (2014). A survey on device-to-device communication in cellular networks. https://core.ac.uk/download/659647813.pdf (accessed on 10th September 2025).
7. Bacciu, D., Bahr, R., Bröring, A., Cavallo, A., et al. (2017). Internet of robotic things: Converging sensing/actuating, hypoconnectivity, artificial intelligence and IoT platforms. https://core.ac.uk/download/84045908.pdf (accessed on 10th September 2025).
8. Balasubramanian, A., Chattopadhyay, A., Fida, A., Gasparri, A., Gasparri, A., Ghaffarkhah, A., Gonzalez-Ruiz, A., et al. (2018). Robotic wireless sensor networks. http://arxiv.org/abs/1705.05415 (accessed on 10th September 2025).
9. Bloomfield, R. E., Fletcher, G., Imai, K., Khlaaf, H., et al. (2020). Towards identifying and closing gaps in assurance of autonomous road vehicles–a collection of technical notes part 1. https://core.ac.uk/download/305120054.pdf (accessed on 10th September 2025).
10. Bolaji-Adetoro, D. F., Adedotun, K. J., Shola, I., & Raji, A. K. (2025). Real-time object detection and collision avoidance in IoT-enabled autonomous vehicles. *Harvard International Journal of Engineering Research and Technology*. https://www.semanticscholar.org/paper/0387a3b34ab1ac17313a6057c4fbacbf031dfc81 (accessed on 10th September 2025).

11. Buyya, R., Kotagiri, R., & Mahmud, R. (2017). Fog computing: A taxonomy, survey and future directions. http://arxiv.org/abs/1611.05539 (accessed on 10th September 2025).
12. Cartolano, F., Filali, F., Gozalvez, J., Krajzewicz, D., et al. (2008). iTETRIS: An integrated wireless and traffic platform for real-time road traffic management solutions. https://core.ac.uk/download/11140413.pdf (accessed on 10th September 2025).
13. Chin, W. H., Efthymiou, C., Fan, Z., Gormus, S., et al. (2011). Smart grid communications: Overview of research challenges, solutions, and standardization activities. http://arxiv.org/abs/1112.3516 (accessed on 10th September 2025).
14. Chowdhury, M. Z., Hossan, M. T., Islam, A., Jang, Y. M., et al. (2018). A comparative survey of optical wireless technologies: Architectures and applications. http://arxiv.org/abs/1810.02594 (accessed on 10th September 2025).
15. Copeland, R., Crespi, N., Rehmani, M. H., Saleem, Y., et al. (2019). Internet of Things-aided smart grid: Technologies, architectures, applications, prototypes, and future research directions. http://arxiv.org/abs/1704.08977 (accessed on 10th September 2025).
16. Denis, R., Dillschneider, J., Lèbre, M. A., & Mouël, F. L., et al. (2014). VANET applications: Hot use cases. https://core.ac.uk/download/49610995.pdf (accessed on 10th September 2025).
17. Dwivedi, Y. K., Hughes, L., Baabdullah, A. M., Ribeiro-Navarrete, S., Giannakis, M., Al-Debei, M. M., Dennehy, D., et al. (2022). Metaverse beyond the hype: Multidisciplinary perspectives on emerging challenges, opportunities, and agenda for research, practice and policy. *International Journal of Information Management, 66*, 102542–102542. https://doi.org/10.1016/j.ijinfomgt.2022.102542 (accessed on 10th September 2025).
18. Farahbakhsh, R., Farahzadia, A., Rezazadeh, J., Shams, Z., et al. (2017). Middleware technologies for cloud of things–A survey. http://arxiv.org/abs/1705.00387 (accessed on 10th September 2025).
19. Fernández-Caramés, T. M., & Fraga-Lamas, P. (2019). Towards next generation teaching, learning, and context-aware applications for higher education: A review on blockchain, IoT, fog and edge computing enabled smart campuses and universities. https://core.ac.uk/download/250404797.pdf (accessed on 10th September 2025).
20. Fomichev, M., Gardner-Stephen, P., Hollick, M., Steinmetzer, S., et al. (2017). Survey and systematization of secure device pairing. http://arxiv.org/abs/1709.02690 (accessed on 10th September 2025).

21. Gill, E., Monna, G. L. E., Scherpen, M., et al. (2007). MISAT: Designing a series of powerful small satellites based upon micro systems technology. https://core.ac.uk/download/pdf/12941826.pdf (accessed on 10th September 2025).
22. Guizani, N., Sharma, V., & You, I. (2019). Security of 5G-V2X: Technologies, standardization and research directions. http://arxiv.org/abs/1905.09555 (accessed on 10th September 2025).
23. Havinga, P. J. M., & Smit, G. J. M. (2000). Lessons learned from the design of a mobile multimedia system in the Moby Dick project. https://core.ac.uk/download/pdf/11454372.pdf (accessed on 10th September 2025).
24. Hong, C. S., Lee, H. W., & Pathan, A. S. K. (2006). Smartening the environment using wireless sensor networks in a developing country. https://core.ac.uk/download/300359207.pdf (accessed on 10th September 2025).
25. Imran, A., Imran, M. A., Mohamed, A., Onireti, O., et al. (2016). Control-data separation architecture for cellular radio access networks: A survey and outlook. https://core.ac.uk/download/74231056.pdf (accessed on 10th September 2025).
26. Jukan, A., Marín Tordera, E., Masip Bruin, X., Ren, G., et al. (2016). Foggy clouds and cloudy fogs: A real need for coordinated management of fog-to-cloud computing systems. https://core.ac.uk/download/81582274.pdf (accessed on 10th September 2025).
27. Kargl, F., Knapik, P., Petit, J., Schoch, E., et al. (2014). Cooperative home light: Assessment of a security function for the automotive field. https://core.ac.uk/download/pdf/31148758.pdf (accessed on 10th September 2025).
28. Liu, F., Cui, Y., Masouros, C., Xu, J., Han, T. X., Eldar, Y. C., & Buzzi, S. (2022). Integrated sensing and communications: Toward dual-functional wireless networks for 6G and beyond. *IEEE Journal on Selected Areas in Communications, 40*, 1728–1767. https://doi.org/10.1109/jsac.2022.3156632 (accessed on 10th September 2025).
29. Nahavandi, S. (2019). Industry 5.0—A human-centric solution. *Sustainability, 11*, 4371–4371. https://doi.org/10.3390/su11164371 (accessed on 10th September 2025).
30. Pujol, F. (2025). Regulatory and policy implications of emerging technologies to spectrum management. https://core.ac.uk/download/pdf/7304707.pdf (accessed on 10th September 2025).

31. Sabeer, S., Ali, S., Hussain, M. A., Begum, A. U., Siddiqua, A., & Arif, M. (2025). Deep learning for secure communication in resource-limited IoT devices. *Communications on Applied Nonlinear Analysis*. https://www.semanticscholar.org/paper/d205aeb683f62247ceaeb6f5e48e9375b7b94c5f (accessed on 10th September 2025).

32. Tao, F., Zhang, H., Liu, A., & Nee, A. Y. C. (2019). Digital twin in industry: State-of-the-art. *IEEE Transactions on Industrial Informatics, 15*, 2405–2415. https://doi.org/10.1109/tii.2018.2873186 (accessed on 10th September 2025).

33. Tassi, A., Ameixieira, C., Demestichas, P., Fazio, P., & Tong, Z. (2018). A city-scale ITS-G5 network for next-generation intelligent transportation systems: Design insights and challenges. https://core.ac.uk/download/195283378.pdf (accessed on 10th September 2025).

34. Tiwari, M. K., Bidanda, B., Geunes, J., Fernandes, K., & Dolgui, A. (2024). Supply chain digitization and management. *International Journal of Production Research, 62*, 2918–2926. https://doi.org/10.1080/00207543.2024.2316476 (accessed on 10th September 2025).

35. Verhoef, P. C., Broekhuizen, T., Bart, Y., Bhattacharya, A., Dong, J. Q., Fabian, N. E., & Haenlein, M. (2019). Digital transformation: A multidisciplinary reflection and research agenda. *Journal of Business Research, 122*, 889–901. https://doi.org/10.1016/j.jbusres.2019.09.022 (accessed on 10th September 2025).

CHAPTER 8

ENERGY MANAGEMENT AND EFFICIENCY

Contents

Introduction.......... 291
8.1. Power Consumption in Embedded Devices.......... 297
8.2. Battery Technologies.......... 302
8.3. Energy Harvesting Techniques.......... 307
8.4. Low-Power Design Strategies.......... 312
8.5. The Role of IoT in Energy Management.......... 318
8.6. Case Studies in Energy Efficiency.......... 322
Summary.......... 327
References.......... 329

INTRODUCTION

The rapid growth of technology has resulted in an urgent need for effective energy management and efficiency, particularly when it comes to embedded devices, battery technologies, and low-power design methodologies. The spiraling increase in the number of electronic devices leads to the demand for energy-efficient solutions never before so pressing. Embedded systems that are intrinsic to present-day devices are prone to face serious challenges due to their low power resources and longevity requirements. This chapter shall aim to examine various strategies and innovations in energy management and highlight the importance of sustainable energy solutions. Keeping in mind the issue of power consumption of such devices, we are in a position to provide insights on other approaches that not only promote efficiency but also limit the environmental impact on energy consumption in the increasingly digital age. From the point of view of power consumption, battery technology is central to determining the run time of embedded devices.

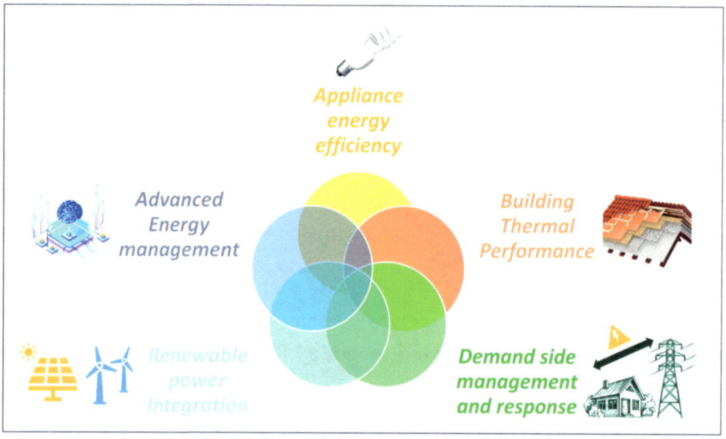

Figure 8.1. *Energy management and efficiency.*

Source: *https://www.mdpi.com/energies/energies-15-08663/article_deploy/html/images/energies-15-08663-g001.png.*

Conventional battery technologies, while widespread, are often found wanting in terms of sustainability, optimized size, and the necessity to replace them at the right time. New technologies such as energy harvesting technologies are gaining popularity, where sources of power are drawn from the environment or from the device's own operational processes. Techniques such as electromagnetic energy harvesting and thermoelectric power generation are potential means of creating self-sustaining systems that improve the usability and efficiency of devices. For instance, Chen et al. (2025) points towards the potential of energy harvesting technologies in driving wearable and implantable medical devices, mitigating the limitation created by traditional battery technologies and offering perspective on the potential for the future to revolutionize the health environment significantly. Lastly, the use of low-power design techniques puts yet another level on complexity and potential regarding increasing energy efficiency. Various techniques can be utilized in optimizing the subsystems of the embedded devices so that minimal power is utilized but the performance is optimized. Techniques involve effective communication protocols to adaptive algorithms that respond to real-time demands on energy resources. Machine-Type Communication (MTC) architectures' growth presents the demand for energy-saving principles built into these systems' functionality, combining various applications across different domains. As quoted by Alves et al. (2024), the optimization of core subsystems has a direct impact on overall energy consumption, with the demand for cross-disciplinary approaches for developing sustainable solutions. Together, these points underscore the need for comprehensive energy management to spur the development of next-generation, green embedded technologies (Figure 8.1).

Definition of Energy Management and Efficiency

Energy management and energy efficiency are central to contemporary technological systems, particularly for resource and energy consumption and optimization. Energy management can be generally described as the systematic monitoring and control of energy utilization within systems to make them sustainable and conserve resources. Efficiency, on the other hand, is a measure used to determine how energy is utilized to deliver useful work without loss, but it should not be unnecessary. In embedded systems, where execution is often under significant constraints such as low processing power and battery, the energy management versus efficiency becomes more crucial. Effective energy management methods not only reduce the operational expense but also extend the device lifespan, thereby enabling increased technology deployment in applications such as the Internet of Things (IoT) and smart spaces. Embedded systems architecture also poses some specific challenges from the perspective of energy consumption, mostly due to the fact that they are multi-functional and need permanent connectivity. For instance, energy-efficient communication protocol design is an important factor in the guarantee that the devices do not disproportionately skew power during interaction within a network. Most functions tend to draw their energy from components around radio transceivers, so there is a need for designers to be focused on low-power circuits and efficient physical layer designs that minimize average energy per information bit for communication. Nonetheless, ineffective network utilization and resource allocation can reverse the efficiency gains of individual devices, leading to throughput and responsiveness losses, which points to the necessity to integrate energy-oriented optimization frameworks into the system's overall design (Biason et al., 2017). Researching context-aware technologies clearly shows the necessity to create smart systems responding to their operating context efficiently while being mindful of energy use. Context-aware applications employ sensors to collect information regarding user behavior and environmental factors to present proactive feedback that optimizes energy utilization. Such dynamic behavior not only enhances the experience but also conserves energy by eliminating redundant operations. Nevertheless, some applications are adversely affected by factors such as restricted power and bandwidth in mobile devices, which hinder the ability to apply such applications. Therefore, the need to identify and surpass these constraints becomes a must for research development in context-aware energy management. To continue research aimed at optimizing middleware services and developing smarter algorithms is capable of impacting the dynamics of energy-conserving technology in a revolutionary manner, leading to a shift towards greener consumption habits (Leung et al., 2016).

Importance of Energy Efficiency in Modern Technology

Energy efficiency has emerged as the foundation of modern technological innovation, especially in the field of embedded devices and battery technology. The growing demand for battery-powered, energy-saving devices calls for sophisticated power management strategies to emphasize power conservation. Embedded systems that are the backbone of applications ranging from consumer electronics to machinery leverage tremendous benefits from low-power design methods. These techniques not only reduce energy consumption but also enhance device performance and lifespan. For instance, technology for energy harvesting is increasingly being integrated into these devices so that devices can harvest power from surrounding energy sources and hence reduce the reliance on conventional batteries. This focus on energy efficiency is very much in line with the requirements of sustainability and cost viability, and thus it is vital that developers integrate energy-conserving considerations into their projects to stay abreast of today's customer needs and regulatory standards. Advances in battery technology also put the focus on energy efficiency in modern technology. For example, research in Vanadium Redox Flow Batteries has demonstrated an increasing trend towards more efficient energy storage systems with scalability and efficiency in renewable energy usage. Such batteries, by virtue of their capability to cost-effectively store and release energy, are particularly well-suited for application in smart grids, where management of variable renewable resources is critical. Through energy efficiency optimization, these technologies allow not only increased utilization of renewables but also satisfaction of energy needs with the least environmental fall-out. With renewable power sources exploding all over the globe, efficient energy storage is the next quest; advancements in battery systems are at the core of outlining a sustainable energy future, finally coordinating technological advancement with environmental stewardship (Guarnieri et al., 2016). Energy efficiency is more than isolated devices and technologies. The importance of energy efficiency reaches out to encompass broader economic and social systems. Research emphasizes that energy-related research is central to understanding contemporary issues faced in the naval business and beyond, facilitating the manner in which energy management methods can optimize operational efficiency (N/A, 2016). By leveraging holistic energy management strategies, industries and businesses can elevate their energy profile, meaning huge cost benefits and increased competitiveness. Furthermore, energy efficiency can also spur the generation of employment opportunities in the green technology sector, as the demand for skilled workers in clean energy solutions continues to exist. The pursuit of energy efficiency is thus not only essential in the development of technology but also a force that drives economic growth and sustainability, which is critical in 21st centuries socio-economic cycles.

Overview of Embedded Devices, Battery Technologies, and Low-Power Design

The explosion of embedded devices in numerous industries has reshaped the method of energy management, particularly considering the key role played by battery technologies. Devices ranging from intelligent household appliances to medical devices depend heavily on efficient power usage to maximize their functionality. With increasing demand for more features in small packages, low-power state management has to be incorporated into embedded systems design so that the devices can limit energy use when idle. This trend is also observed in the literature, where it is emphasized that retrofitting the existing electricity infrastructure to accommodate contemporary needs is necessary, as various applications rely on both renewable sources and energy-conservation legislations that allow the embedded devices to deliver their best in the context of a smart grid (Borghetti et al., 2013). Besides, real-time power management techniques are necessary for continuance of operation without compromising battery life, thus maximizing the boundaries of what can be achieved in intelligent device technology. Battery technology is also an industry stalwart in making efficient embedded systems with a wide variety of solutions being available to facilitate low-power design. Standard batteries are often short in terms of capacity and number of charging cycles, so substitute sources of energy are needed to guarantee the long-term operation capability. Recent technological innovations in the field of battery chemistry, such as lithium-ion and more recent solid-state technologies, promise greater performance in less weight and size. Moreover, harnessing energy from technologies such as solar cells or ambient sources has changed the very nature of these systems. As hinted in wireless sensor network literature, power harvesting techniques are crucial in enabling nodes to tap into energy from their surroundings, yet their nature of volatility poses difficulties to reliability and quality of service guarantees (Berzosa et al., 2012). Accordingly, an in-depth knowledge of the technologies informs the methods used in constructing next-generation embedded systems. Use of low-power design methodologies is essential to enhance energy efficiency in embedded devices, and this demands a multidisciplinary design and development methodology. Low-power architecture not only caters to the hardware but also demands software optimization to ensure that the consumed energy is minimized for all operation states. Technologies such as dynamic voltage and frequency scaling (DVFS) allow devices to adjust power usage in real time based on workload requirements, cutting energy use even lower. More advanced communication protocols are also highly essential for managing energy usage among connected devices, particularly in environments such as the smart grid. As embedded systems are becoming more and more integral to our daily life, the convergence of these varied factors—advanced battery technologies, energy harvesting methods, and innovative low-power designs—prepares the grounds for innovations in the future that aim not only for efficiency but also sustainability in energy management practices (Berzosa et al., 2012; Borghetti et al., 2013) (Figure 8.2).

Figure 8.2. *Low-power embedded systems design and power management.*

Source: https://external-content.duckduckgo.com/iu/?u=https%3A%2F%2Fwww.nwengineeringllc.com%2Farticle%2Fimg%2Flpembed-0.png&f=1&nofb=1&ipt=d9b02b7da569b6307c9e97fcc7973c043ef022a5bbf089e64cd764f4bfcba9c6.

8.1. POWER CONSUMPTION IN EMBEDDED DEVICES

The expanding application of embedded devices across various sectors emphasizes the pressing need to solve power consumption issues inherent in such technologies. Embedded devices, typically applied where operation must be uninterrupted, tend to energy inefficiency that leads to decreased performance and increased operating cost. This comes into operation in pervasive systems, where Wireless Sensor Networks (WSNs) are employed for facilitating functionality. In all such applications, power harvesting emerges as a viable option where devices can collect energy from local environmental sources. While being promising, the unreliable nature of availability of energy in such forms of harvesting can negatively impact user experiences, generating perceptions of unreliability of service delivery. Thus, engineers and researchers have to innovate in embedded device design structures to maximize energy management, not just meeting operational needs but also effective energy consumption levels to ensure user satisfaction (Berzosa et al., 2012). Apart from energy harvesting techniques, contemporary embedded systems also intend to integrate low-power design philosophies to minimize energy usage. For instance, driving hardware design and taking advantage of software optimizations enhance the energy efficiency of embedded systems without disrupting functionality. Specialized low-energy elements, for instance, microcontrollers created with energy-efficient architectures, contribute meaningfully to such a purpose. Moreover, applying techniques like Dynamic Voltage and Frequency Scaling (DVFS) and sleep mode enables systems to scale power consumption based on active workloads and operating requirements. Such energy management dynamically not only lengthens the battery duration but also reduces the generation of heat, which is particularly precious in highly integrated embedded systems. Research has shown that such techniques can lead to vast energy savings as well as improved sustainability for various embedded applications (Duarte et al., 2018). Besides, energy management in embedded systems must consider the growing demands for greater connectivity, i.e., those included in Internet of Things (IoT) ecosystems. Each connected device contributes to a collective boost in the total energy footprint, quoting the necessity for intelligent energy management architectures that best balance function and power consumption. This has created interest in the development of more sophisticated Model-Based Predictive Control (MBPC) systems that allow dynamic energy use adjustments based on real-time inputs and prediction. Recent advancements show that incorporating self-powered wireless sensors into MBPC is able to significantly reduce the energy costs of climate control systems in buildings without reducing comfort levels for building users. Accordingly, there is still plenty of room to push the power consumption strategies in embedded systems ahead, enhancing efficiency in operation alongside addressing the imperative of sustainability within this rapidly developing technological field (Duarte et al., 2018) (Figure 8.3).

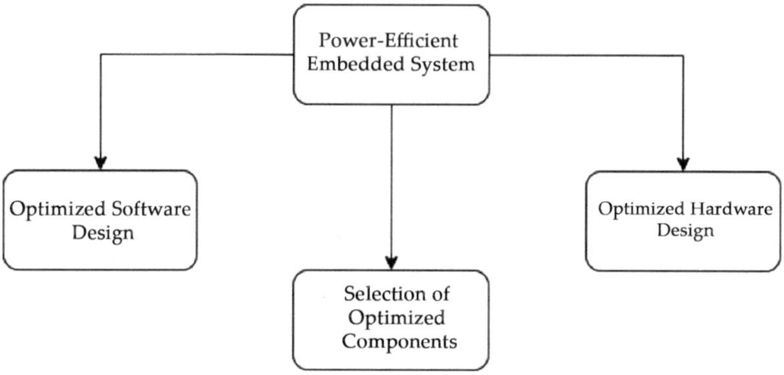

Figure 8.3. *Key factors for optimizing power consumption in an embedded device.*

Source: https://external-content.duckduckgo.com/iu/?u=https%3A%2F%2Fwww.engineersgarage.com%2Fwp-content%2Fuploads%2F2022%2F07%2FImage-1.png&f=1&nofb=1&ipt=ba906f17afc26e8b2a2ae597c954cd9364b38191f3358c6eec8e24df38c4552e.

8.1.1. Factors Influencing Power Consumption

The determinants of power usage in embedded systems are multifaceted and include both hardware characteristics and operational dynamics. Dominant hardware characteristics such as microcontrollers, sensors, and communication interfaces majorly contribute to total energy efficiency. For instance, low-power modes in microcontrollers can greatly reduce energy during idle periods. Additionally, the level of consumption is impacted by the selection of sensor technology; varying sensors consume varying levels of power based on their requirements while operating and processing the signal. These technologies are consistent with facts in recent energy studies, highlighting the importance of hardware and software optimization to achieve recordable reductions in power consumption (N/A, 2016). This makes it essential that designers make thoughtful decisions about their component selection with a priority given to energy-efficient technologies to help facilitate the overall agenda of sustainable energy management. Embedded device operating conditions also influence power consumption. Environmental condition fluctuations, such as temperature or humidity, may alter the operational efficiency of electronic components.

For example, high temperatures may lead to higher leakage currents in semiconductors, which consume more energy overall. Furthermore, device duty cycle—how busy and idle the device is—is proven to be a crucial factor in determining energy consumption. These operating parameters can be tuned to enable one to tailor energy consumption profiles for specific applications, hence improving overall efficiency. Additionally, real-time monitoring of the devices operating conditions may provide information on how external factors influence power demand, with the potential for further tuning (Aslam et al., 2018). Such information emphasizes the need for adaptive methods in deploying and designing energy-efficient embedded systems. Integration of energy harvesting techniques is a primary facilitator towards managing power consumption, particularly for remote or mobile embedded devices. These technologies allow the devices to harvest ambient sources of energy—solar, heat, or kinetic energy—by converting them to useful electrical energy. By reducing the utilization of conventional battery technologies, energy harvesting can significantly lower the frequency of battery replacements and the resulting environmental effects. Energy harvesting offers great advantages but singular challenges in terms of the availability of energy and storage capacity. These limitations necessitate careful design trade-offs to ensure adequate energies captured are enough to satisfy the device requirements without overly increasing complexity or cost (Aslam et al., 2018). Hence, energy harvesting innovation is important in boosting low-power design strategies and operating efficiency, resulting in a sustainable energy future.

8.1.2. Impact of Power Consumption on Device Performance

The relationship between power consumption and device performance is of paramount importance, especially in areas of embedded systems and the Internet of Things (IoT). Overpower consumption can result in serious issues, which can impact the performance and lifespan of devices. High energy consumption typically translates to lower durations of operation, which puts restrictions on these devices' performance. In addition, overloading may lead to over-heating and excessive wear on electronic devices, thus lowering overall performance. As the demand for ubiquitous computing grows, optimizing power usage becomes increasingly clear. Low power design techniques are required in an effort to ensure embedded devices can operate efficiently while never sacrificing functionality or reliability. Evolution of such strategies demands a subtle understanding of system design and the precise functions that have been delegated to every component of the device in order to maximize output with minimum input energy requirements. Moreover, the evolution of energy harvesting technology invokes a pathbreaking approach to combating the derogatory effect of power consumption on device performance. With the utilization of ambient environmental energy sources such as solar, thermal, or kinetic energy, it is possible that devices have longer cycle operations before they require an instant recharge or replacement of batteries. This is especially applicable in the case of installing devices in remote locations. The limitations, however, include the inconsistency and uncertainty of these energy sources, which can impact continuity and quality of delivered services. Consumers may perceive systems that use energy harvesting as being unreliable when there is performance failure, which may ultimately render the functionality of the technology useless (Berzosa et al., 2012). To mitigate this, an equilibrium approach to weigh the advantages of energy harvesting against its disadvantages is required in deploying firm, user-friendly systems. Further, power management issues are more than just mere energy sourcing; they also entail maximizing communication protocols for IoT networks. Communication interface, and the radio transceiver in particular, typically is a major power consumer of such devices. Research has revealed that improving energy efficiency in the physical layer is not a sufficient condition for an overall system performance improvement if other system resources are handled inadequately (Biason et al., 2017). This provides evidence that energy management should be treated holistically in order to integrate signal processing, channel access, and routing protocols. Using an energy-centered and context-aware optimization model, engineers can design systems that resource allocation is adaptive, allowing them to respond dynamically to varying operating conditions. This end-to-end model not only maximizes energy efficiency but also optimizes device performance overall to ensure a brighter and more sustainable future for embedded technology (Figure 8.4).

Figure 8.4. *Low-power consumption as a critical factor in smart devices.*

Source: https://www.designing-electronics.com/wp-content/uploads/2022/05/Anders2-scaled.jpg.

8.1.3. Strategies for Measuring and Analyzing Power Usage

In order to effectively manage energy consumption in embedded systems, the use of aggressive measurement and analysis strategies for power consumption is crucial. One prominent method involves the use of energy management frameworks that provide real-time monitoring of power consumption metrics. Utilizing cutting-edge sensors and analytics, these frameworks can acquire detailed insights into energy consumption patterns across various operation states. The level of detail offers an option for identifying power-hungry components, paving the way for targeted optimizations. For instance, as indicated in the recent research, low-power circuit design and the physical layer can significantly impact average energy used in the process of communication activities. However, actual improvements in efficiency can be countered by resource management solutions designed wastefully, hence the need for end-to-end energy-oriented plans that merge hardware design methods with performance and resource management (Biason et al., 2017). Along with this, context-aware applications have been shown to be at the heart of upgrading energy measurement techniques. Leveraging contextual information on device usage and surroundings, designers have been able to create systems that learn to minimize power usage based on real-time feedback. Context-aware environments enable more in-depth understanding of power consumption, and this is particularly beneficial for smartphones and other portable devices. These programs are able to sensibly adjust their use of resources in relation to user activity and environmental surroundings, and thus ultimately maximize battery life and overall effectiveness. In spite of the benefits, challenges still exist, mainly concerning the extent of resources mobile devices have at their disposal, such as power and memory. Challenges' overcoming can provide enhancements in energy management practices in mobile computing (Leung et al., 2016). In addition to advanced measurement techniques, incorporating innovative signal processing methodology is also a significant contributor to managing power consumption in embedded systems. Techniques such as adaptive signal processing can reduce the amount of energy used while ensuring data transmission integrity. By dynamically adapting to various conditions such as channel access and routing protocols, networks are capable of optimizing network longevity without wasting energy. Policies that balance processing and communication responsibilities along with taking into account energy impacts can enhance the performance of IoT devices. This ensures that energy efficiency is taken into account at all levels of design and operation, eventually leading to improved device lifetimes and more energy-efficient habits in the broader energy management and efficiency context (Biason et al., 2017).

8.2. BATTERY TECHNOLOGIES

The contribution of battery technologies to modern arrangements of energy management cannot be overstressed, particularly in light of the increasing need for efficient power utilization in embedded devices. With smart technologies gaining ground, advancing battery performance is critical to ensuring device longevity and customer satisfaction. New battery systems like lithium-ion and solid-state batteries offer improved energy densities and safety profiles for more reliable and longer life power sources. In addition, as energy management systems change, the use of batteries with renewable power sources—photovoltaic arrays, for instance—can help in providing an improved response to changing energy loads, thereby maintaining an optimal balance between generation and consumption. The very combination of energy generation and storage capacity is also observed in the design aspects of modern systems, where inherent energy storage is coupled with sophisticated demand management techniques to maximize energy utilization in applications that are diverse (Counsell et al., 2011). In hybrid energy system discussions, the combination of incompatible battery technologies is one such achievement in the field. For instance, a combination of traditional batteries, flywheels, and ultracapacitors can enhance energy density as well as provide short bursts of energy when required. Research in these hybrid systems is growing, especially in rail transport, where efficiency in energy is imperative. One such notable example is that of hybrid locomotives, where the integration of a flywheel with accumulators and a diesel engine has exhibited significant energy management improvements with optimized power flow strategies. Such examples demonstrate the versatility of battery technologies to match their own to special energy requirements (Akli et al., 2009). With new battery technologies, they offer exciting solutions to the multifold issues of energy use in transportation and embedded systems. Besides that, environmental effects of battery technologies remain on the energy management agenda. As batteries find their way into more consumer electronics and electric vehicles, lifecycle, recyclability, and sustainability issues get more contentious. New battery chemistries such as sodium-ion and metal-air batteries are not just providing more energy capabilities but are also seeking to minimize environmental impacts. Constructing regulatory standards for better recycling and handling of hazardous materials is increasingly significant. Therefore, the future of battery technology does not just hinge on enhanced performance but on producing greener designs as well. This synchronization of technological advancement with environmental sustainability principles is key to facilitating a responsible energy environment, summarizing the overall objective of energy efficiency and management within embedded systems and beyond (Figure 8.5).

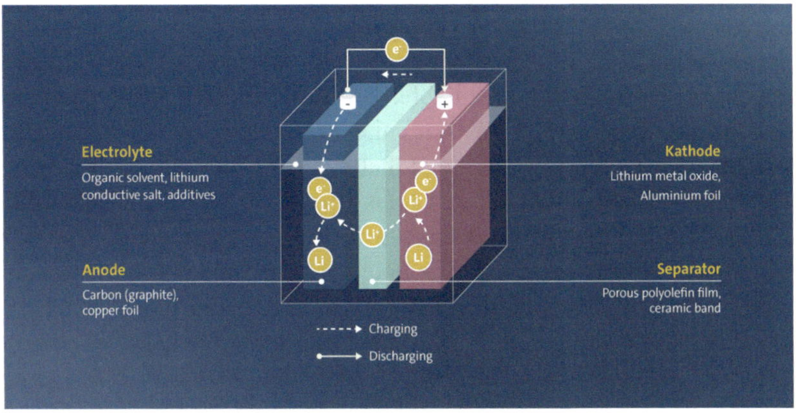

Figure 8.5. *Three battery technologies that could power the future.*

Source: https://media-exp1.licdn.com/dms/image/C5612AQFg5YWL7pShgw/article-cover_image-shrink_720_1280/0/1622472355233?e=2147483647&v=beta&t=FWzVEvvf6HEgnAmDYCAhD2I4PiOMeTqm9vNU5ilF8YI.

8.2.1. Types of Batteries Used in Embedded Devices

The growth of embedded devices has called upon the development of various battery technologies uniquely suited to their individual energy needs. Lithium-ion batteries are one of the most utilized battery types due to their reusability and high energy density, which makes them suitable for applications needing long power retention. These batteries can handle the spurious working loads of embedded systems efficiently without significant degradation in performance over time, thus enabling devices to offer functionality irrespective of working loads. Moreover, advances in lithium technologies, such as lithium polymer batteries, provide greater flexibility and form factor allowance, maximizing space in compact designs. But because these gadgets are typically used in environments where recharging will not be feasible, it is extremely important to consistently pursue innovations in energy harvesting and battery management to lengthen overall functional lifespan, a capability that is entirely in line with the tenets advocated by embedded systems energy management methods (Counsell et al., 2011). Furthermore, the incorporation of energy harvesting approaches within embedded devices is also a very important breakthrough in green power solutions. Energy harvesting enables devices to tap power from environmental sources of energy—solar, heat, or motion energy—to supplement their batteries, thus increasing life significantly. One such example is the development of photovoltaic cells for portable devices that enable self-sustaining energy systems by converting sunlight into electrical energy directly. This is particularly helpful in remote or wearables technology where other charging methods might be unsuitable. Besides, research in projects such as the MOBY DICK project emphasizes the importance of this integration by focusing on adaptive energy solutions that can support multimedia computing in battery-constrained environments (Havinga et al., 2000). By coordinating energy harvesting and advanced battery technology, embedded systems can achieve unprecedented levels of efficiency and flexibility. Apart from lithium and energy harvesting technology, efforts are also being made in the area of embedded devices through the development of alternative battery technologies like zinc-air and solid-state batteries. Solid-state batteries offer improved safety as well as energy density, and the potential to revolutionize power supplies in mission-critical applications like medical devices and autonomous systems. They eliminate the risk of flammability of liquid electrolytes and assist in improved thermal stability, making them ideal for devices under strict conditions. Zinc-air batteries also have high energy density at lowered cost compared to standard alternatives, utilizing ambient oxygen as a reactant. Their minimal weight and environmental friendly constituents make them ideal for application where environmental friendliness dominates without compromising on functional effectiveness. In the end, developing an overall understanding of these types of batteries is needed to advance energy management and advance embedded device development in order to tackle modern power consumption issues effectively.

8.2.2. Advancements in Battery Technology and Their Implications

Advancement in the technology of batteries has in recent years introduced revolutionary improvements that enhance efficiency and energy management in various applications, including embedded systems. Lithium-sulfur and solid-state battery technologies have emerged, offering improved energy density and better safety compared to their traditional lithium-ion counterparts. The technologies introduce the possibility for extended lengths of use in devices without the requirement to expand the number of times they need to be recharged, thus promoting a more sustainable usage method for energy. Also, with the ubiquity of handheld electronics, efficient battery systems play a crucial role in eliminating energy wastage and extending embedded system lifespan. Not only does its integration improve overall performance but also it can initiate the development of a circular economy in terms of enhanced recyclability and power longevity, setting a new standard of energy efficiency for digital devices. The effects of these innovations extend beyond raw performance improvements;

they have effects on the space of energy management methods as a whole. For instance, fresh technologies in battery development allow innovation in low-power design methods, including energy harvesting technologies in embedded and portable systems. This integration can result in self-powered devices that harvest ambient energy, such as solar or kinetic energy, and use it to power batteries.

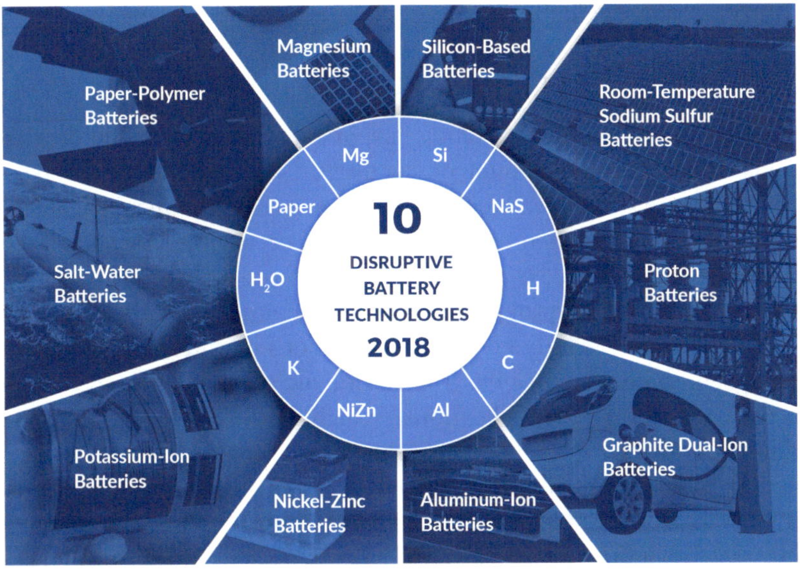

Figure 8.6. *Advancements in battery technology and their implications.*

Source: https://www.solarpowerworldonline.com/wp-content/uploads/2019/01/10-battery-technologies.png.

Hence, such a paradigm shift can immensely reduce reliance on traditional power sources and reduce operational costs in the long term. In addition, such methods facilitate new applications in health care and remote network monitoring, wherein battery lifetime extension is important to ensure continuous operation and enhanced patient care. The continued advancement of battery technology is a vital impetus to bringing about a more efficient and sustainable approach to energy management across all industries. With such advancements, the evolving field of energy harvesting is equally pivotal in rejuvenating the manner power consumption in embedded devices is addressed. By taking advantage of commercially available technology and affordable production methods, as with the recent developments, developers are also capable of creating solutions that are economically feasible as well as efficient. COTS device-based projects attempt to innovate and democratize access to cutting-edge satellite and sensor technologies by bringing down costs, hence making them applicable on a wider scale. This improvement is particularly significant in academia and small-to-medium enterprises with limited budgets. With the integration of battery development technology and energy-efficient harvesting mechanisms, not only is a power-efficient environment created but also the possibility of empowering individuals to incorporate technology more naturally into everyday life (Borri et al., 2010; Correa et al., 2005) (Figure 8.6).

8.2.3. Challenges in Battery Life and Energy Density

The rapid advancement of technology has, in every way, increased the demand for efficient devices and maximum power usage devices. Battery life and battery energy density are two of the most significant concerns on this front. The latter refers to the amount of energy per unit of mass or volume, a critical factor in deciding how long a device can run before needing a recharge. The new battery technologies, although improved, many times are unable to provide the energy requirements of increasingly complex applications in embedded systems. The developers continuously push the frontiers of such technologies as they create innovative context-aware applications, which rely heavily on the constant availability of power. This situation heightens the issue, as native mobile device middleware services are usually subjected to the limitations of constrained power, memory, and bandwidth, making it incumbent upon researchers to propose novel solutions for enhancing battery efficiency and lifespan (Leung et al., 2016). Furthermore, the energy density problem is compounded by the growing adoption of renewable energy sources. While these sources are touted as being sustainable, their intermittent character demands advanced energy storage technologies that can efficiently store and utilize stored power. A technology coming onto the scene that has already been proved suitable for application in smart grid applications is the vanadium redox flow battery. This technology is different from conventional batteries in that it uses liquid electrolytes, thus allowing scalability and even higher energy density. However, with all its advantages, universal utilization of vanadium technology has drawbacks in terms of cost, system complexity, and energy efficiency. These need to be surmounted in order to develop fully the potential of battery technologies seamlessly meshing with renewable sources for effective management of energy for embedded devices (Guarnieri et al., 2016). Lastly, the use of low-power design methods is critical in resolving battery life and energy density problems. Effective power management techniques, such as energy harvesting, enable devices to scavenge and retain limited amounts of energy from the environment. Such techniques are particularly beneficial for devices that are deployed in remote or hard-to-reach areas or far from power sources where recharging is not feasible at regular intervals. In addition, optimization of device components in order to reduce power consumption can significantly extend battery life. However, the performance compromise, especially processing capacity and responsiveness, usually presents further challenges. Thus, an in-depth comprehension of how power consumption interacts with device function is vital towards the creation of battery technologies while capable of effectively managing energy demands in the new paradigm of embedded devices (Leung et al., 2016).

8.3. ENERGY HARVESTING TECHNIQUES

Energy harvesting techniques play a key role in enhancing the efficiency and sustainability of embedded devices, particularly for power management systems. By harvesting energy from environmental sources such as solar, thermal, or kinetic power, these techniques allow devices to operate without traditional power supplies. This is most beneficial for pervasive system uses and Wireless Sensor Networks (WSNs), where the equipment generally has a limit in terms of battery life and energy provision. The adoption of power harvesting techniques not only alleviates the pain points of battery drains, but also ensures a more stable user experience with continuous operation. As emphasized, intermittence of the energy source presents challenges that can affect the perception of the users regarding the dependability of the system and therefore render new methods essential to the success of implementing energy-harvesting WSNs and pervasive systems (Berzosa et al., 2012). Energy-harvesting technologies are also enhanced in efficiency through optimization of system architecture towards energy consumption versus operational needs. Among IoT devices, radio transceiver is a primary source of energy dissipation, for which communication processes are power consuming. Low power circuit design and energy-conscious communication protocols are thus gaining attention from researchers for fulfilling these systems' energy requirements. Devices are able to lower energy expenses associated with data transmission through signal-dependent processing techniques and channel access optimization techniques. The integration of adaptive algorithms is also a crucial role in sustaining the network to adapt dynamically to various conditions for guaranteeing maximum performance and longevity (Biason et al., 2017). The integration of the design and operational efficiency in this emphasis describes the importance of comprehensive measures to energy management. With ongoing research in energy-efficient designs, the practical implications of energy harvesting technologies exceed power generation. They revolutionize the embedded system design, opening up a paradigm shift in which the devices are autarkic and less susceptible to power supply disruptions. However, the operation of power harvesting systems must consider the quality and reliability of the power harvested since power fluctuations can impact service continuity. Further research into how these issues can be addressed is unavoidable. Both experimental and theoretical research indicate that new designs can make energy harvesting systems more viable and able to deliver user expectations in a way that reduces the need for conventional centralized power grids. Not only does this guaranteed energy sustainability, but it also provides the potential for more extensive use of renewable energy sources for energy management policies (Berzosa et al., 2012) (Figure 8.7).

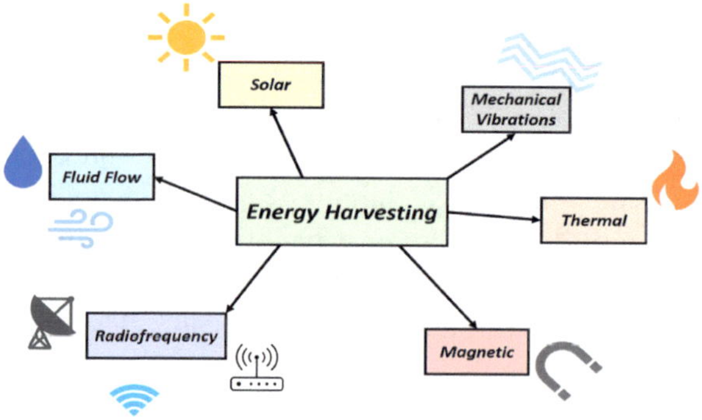

Figure 8.7. *Energy harvesting sources.*

Source: https://www.researchgate.net/publication/372013460/figure/fig1/AS:11431281390242635@1745248855266/Energy-harvesting-sources.tif.

8.3.1. Overview of Energy Harvesting Methods

Energy harvesting methods have become essential components of today's energy management systems, particularly for embedded devices and wireless sensor networks (WSNs). They capture environmental energy from the surrounding environment in the form of solar, thermal, vibration, and electromagnetic energy and convert this into available electrical energy. The continuous requirement for low-power consumption in electronic devices has brought a need to move towards energy harvesting technologies that will support the functioning of such devices without the need to replace batteries repeatedly. Although the possibility of energy harvesting is great, it should be understood that the harvested energy is intermittent in nature and is dependent on the environmental conditions. Hence, energy harvesting systems need to be designed in such a manner that they can manage these variations efficiently so that the reliability and stability of provided services to clients are ensured. This variation can affect perception by users and consider these systems unreliable if proper management is not achieved (Berzosa et al., 2012). Wireless energy transfer (WET) technologies have a long way in increasing the list of energy harvesting methods. By utilizing radio frequency (RF) signals, WET technologies allow devices to be charged without making physical contacts, hence enhancing placement and deployment convenience in devices.

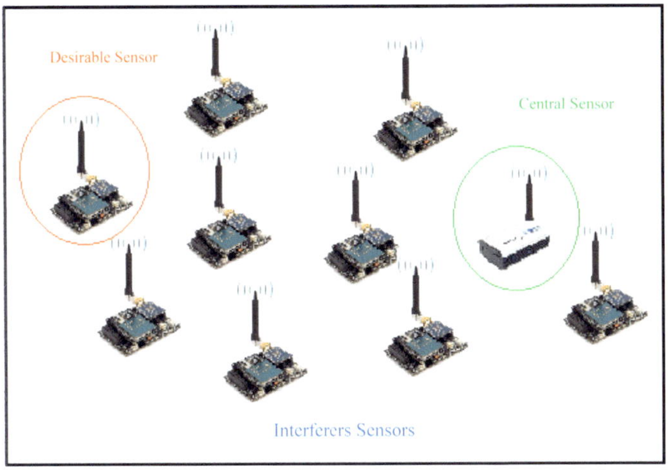

Figure 8.8. *Wireless sensor network (WSN) with one desirable sensor.*

Source: https://www.researchgate.net/publication/281335972/figure/fig1/AS:652951903289354@1532687273660/Wireless-Sensor-Network-WSN-with-one-desirable-sensor-multiple-interferer-sensors-and.png.

This innovative technique addresses one of the greatest limitations of traditional battery-operated systems: the disturbance and inconvenience of operational restriction and battery maintenance needs. With WET, devices can have a more sustained operation, which contributes to overall resilience and lifespan of embedded systems. Under this method, there is the inclusion of new challenges, such as the need for specialized wireless power transmitters as well as the issues related to the simultaneous transmission of energy and information. Therefore, ongoing studies in WET aim at optimizing such systems for effective use in a wide variety of applications, making them more and more popular in energy management (Bi et al., 2014). In summary, the energy harvesting technology landscape is constantly evolving with the integration of state-of-the-art technology such as solar cells, piezoelectric materials, and wireless energy transfer systems. Both methods have their strengths and weaknesses, requiring developers and researchers to balance these concerns when designing effective energy management solutions. The significance of effective energy harvesting is also seen when embedded devices and low-power design approaches are contemplated, where these technologies not only prolong device lifespan but also cut down on waste in the environment through less dependence on disposable batteries. Along with the improving technology, the implementation of more than one energy source will be crucial in overcoming present boundaries and increasing the scalability and acceptability of energy harvesting systems. Ultimately, collaboration between academia and industry will be crucial in realizing the true potential of energy harvesting as a viable way of powering the next generation of intelligent devices (Figure 8.8).

8.3.2. Applications of Energy Harvesting in Embedded Systems

The use of energy harvesting technology in embedded systems is a radical approach aimed at solving the central issue of power consumption, particularly for wireless sensor networks (WSNs). The systems are increasingly utilized in pervasive applications such as smart homes and environmental monitoring. Energy harvesting technologies enable nodes in these networks to tap energy from the surroundings, e.g., solar, thermal, or mechanical, to keep running for extended periods without needing battery replacement every now and then. This shift enhances system reliability without ensuring uninterrupted delivery of service at all times, which is crucial in order to maintain user confidence and satisfaction. In spite of this, the unpredictability of the harvested energy offers challenges that can lead to intermittent supply of energy, affecting system performance and user experience. Efficient energy management systems must thus be created in a bid to optimize energy consumption and counteract perceptions of unreliability of energy harvesting technologies (Berzosa et al., 2012). Another significant field in energy harvesting application areas of embedded systems is device design and efficiency. Specifically, the radio transceiver module usually drains most of an IoT device's energy, so innovative low-power circuit structures are essential. Most of the research has instead focused on designing energy-efficient physical layer schemes to maximize energy expenditure per information bit for the communications process. Even though such advancements are essential, they are undone by poorly designed network protocols and resource management policies and hence impact the performance of embedded systems in general. For improving energy efficiency, a context-aware and energy-centric optimization framework is introduced. The framework involves several key strategies like joint design of routing and channel access protocols, which allow for maximized network lifetime and adaptive flexible operation in different working conditions (Biason et al., 2017). In addition, energy harvesting also facilitates the dissemination of sustainable and self-sustaining embedded systems across all fields of life. Through minimizing dependence on traditional sources of energy, such as batteries, the systems are likely to greatly minimize battery disposal and resource depletion environmental effects. Energy-aware communication protocols and adaptive signal processing are essential design principles which need to be used to optimize the harmony between power consumption and energy harvesting in embedded systems. The use of such a visionary approach not only applies maximum lifetime to embedded systems but also helps in developing strong infrastructures that can actually operate in diverse scenarios. With the advancement of energy harvesting technology, it holds the potential to deliver solutions towards sustaining long-term efficiency and sustainability in embedded systems that fit into the broader scheme of technological advancement (Berzosa et al., 2012; Biason et al., 2017).

8.3.3. Benefits and Limitations of Energy Harvesting Technologies

With more demands on sustainable and efficient energy solutions, energy harvesting technologies have emerged as an important research area. The technologies recover ambient energy from vibration, thermal gradient, and solar irradiation sources and convert them to usable electrical energy, providing an attractive alternative to traditional batteries. Among various energy harvesting techniques, piezoelectric transduction stands out due to its high electromechanical coupling coefficient, which enables devices to capture power from mechanical strain. Additionally, breakthrough research in materials science has made it possible to fabricate stretchable and flexible piezoelectric generators, further increasing their application in various embedded devices. This adaptability makes them ideal to power low-consumption electronics in diverse environments with potential decreased reliance on nonrenewable energy sources while enhancing energy management and efficiency in embedded systems (Sezer et al., 2020). While there are prospective advantages from energy harvesting technologies, there remain boundaries to them that must be endured. The power output from harvesting methods typically remains intermittent and insufficient for high-power applications, particularly where energy demand exceeds what existing transduction technology can offer. For instance, although piezoelectric generators are readily capable of providing sufficient power to low-energy devices or small sensors, their performance under heavy load or under transient mechanical stimulation deteriorates. Also, the in-service life of these kinds of systems can be shorted due to material degradation or fatigue, impacting long-term sustainability. Hybrid systems formed by synergizing energy harvesting with traditional battery technologies can provide a possible solution, balancing between short-term power needs and long-term energy autonomy, and addressing these limitations (Sezer et al., 2020). Moreover, the changing landscape of communication technologies, particularly the transition from 5G to next generation 6G networks, poses challenges and prospects for energy harvesting systems. With wireless devices becoming increasingly connected and bandwidth-demanding, their power requirements will augment, demanding creative mechanisms to efficiently handle power consumption. Energy harvesting could have a critical role to play in powering the forecast growth of the Internet of Things (IoT), especially in situations where conventional sources of power are impossible. Notwithstanding, such reliance on energy harvesting technologies in this context must be weighed against the realities of energy efficiency and how far such systems can stretch in providing the demands of next-generation connectivity. Research in intelligent design ideas, such as low-power circuit designs and energy-aware communication protocols, will be essential in ensuring that energy harvesting systems play an effective part in the evolving energy management situation (Figure 8.9).

Figure 8.9. *Energy harvesting techniques and trends in electronic applications.*

Source: https://media.springernature.com/lw685/springer-static/image/chp%3A10.1007%2F978-3-031-35965-1_11/MediaObjects/531179_1_En_11_Fig1_HTML.png.

8.4. LOW-POWER DESIGN STRATEGIES

Within the strategies of low-power design, integration with renewable power sources and existing power infrastructures is one of the critical considerations to be made. New solutions to how energy efficiency may be established within various applications, particularly in building infrastructures, are being formed. By synchronizing renewable power production, such as through solar panels, with the embedded device's electricity consumption, a cleaner energy system can be designed. This involves minimizing energy usage by optimized demand management utilizing dynamic energy storage systems. For example, a system can be designed to effectively integrate energy-saving technology with embedded storage so that it can serve local energy requirements, seriously minimizing the use of conventional power.

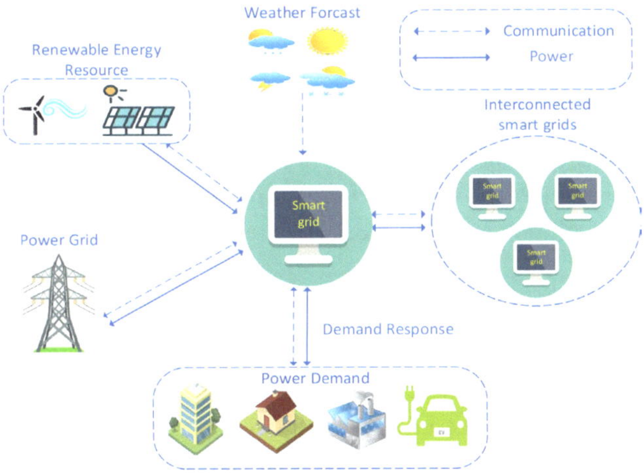

Figure 8.10. *Integration of renewable energy in microgrids and smart grids in deregulated power systems.*

Source: https://advanced.onlinelibrary.wiley.com/cms/asset/5263afa3-da3c-4cfe-8d62-a8d462eda858/aesr202400088-fig-0008-m.jpg.

Evidence for the effectiveness of such methods is in live environment trials that demonstrate their usability and efficiency in real-time energy management, as evidenced by outcomes by Counsell et al. (2011). Low-power design methods also entail innovations in energy storage technologies, which are of prime importance in energy optimization. Systems that combine a mix of storage devices, such as flywheels, ultracapacitors, and batteries, can enable efficient distribution and management of power. A hybrid locomotive developed by the French National Railways provides an illustration of how a diesel engine may be employed in conjunction with a flywheel and accumulators to power requirements. Design and deployment of energy management strategies that control the power exchange among these devices are very important in achieving operational efficiency. Focus on maintaining the best balance between generation and storage maximizes performance overall. The strategies are also non-compromising in curbing energy wastage, as observed by Akli et al. (2009), with low-power designs capable of revolutionizing vehicular and embedded systems. Additionally, the analysis of low-power approaches also mirrors an overall movement toward energy-conscious designs in other areas of technology. With embedded devices spreading into consumer electronics, medical care, and industrial control systems, the need for creative power conservation efforts only grows more urgent. These approaches focus on reducing operational power consumption while increasing functionality, typically leading to the utilization of sophisticated semiconductor technologies that reduce voltage levels and minimize power leakage. Essentially, the integration of smart energy management systems allows real-time variability based on usage patterns, ultimately creating more efficient and longer-lasting batteries. With the ownership of such ideals as adaptability and energy harvesting, low-power design methods are leading the way toward a sustainable future for energy, proving their essential role in future technological development (Figure 8.10).

8.4.1. Principles of Low-Power Design

Low-power design techniques are responsible for reducing power consumption while maximizing performance, particularly for embedded systems. Effective low-power design begins with understanding the energy requirements of various parts and using controls that reduce power consumption through every stage of operation. Techniques such as dynamic voltage and frequency scaling (DVFS) allow systems to control their performance in relation to workload demands, not only decreasing energy expenditure but also enhancing thermal management. By workload profiling and applying adaptive interventions, designers are able to achieve significant energy savings. Deployment of sleep modes and power gating can also reduce idle power consumption, leading to more eco-friendly device operation. By these rules, low-power design not only facilitates the management of energy but also contributes to the reliability and lifetime of embedded devices. Another significant topic in low-power design is the use of energy harvesting technologies, which enable devices to capture ambient power and transform it into useful energy. By harnessing environmental energy such as light, heat, or motion, energy harvesting can potentially extend battery life or even allow for constant operation of electronic devices. This change in source dependence reconfigures the design paradigm; rather than having to design on the basis of limited battery capacities alone, engineers can develop innovative designs considering intermittent energy availability. For instance, protocols need to be designed for dealing with the randomness of harvested energy in an efficient manner. The merge of the energy harvesting circuit with smart power management systems can lead to a paradigm shift in energy budget perception and therefore optimize the overall operation of embedded systems (Alizai et al., 2016). Finally, low-power design methodologies should also consider the impact of component choice and system architecture on energy efficiency. Microelectronic advancement is vital as there are more power-efficient components that are being developed continuously. For instance, the use of low-power sensors and microcontrollers diminishes energy use at the source, then enhances system performance. Furthermore, computation algorithms optimized can significantly reduce computation energy required, especially in wireless sensor networks. Merging these advanced components with the effective energy management strategies, such as in the PowerTag project, illustrates how complex systems can achieve a 44% increase in functioning lifetime through careful design (Janek et al., 2007). Consequently, implementing these low-power design practices not only supports efficiency but also encourages innovation in green technology solutions.

8.4.2. Techniques for Optimizing Power Usage in Hardware

Power consumption in hardware must be optimized to enhance the energy efficiency of embedded devices, particularly for the Internet of Things (IoT). The radio transceivers of the devices consume the majority of the energy and hence efficient physical layer design is imperative. Against this usage, designers have been working hard to highlight low power designs for circuits that minimize the average energy used per information bit in data communication. However, if the network and processing structure of the entire network is not properly planned, these gains would be eliminated with severe throughput and responsiveness degradation.

A holistic solution involves an energy-centric optimization paradigm that optimizes communication tasks against signal-dependent processing approaches, such as compression and feature extraction. A holistic solution guarantees power efficiency takes into account not just hardware design but also software and network protocols, opening the door to effective energy management in embedded systems (Biason et al., 2017). Besides, the emergence of LPWANs has revolutionized the energy consumption pattern of IoT devices, facilitating a new wireless communication paradigm focusing on energy saving in conjunction with QoS enhancement. With the network base supported by billions of interconnected devices, network management mechanisms are crucial in addressing the power consumption issue. A sample is a dynamic file synchronization protocol that is specifically developed for low-power IoT usage. This protocol optimizes the amount of data that is exchanged, hence minimizing the degree of interaction between devices in the network. By reducing the quantity and frequency of communication, it essentially prolongs battery life, which is essential in the majority of IoT applications. The adaptability present in this protocol allows it to function well in conjunction with many Layer 2 technologies, thus enabling broad applicability in terms of situations and total energy efficiency in IoT devices (Biagi et al., 2018). Application of adaptive algorithms and reconfigurable architectures further enhances power optimization techniques in hardware.

These techniques offer self-adaptability across a broad group of operating conditions, allowing systems to modify their behavior dynamically according to energy limitations. They enable intelligent resource handling, whereby devices learn to adjust processing jobs and channel access protocols based on the current circumstances. For instance, their adaptative features can extend the lifespan of an IoT network by conducting energy efficiency versus responsive data communication requirements in a balanced manner. Secondly, using learning architectures enables continuous improvement and optimization of power handling methods with time, such that energy consumption is always maximally efficient. As these technologies advance, the integration of energy-aware design concepts and adaptive processing will increasingly enhance the sustainability and longevity of embedded devices in the context of current energy management (Biason et al., 2017) (Figure 8.11).

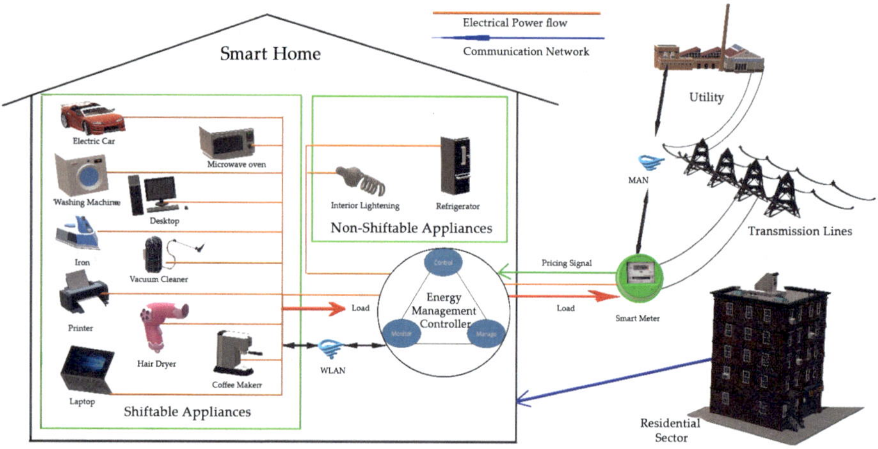

Figure 8.11. *Proposed system model.*

Source: https://www.mdpi.com/electronics/electronics-09-00406/article_deploy/html/images/electronics-09-00406-g001.png.

8.4.3. Software Strategies for Reducing Energy Consumption

The study of software mechanisms for reducing energy consumption in embedded systems is a key area of energy management and efficiency. Context-aware applications present unique challenges, as they would likely depend on collecting continuous data from a set of sensors, leading to higher energy requirements. Employing analysis techniques focused on battery usage behavior can allow the developer to reduce the runtime costs associated with context-sensitive functionality by significant extents. For example, policies can be followed in order to lower the number and amount of context information published that triggers energy savings directly. Additionally, tools such as the Nokia Energy Profiler can contribute to identifying inefficiencies as well as validating the effects of proposed adjustments in various operating states, thereby ensuring a more sustainable energy management policy in embedded systems (Ardito et al., 2010).

Also noteworthy is the underlying computational architecture that powers embedded devices, wherein optimizations to software can assist in enhancing energy efficiency. The recent trend among graphics processing units (GPUs) confirms this relationship; as computational demands increase, so does power consumption.

But what research focuses on to enhance the energy efficiency of GPUs are observations that can be extended to other platforms. By classifying techniques based on their research concepts, for instance, dynamic voltage scaling or parallel computing strategies, scientists will be in a position to spot techniques that enormously reduce power usage. By comparative studies of GPU energy efficiency versus other processing units like FPGAs and CPUs, the world can promote collective effort towards realizing sustainable energy solutions in embedded systems (Mittal et al., 2014).

In addition to the above steps, software development practices must accord top priority to low-power design concepts in the product life cycle. The principles are optimizing algorithm efficiency, programming with energy-aware languages, and applying power management practices at the software level. Proper coding can lead to lower operating cycles and less energy expenditure in general, while programming using energy-aware languages can efficiently limit energy-intensive operations at runtime. Besides this, employing sleep or low-power states in times of idleness can save even more energy. These software-oriented mechanisms can be integrated into design and deployment to assist developers in realizing substantial energy savings, which are consistent with broader efforts at green adoption of technology in embedded systems (Ardito et al., 2010).

8.5. THE ROLE OF IOT IN ENERGY MANAGEMENT

The expansion of the Internet of Things (IoT) has transformed energy management, particularly with focus on power consumption optimization in various systems. Through the utilization of sensors and smart devices, IoT provides the capability to monitor and control energy consumption in real time, a crucial aspect for efficiency improvement. For instance, in building automation systems, the utilization of Model-Based Predictive Control (MBPC) has been significantly promising energy conservation. One of the key innovations is the IMBPC HVAC system, where self-powered wireless sensors and an IoT platform are used to ensure thermal comfort and reduce electricity bills in real-world applications. This integrated process is an exemplar of how IoT can enable the transition to cleaner energy practices, with economic benefits, as well as environmental gains (Duarte et al., 2018).

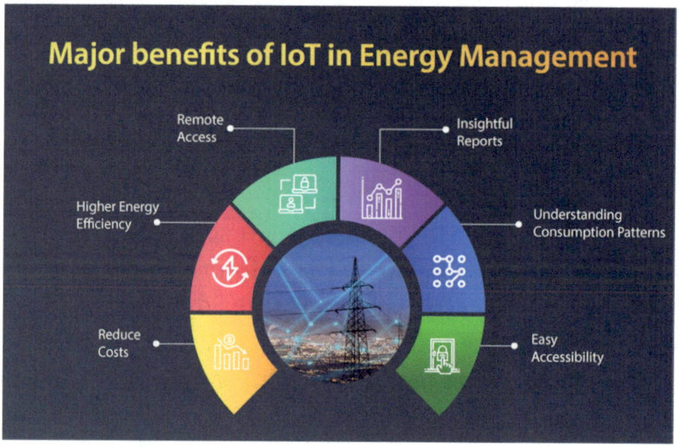

Figure 8.12. *IoT in energy management.*

Source: https://www.rejigdigital.com/wp-content/uploads/2023/03/13-03-01.jpg.

Besides, IoT devices are also significant in the evolution of low-power design methods, as necessitated by growing energy consumption of devices connected. A radio transceiver of an IoT device is a significant source of energy consumption, which has prompted much research into power-efficient circuit design. Lacking consideration of the system architecture as a whole, however, it is possible for hardware advancements to be able to only so much. Optimized network protocols, processing, and resource management solutions are essential for efficient energy management. With an energy-focused and context-aware optimization system, IoT systems will be able to enhance efficiency in operations while extending the lifespan of their functionality. The combined approach will guarantee that energy usage corresponds to usage patterns and system demands in a bid to optimize overall efficiency (Biason et al., 2017). As the world continues to focus on green energy sources, IoT will play a critical role in facilitating the infusion of such technologies into current systems. IoT smart grids can dynamically manage the supply and consumption of electricity, which can allow for a more resilient energy grid. This becomes crucial in the integration of non-predictive energy sources like wind and solar, which require sophisticated management to deliver balanced supply and demand. IoT has the ability to optimize energy flow through data analysis from many endpoints, thereby stabilizing the grid and promoting energy conservation. The intersection of IoT and energy management, therefore, presents historic opportunities not just to maximize efficiency but to achieve a sustainable energy future using advanced technology for stewardship of the environment (Duarte et al., 2018) (Figure 8.12).

8.5.1. Integration of IoT Devices in Energy Efficiency

IoT devices have become more and more central to energy efficiency initiatives, particularly with regard to building management systems. With approximately 40% of electricity used globally sourced from buildings, energy consumption needs to be maximized using smart technologies in the interest of sustainability (Duarte et al., 2018). IoT devices enable the acquisition of data and analytics in real-time, which can further optimize predictive control mechanisms for larger energy management. Of these technologies, Model-Based Predictive Control (MBPC) has been particularly promising. With the use of MBPC, phenomenal energy savings can be realized, especially in heating, ventilation, and air conditioning (HVAC) systems, without compromising occupant comfort. However, the field application of these technologies in existing buildings remains limited, and more need exists for further commercial products that enable theoretical models to converge into real-world integration in order to maximize energy efficiency on all sides. The problem of energizing IoT devices in a sustainable manner cannot be ignored when considering energy efficiency. Conventional battery-powered devices are far from being ideal, largely because of the limitations of battery lifetimes and the holy grail of operating continuously. Recent developments in wireless energy transfer (WET) technology provide exciting possibilities for overcoming these challenges. With wireless powered communication (WPC), devices can be powered by specialized wireless power transmitters and hence enjoy an uninterruptible energy source. This approach not only extends the operating lifespan of IoT devices but also enables better functionalities in terms of energy management system communications. WPC system design and integration, however, involves addressing new energy transmission and device synchronism problems. The reconciliation of wireless infrastructure benefits with the inborn complexities thus becomes crucial for optimizing energy-efficient IoT solutions (Bi et al., 2014). Also, the use of low-power design techniques is an essential aspect of making integrated systems of IoT more energy-efficient. By focusing on energy-saving components and usage patterns optimization, developers can create embedded devices that consume minimal power while delivering maximum functionality. This approach instantly leverages the strengths of IoT technologies, which derive power from efficient data processing and communication. For instance, the integration of next-generation sensor technologies with low-power communication protocols can lead to substantial overall power reductions. Additionally, complementariness of these low-power designs with energy harvesting techniques also guarantees sustainability through the utilization of ambient sources of energy for powering devices. By bringing such innovative solutions together, the inclusion of IoT devices into energy management becomes the template for making energy systems more efficient and sustainable in today's world, drawing a blueprint for the future development of smart technology (Figure 8.13).

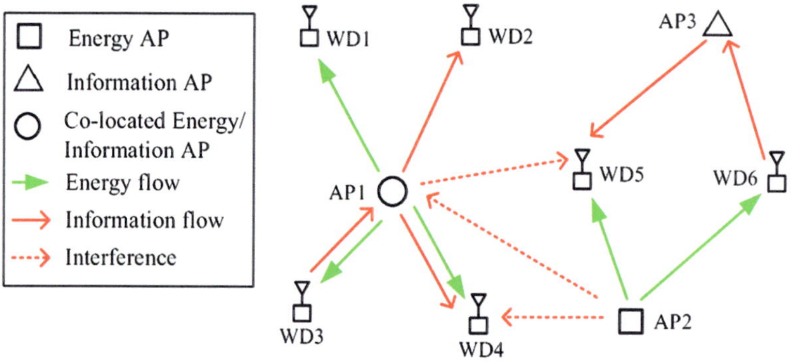

Figure 8.13. *A network model for wireless powered communication.*

Source: https://www.researchgate.net/profile/Chin-Keong-Ho/publication/264673637/figure/fig1/AS:392534846132241@1470599008815/A-network-model-for-wireless-powered-communication.png.

8.5.2. Data Analytics for Optimizing Energy Consumption

In the evolving energy management scenario, data analysis is leading the charge of achieving maximum energy utilization across the board from embedded devices to renewable energy sources. By leveraging advanced analytical methods, companies are able to harness meaningful observations from big data gathered through smart meters, sensors, and IoT devices. This approach to analysis allows energy managers to identify trends, forecast demand, and implement accurate strategies that recover a great deal of energy otherwise wasted. For instance, smart grids utilize data analytics to facilitate bi-directional energy flow, optimizing power generation while maintaining system reliability and security. Techniques such as predictive analytics and machine learning have the ability to refine these processes even more so that the users have the ability to make well-informed choices based on the principles of sustainable energy management as outlined by Copeland et al. (2019). Additionally, the use of IoT devices in renewable energy systems has the ability to revolutionize data collection and real-time monitoring capabilities. When such devices saturate the electrical network, they are generating large sets of data which are critical to system optimization. Battery technologies and energy harvesting solutions can especially gain maximum value from analytics-based approaches that allow them to maximize operational efficiency. Studying the difference in generation and consumption patterns can help stakeholders optimize battery discharge cycles and enhance load management. This systemic approach not only maximizes the effectiveness of energy storage systems but also grid stabilization via enabling a more predictable utilization of variable renewable sources, a feature thoroughly analyzed by Ahmed et al. (2024). Ultimately, the use of data analytics in this context creates a more resilient and responsive energy landscape. The benefits of energy optimization via data analytics go beyond the operational efficiency; they are accompanied by broader implications for sustainability and energy resilience. By adopting such analytics-driven strategies, organizations contribute to a more sustainable energy system that reverses the pressure of climate change and rising energy demand. By providing energy stakeholders with information regarding user behavior and operation inefficiencies, data analytics allows them to implement conservation practices aligned with environmental goals. Moreover, this convergence of technologies not only raises customer engagement through monitored energy management solutions but also instills a culture of responsibility and openness regarding energy usage. With the continued revolution of the energy sector, the relevance of the intersection of data analytics and energy optimization will be central to shaping energy policies and practices for the future, making it imperative that research and innovation in this area continue.

8.5.3. Future Trends in IoT and Energy Management

As the Internet of Things (IoT) technology continues to advance at a rapid rate, it is now more critical than ever before to deal with energy management within this ecosystem in order to support the broad spectrum of demands in varied applications. The future will witness sophisticated cloud-based platforms that unify heterogeneous IoT devices and systems. These frameworks will allow service developers to combine an array of connectivity options, communication patterns, and protocols to make it easier for energy-efficient services to be developed across the different IoT spaces. This integration is key to facilitating the highest efficiency paybacks achievable from IoT technology. Last but not least, the ability to process low-resource IoT devices cost-effectively will drive the development of innovative energy management solutions that utilize advanced data analytics and connectivity to realize optimal power efficiency (Demeester et al., 2015). Wireless sensor networks, one of the cornerstones of IoT applications, entailing design and deployment further emphasize the need for energy-efficient solutions. As such sensors are typically battery-powered and resource-limited, studying advances in low-energy design techniques as well as energy harvesting is very important. Future work is likely to focus on optimizing energy efficiency through the use of multiple Media Access Control (MAC) protocols, which have a large influence on power usage in sensor networks. Methods such as duty cycling, along with sophisticated MAC protocols, offer ways to conserve energy while maintaining communication reliability among devices. With security mechanisms becoming more integrated into the networks, an energy load versus security measures balance for countering potential attacks will remain a significant consideration for researchers and practitioners (Udoh et al., 2019). Furthermore, energy storage technology will drive the development of the future of IoT and energy management. Improved battery technologies, such as lithium-sulfur and solid-state batteries, can potentially provide greater energy density, longevity, and safety, and improve the operating lifetime of IoT devices. Energy-harvesting systems with the ability to draw from the ambient sources of power, such as solar, thermal, or kinetic energy, are also possible. With the integration of these technologies and low-power design practices, the innovators are able to deploy self-sustaining systems that significantly reduce reliance on traditional power sources. This holistic energy management approach not only aids sustainability but also aligns with the increase in green solutions in the IoT landscape. Through this, innovation in energy harvesting and storage technologies will play an important role in the attainment of a more scalable and efficient IoT (Demeester et al., 2015).

8.6. CASE STUDIES IN ENERGY EFFICIENCY

The push towards energy efficiency in embedded systems has seen some remarkable innovations that speak to the nuances of power consumption. Of note is the inclusion of wireless energy transfer technology, which has been touted as a solution to one of the biggest challenges facing battery-powered devices: battery drain. The traditional devices are typically riddled with operational disruptions due to the need to constantly switch and replace batteries. With the advancements in the radio frequency (RF) enabled WET, wireless powered communication (WPC) is recognized as a suitable alternative, enabling constant operations and extending the life of device operations. Through enabling infinite energy supply, WPC not only enhances the performance and reliability of wireless communications but also presents new research challenges that require a thorough study of design strategies and the viable implementation strategy, displaying its revolutionary capability in the area of energy efficiency (Bi et al., 2014).

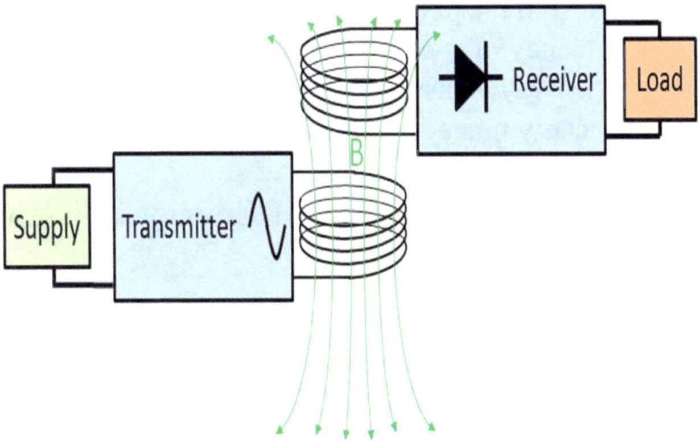

Figure 8.14. *Wireless power transfer circuit working.*

Source: https://www.watelectronics.com/wp-content/uploads/Wireless-Power-Transfer-System.jpg.

In addition to WET, ambient energy harvesting is another important method for enhancing energy efficiency for embedded systems. As the micro-electronics environment is constantly evolving, innovations that are able to harvest environmental sources of energy, such as solar, thermal, and vibration energy, are redesigning the function of wireless sensor networks (WSNs). Ambient energy harvesting enables the systems to prolong their working lifetime or even achieve eternal functionality, eliminating the limitations of finite energy budgets. This development significantly alters paradigms of WSNs design and implementation from battery-limited models to architectures for potentially unbounded but intermittent power supplies. Through research on real-world applications and case studies, researchers have documented the success of these technologies in the field and provided decisive insight into implications for energy management in a wide variety of embedded situations (Alizai et al., 2016). Low-power design methodologies also contribute further to improvement in energy efficiency. Hardware architecture and software algorithms are optimized to reduce energy consumption with little degradation in performance. Techniques such as power gating and dynamic voltage and frequency scaling (DVFS) are important techniques for optimizing high degrees of energy savings in embedded systems. Case studies illustrate practical application of these strategies, with lowered levels of power consumption across different contexts—spanning IoT devices to wearable technology. Additionally, ongoing research aims at developing better methodologies that synergistically bring together hardware capacity and energy management systems in order to develop more responsive and efficient devices. These case studies attest to the potential of low-power design techniques to not only enhance the efficiency of specific devices but also to benefit society more by reducing overall energy consumption across industries, solidifying their place in sustainable energy management policy (Figure 8.14).

8.6.1. Successful Implementations of Energy-Efficient Embedded Systems

The successful deployment of energy-efficient embedded systems largely hinges on the harnessing of new design paradigms to maximize low power consumption without affecting performance. One such methodology is the development of pervasive systems integrated unobtrusively into a network of Wireless Sensor Networks (WSNs). Such systems are particularly beneficial in maximizing user experience by offering services enhancing quality of life through monitoring and automation. However, the sustainability of these services relies on proper battery management along with the integration of Power Harvesting (PH) techniques that allow individual nodes of the WSN to harvest power from the surrounding atmosphere. This capability not only sustains operation time but also minimizes reliance on traditional power supplies, thus becoming an incredible innovation in embedded systems intended to meet the demands of today's applications (Berzosa et al., 2012). Besides, although energy-harvesting embedded systems bring enormous benefits, there are potential challenges that should be tackled, particularly related to PH source dependability. The intermittency of energy harvesting can lead to different degrees of energy availability, which might be detrimental to user attitudes as well as system reliability. Implementation of robust energy management policies and adaptive control algorithms is essential in minimizing these risks as they can incessantly modify system operations based on current power availability. This function improves user confidence in system operations, therefore leading to broader adoptions and application of energy-saving technologies. System developers must focus on creating versatile interfaces and feedback mechanisms that allow users to interpret system states, thus bridging the technical performance – user expectations gap (Berzosa et al., 2012). Additionally, the research of alternative battery technologies and energy harvesting methods becomes imperative in advancing embedded systems functionality. Lithium-ion batteries, which feature high energy density, have dominated the industry for decades but are increasingly being supplemented by other technologies such as supercapacitors and novel battery chemistries with greater charging rates and cycle stability. Further, integrating solar panels or piezoelectric generators into embedded systems provides new opportunities for green energy harvesting, most notably in those applications where long autonomy is significant. The effective application of such power-saving technologies and measures proves the synergy involved in applying design intelligence, advanced material science, and power management techniques. The multilayered mechanism not only surmounts the current limits of employing power but also provides a foundation for future breakthroughs in embedded systems capable of unprecedented efficiency.

8.6.2. Comparative Analysis of Energy Consumption in Various Devices

When comparing energy consumption across devices, it is significant to consider the different operational architectures and technologies employed. Embedded devices, generally committed to specific, specialized tasks such as IoT sensors or medical wearables, exhibit unique power profiles according to design and purpose. As these devices spend significant amounts of time in low-power states between computing or transmission, their energy consumption is measured based on both their cycle of operation and the particular situation in which they are operating. It has been shown that energy usage can vary significantly based on communication protocol and hardware design efficiency, which means that technology selection is critical as much as the intended purpose of the technology. Consequently, an exhaustive comparative study must also consider such intricacies in order to accurately assess energy performance. The mutually interdependent nature of energy management and the performance of embedded devices emphasize the significance of low-power design techniques. Power consumption does not solely depend on the device; it is significantly influenced by the determinations taken in its physical layer design as well as processing capability. As has been noted through research, most of the energy is consumed in the radio transceiver components of IoT devices, making it necessary to adopt quality design methods that save energy without compromising performance. Techniques such as adaptive signal processing and channel access optimization can go a long way towards enhancing energy efficiency, which allows for increased lifetimes of operation in battery-powered devices. With improvements in design methods, ensuring that emerging technologies are developed with energy-centric practices will be critical for the sustainability of embedded systems in the more networked world. Some of this comparative study involves learning about battery technologies and energy harvesting methods that are the break points for the devices' longevity and efficiency. Understanding the energy density, cycle life, and charging time of various battery technologies is critical to enable well-informed embedded system design choices. At the same time, energy harvesting methods such as solar or thermoelectric power introduce new solutions to minimize reliance on conventional batteries. Resources like the NPS report of energy research identify the importance of integrating energy solutions to the larger Naval enterprise, even as the relevance of energy efficiency in contemporary applications is mooted (N/A, 2016). An analysis of these technologies against conventional measures of energy consumption makes possible a comprehensive view that reflects directions towards improved energy management across different applications, contributing to theory and practice advancement.

8.6.3. Lessons Learned from Industry Leaders in Energy Management

Analysis of energy management practices provides important insights based on industry leaders' experiences in implementing energy-saving measures. Notably, the Building-to-Grid testbed project case studies at UC Berkeley emphasize the significance of sound data aggregation and analysis. The collaboration between university researchers and industry companies enabled the in-depth mapping of energy flows in a building, the value of which in terms of understanding energy consumption patterns is the main finding. This intervention strategy not only makes it possible to identify inefficiencies but also enables the creation of customized solutions that are aligned with specific energy needs. These projects highlight the necessity of interdisciplinary collaboration, as observed in the cooperative dynamics put in place in this project, that are crucial in driving innovation and optimizing the utilization of energy in diverse settings, a lesson that is valuable to stakeholders in the fields of embedded devices, battery technology, and low-power design practices (Peffer et al., 2010). Moreover, the successful implementation of energy management systems relies in most instances on the application of smart metering technologies, as can be seen from the findings of the UC Berkeley program. The installation of sophisticated sensors and submeters played a pivotal role in the gathering of real-time data, which informed decision-making initiatives and enabled timely adjustments to energy consumption. These technologies allow organizations to monitor their energy usage closely, leading to greater energy conservation and efficiency. This information obtained from these technologies not only leads to improved operational efficiency but also serves as a blueprint for integrating energy harvesting methods into embedded systems. Industry pioneers demonstrate that investing in such technologies can reap rewards over the long haul, allowing organizations to respond rapidly to shifting energy situations and apply low-power design techniques that reduce overall consumption (Peffer et al., 2010). Finally, industry pioneers' experiences highlight the necessity of a roadmap in driving the rollout of energy management systems. The methodical approach the UC Berkeley group took, which included developing a thorough methodology and evaluating the process, validates the need for a methodical framework for energy efficiency projects. This plan may serve as a template for other projects with the same goal that would potentially close the divide between energy use and sustainability goals. Additionally, working with local utilities, as illustrated in the testbed project, identifies the value of robust partnerships that facilitate data exchange and coordination. These synergies enhance the capability for energy communication and effectiveness in management. Emphasizing these findings can empower future projects to spearhead cutting-edge solutions in energy management, particularly in the fields of battery technologies, energy harvesting, and low-power efficient design (Peffer et al., 2010).

SUMMARY

The intricate relationship between energy management and efficiency in embedded devices underscores the dire need for new solutions in the domains of battery technologies and energy harvesting. The constantly changing nature of embedded systems calls for a finely honed sensitivity to power consumption trends, particularly as devices increasingly infiltrate everyday life. As has been pointed out in N/A (2016), expanding research on energy indicates a growing acknowledgment of energy as a core aspect of modern technology. The conversion is not only a requirement of technology but an intrinsic aspect of sustainable development, substantiating the point that effective energy methods can both enhance performance and prolong device life. Moreover, adopting low-power design techniques is necessary for power-efficient system integration, particularly in resource-constrained environments such as wireless sensor networks. The findings of Udoh et al. (2019) support this perspective by demonstrating how security factors, while necessary, may also impact energy loads on the devices. This is significant to understand because it indicates that effective energy management must balance both efficiency and security. With low-power solutions at the forefront, it is now possible to cut down on energy drains and retain vital functionalities in fully functional state, in order to foster the operational life cycle of embedded devices. Finally, with energy management technologies, low-power design methods, and advanced battery products, there is a window towards the realization of sustainable energy consumption among embedded devices. With the evolution of this field, ongoing R&D will continue to lead the way in addressing the challenges posed by power consumption and efficiency. By creating coordination among academia, industry, and government departments, as identified in N/A (2016), the vision for a more energy-efficient technological world is possible. This commitment to research and application will not just enhance performance but also towards a more sustainable future in a variety of technological fields. The research on energy efficiency and management in embedded systems encompasses a number of significant topics with a dominant focus on power consumption, battery technology, energy harvesting, and low-power design methodologies. The growing pervasiveness of embedded systems, especially in areas like healthcare, necessitates innovative approaches to optimizing energy utilization. One fundamental aspect of this is the investigation of the limitations of traditional battery technologies, namely in implantable medical devices where longevity, miniaturization, and the need for periodic maintenance may hold back usability and patient safety. This evokes the strong need for new energy sources, highlighting the revolutionary power of energy harvesting techniques harnessing the human body's natural sources of energy as likely power alternatives for the likes of pacemakers and wearables (Chen et al., 2025). And as emerging communications technologies advance, so too does the need for efficient use of energy standards, especially when it comes to Machine-Type Communication (MTC) devices. The study finds that the optimization of subsystems such as radio communication, processing, and sensing technologies is essential in order to improve overall energy efficiency. Each subsystem has its own challenges and trade-offs that should be addressed in order to deliver optimal performance at minimal power cost. By delineating approaches to improve these subsystems, it is apparent that communication framework innovation is central to the design of sustainable energy solutions that meet the mounting demands of networked devices in various industries (Alves et al., 2024). In summary, low-power design methodologies and good energy management practices are essential in redefining the operational model for embedded devices. With increased research, an inter-disciplinary model becomes unavoidable, where integration of multiple disciplines is emphasized to meet today's energy demands. Synergy among energy harvesting technologies and advanced battery technologies is an essential component in achieving a paradigm shift towards autonomous biomedical devices and communication platforms. With innovation and continuous improvement in such areas, it is possible to build the groundwork for intelligent, energy-saving devices that are ahead of the demands of modern society while maintaining maximum operational efficiency (Alves et al., 2024; Chen et al., 2025).

REFERENCES

1. Abdel Gelil, I., Aguayo, F., Boyle, G., Breukers, M., et al. (2012). Policies for capacity development. https://core.ac.uk/download/9836421.pdf (accessed on 10th September 2025).
2. Aguado Correa, F., Barcos, L., & Álvarez Gil, M. J. (2005). Benefits of connecting RFID and lean principles in health care. https://core.ac.uk/download/29427288.pdf (accessed on 10th September 2025).
3. Ahmed, H., Al Khatib, O., Barbulescu, E. D., Nassereddine, A., et al. (2024). Internet of things important roles in hybrid photovoltaic and energy storage system: A review. https://core.ac.uk/download/622528700.pdf (accessed on 10th September 2025).
4. Akli, C. R., Jaafar, A., Jeunesse, A., Roboam, X., et al. (2009). Sizing and energy management of a hybrid locomotive based on flywheel and accumulators. https://core.ac.uk/download/12043876.pdf (accessed on 10th September 2025).
5. Alizai, M. H., Bhatti, N. A., Mottola, L., Syed, A., et al. (2016). Energy harvesting and wireless transfer in sensor network applications: Concepts and experiences. https://core.ac.uk/download/84891109.pdf (accessed on 10th September 2025).
6. Alves, H., Basutli, B., Chuma, J., Ntabeni, P., et al. (2024). Device-level energy efficient strategies in machine type communications: Power, processing, sensing, and RF perspectives. https://core.ac.uk/download/628876665.pdf (accessed on 10th September 2025).
7. Anthony, D. W., Kroener, M., Nazir, S., Newey, S., Peijiang, C., & Samuel, M. (2018). Autonomous monitoring framework for resource-constrained environments. https://core.ac.uk/download/293885447.pdf (accessed on 10th September 2025).
8. Ardito, L., Falcarin, P., Mejia Bernal, J. F., Morisio, M., et al. (2010). Towards an efficient context-aware system: Problems and suggestions to reduce energy consumption in mobile devices. https://core.ac.uk/download/pdf/11418221.pdf (accessed on 10th September 2025).
9. Aslam, N., Cao, Y., Hussain, S., Khan, M., et al. (2018). A critical analysis of research potential, challenges and future directives in industrial wireless sensor networks. https://core.ac.uk/download/96885354.pdf (accessed on 10th September 2025).
10. Berzosa, J., Grabham, N. J., Mabe, J., Tudor, J., et al. (2012). Social issues of power harvesting as key enablers of WSN in pervasive computing.

11. Bi, S., Ho, C. K., & Zhang, R. (2014). Wireless powered communication: Opportunities and challenges. http://arxiv.org/abs/1408.2335 (accessed on 10th September 2025).
12. Biagi, M., Cuomo, F., Listanti, M., Petroni, S., et al. (2018). Adaptive data synchronization algorithm for IoT-oriented low-power wide-area networks. https://core.ac.uk/download/188829070.pdf (accessed on 10th September 2025).
13. Biason, A., Kelly, M., Pielli, C., Rossi, M., et al. (2017). EC-CENTRIC: An energy- and context-centric perspective on IoT systems and protocol design. https://core.ac.uk/download/84922330.pdf (accessed on 10th September 2025).
14. Borghetti, G., Menga, G., Procaccianti, G., Kok, L., Ardito, M., Morisio, M., Mo, Y., et al. (2013). Smart grid technologies in Europe: An overview. https://core.ac.uk/download/pdf/11431923.pdf (accessed on 10th September 2025).
15. Borri, M., Del Corso, D., Passerone, C., Reyneri, L., et al. (2010). Design solutions for modular satellite architectures. https://core.ac.uk/download/pdf/11421001.pdf (accessed on 10th September 2025).
16. Chen, J., Liang, Y., Lin, J., Lin, S., et al. (2025). Sustainable power solutions for next-generation medical devices. https://core.ac.uk/download/661941307.pdf (accessed on 10th September 2025).
17. Copeland, R., Crespi, N., Rehmani, M. H., Saleem, Y., et al. (2019). Internet of Things-aided smart grid: Technologies, architectures, applications, prototypes, and future research directions. http://arxiv.org/abs/1704.08977 (accessed on 10th September 2025).
18. Counsell, J., Doran, S., Logan, R., Stewart, R., et al. (2011). A novel DSM philosophy for building integrated renewable systems. https://core.ac.uk/download/9036219.pdf (accessed on 10th September 2025).
19. Demeester, P., Hoebeke, J., Moerman, I., Van den Abeele, F., et al. (2015). Integration of heterogeneous devices and communication models via the cloud in the constrained Internet of Things. https://core.ac.uk/download/55746971.pdf (accessed on 10th September 2025).
20. Duarte, H., Ferreira, P. M., Ruano, A., Silva, S., et al. (2018). Wireless sensors and IoT platform for intelligent HVAC control. https://core.ac.uk/download/187434288.pdf (accessed on 10th September 2025).
21. Guarnieri, M., Mattavelli, P., Petrone, G., Spagnuolo, G., et al. (2016). Vanadium redox flow batteries: Potentials and challenges of an emerging storage technology. https://core.ac.uk/download/80155277.pdf (accessed

on 10th September 2025).

22. Havinga, P. J. M., & Smit, G. J. M. (2000). Lessons learned from the design of a mobile multimedia system in the Moby Dick project. https://core.ac.uk/download/pdf/11454372.pdf (accessed on 10th September 2025).
23. Janek, A., Pistauer, M., Preishuber-Pfluegl, J., Steger, T., et al. (2007). Lifetime extension of higher-class UHF RFID tags using special power management techniques and energy harvesting devices. https://core.ac.uk/download/62912624.pdf (accessed on 10th September 2025).
24. Leung, K. K., Leung, V. C. M., Liu, C. H., Moreno, W., et al. (2016). Context-awareness for mobile sensing: A survey and future directions. https://core.ac.uk/download/30610426.pdf (accessed on 10th September 2025).
25. Mittal, S., & Vetter, J. S. (2014). A survey of methods for analyzing and improving GPU energy efficiency. http://arxiv.org/abs/1404.4629 (accessed on 10th September 2025).
26. Peffer, T. (2010). UC Berkeley's Cory Hall: Evaluation of challenges and potential applications of building-to-grid implementation. https://core.ac.uk/download/288432727.pdf (accessed on 10th September 2025).
27. Sezer, N., & Koç, M. (2020). A comprehensive review on the state-of-the-art of piezoelectric energy harvesting. *Nano Energy, 80*, 105567–105567. https://doi.org/10.1016/j.nanoen.2020.105567 (accessed on 10th September 2025).
28. Tariq, F., Khandaker, M. R. A., Wong, K. K., Imran, M. A., Bennis, M., & Debbah, M. (2020). A speculative study on 6G. *IEEE Wireless Communications, 27*, 118–125. https://doi.org/10.1109/mwc.001.1900488 (accessed on 10th September 2025).
29. Udoh, E., & Udoh, E. (2019). An energy aware and secure MAC protocol for tackling denial of sleep attacks in wireless sensor networks. https://core.ac.uk/download/305119264.pdf (accessed on 10th September 2025).

CHAPTER 9

HUMAN-MACHINE INTERACTION (HMI)

Contents

Introduction.......... 333
9.1. Historical Development of HMI.......... 337
9.2. Types of Human-Machine Interfaces.......... 342
9.3. The Role of Artificial Intelligence in HMI.......... 346
9.4. Usability and User Experience in HMI.......... 351
9.5. Challenges in Human-Machine Interaction.......... 356
9.6. Future Trends in HMI.......... 361
Summary.......... 365
References.......... 368

INTRODUCTION

The vibrant field of Human-Machine Interaction (HMI) is a crossroads at which technology, cognitive science, and social interaction meet. It is interested in the question of what human-machine interaction is and how these interactions can best be optimized for optimal user experience and useability. As machines become more part of daily life, understanding HMI becomes imperative. Effective interactions depend on the cognitive character of user participation, which must be fully comprehended as regards how humans process information, employ language, and how emotional processes function in these interactions. The extensive implications of HMI cross numerous domains, including health, education, and industry, and therefore require a multidisciplinary approach in being able to entirely grasp human users and machine interfaces (O'Brien et al., 2013). In the HMI paradigm, the development of artificial intelligence (AI) has profoundly impacted our conceptualization of human-computer interactions. While AI technologies have the potential to mechanize extensive amounts of human work, they also pose important questions about sociocultural effects. One particularly significant issue is the representation of AI systems, as the depiction of these systems tends to mirror and ratify gender stereotypes, making these technologies more difficult to conceptualize in society. Feminist theory has long understood technology as something employed to counteract what society expects, but many applications of AI today tend to reinforce current dichotomies.

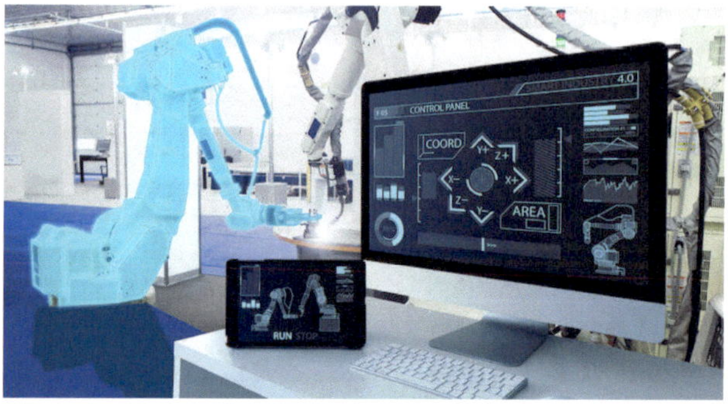

Figure 9.1. *Human-machine interaction (HMI).*

Source: https://www.automationreadypanels.com/wp-content/uploads/2024/01/Human-Machine-Interface-HMI-750x400-1.jpeg.

AI handled through chat, for example, acts on what have been traditionally feminine traits and thereby restricts their ability to perform emotional labor or even occupy jobs coded as feminine (Vega et al., 2019). This evolution heralds a paradox in HMI, requiring the critique of the tacit norms that underlie AI design and performance. Interdisciplinary research on HMI reveals the importance of cognitive science, linguistics, and social theory, unearthing the diversity of determinants subsidizing effective human-machine interfaces. Translatology on the cognitive level illustrates this juncture in the sense that it investigates how the processes of cognition influence translation processes and how conclusions could translate to informing HMI design. Through the study of the processes of cognition that underlie communication, scholars can better understand users' behavior and attain communication in accordance with users' mental models. Additionally, the potential that HMI has towards influencing future technological innovation calls for ongoing debate regarding ethical concerns as well as economic implications. As HMI continues to evolve, it becomes more critical to critically evaluate not only technological potential but also the wider social narratives to which they contribute or counter, opening up more inclusive and effective human-machine interaction (O'Brien et al., 2013) (Figure 9.1).

Definition of Human-Machine Interaction

The definition of human-machine interaction (HMI) encompasses the intricate interaction between humans and technical systems with user experience, interface design, and cognitive ergonomics taken as a priority. In essence, HMI seeks to facilitate effective communication and functionality between the user and the machine with a focus on enhancing usability and productivity. The role of HMI is increasingly gaining prominence in various sectors like automobile, healthcare, and consumer electronics. The growing complexity of technology demands more understanding of the impact of human senses, feelings, and mental processes on dealing with machines. Researchers have, therefore, emphasized the importance of designing systems that not just function effectively but also talk to people emotionally and aesthetically, thereby defining a combined user experience. A fundamental aspect of defining HMI is the investigation of the perceptual and cognitive influences on user actions with technology. An example is that automotive studies of HMI have demonstrated how consumers assess switch haptics as a critical element in their perception of vehicles. A test with premium sports utility vehicles confirmed that customers evaluated the haptic feel of controls under different conditions and gained useful insights into users' preferences and expectations. Factors like Affective, Robustness and Precision, and Silkiness emerged as key drivers in customer acceptance and satisfaction. This information highlights the requirement to incorporate user feedback throughout the design cycle, as well as ensuring that human factors are taken into account to improve the safety, comfort, and overall usability of HMI systems (Pitts et al., 2008). Furthermore, cognitive ergonomics has a key contribution to make in understanding the definition and implications of HMI through a focus on user performance and prediction of error with technology. The development of techniques for human error prediction supports the design of user-friendlier systems and the improved experience of interaction. The idea of 'rewritable routines' emphasizes the malleability of human cognition in response to technology, suggesting a dialectic between user and system design. With the study of witnessed behavior coupled with predicted errors, cognitive models may be formulated by researchers that will inform the creation of HMI to maximize efficiency and minimize the likelihood of errors in interactions. As technology systems increasingly become integrated into everyday life, studying these cognitive factors will be a vital component in spearheading the understanding and efficacy of human-machine interaction (Baber et al., 1997).

Importance of HMI in Modern Technology

The significance of Human-Machine Interaction (HMI) in today's technology cannot be overemphasized, as it practically determines the manner in which individuals engage with digital environments and automated machines. HMI serves as a key mediator that makes it easier for humans to communicate with machines, promoting an enriching user experience through facilitating intuitive and accessible interfaces. From healthcare to entertainment, effective HMI can enhance productivity and emotional well-being, towards a partnership-centered relationship between humans and machines. The application of artificial intelligence in arts programs, for example, not only illustrates high-level computing capability, but also confers mental health benefits, particularly on at-risk populations like individuals with dementia. By tapping into the capability of AI to elicit emotional reactions, HMI has the ability to allow therapeutic interactions to enhance cognitive stimulation and social interaction in users, emphasizing its critical place in contemporary technological uses (Harwood et al., 2019). Contemporary technologies under the umbrella of HMI, including *cyborgian* integration, further emphasize the transformative power of the human–machine union. The integration of human traits with advanced technologies, or melding, is a paradigm shift in how service sectors conduct operations. Such an integration enables frontline service staff to leverage bio- and psychophysical enhancements that enable more intuitive and empathetic interactions with clients, hence contributing to enhanced service quality. These cyborgian applications accentuate not only the potential for increased operational efficiency but also the capacity to address perplexing service issues in real time. As more organizations embrace the use of such technologies, there is a need to appreciate the implications of cyborg technologies, particularly those on ethical considerations and how they influence human agency in service contexts (Garry et al., 2019). The worth of HMI is greater than mere efficiency—it significantly redefines social relationships and personal lives. With increasingly advanced HMI technologies come new challenges demanding ongoing research and ethical discussion. Advances in the area of neural interfaces and robotics necessitate the complete picture of how these technologies alter human interaction. The implications of privacy, autonomy, and the human condition of HMI need to be examined with caution in a manner that supports the establishment of frameworks that ensure a high regard for human well-being. This is crucial in establishing regulatory frameworks that strike a balance between innovation and ethical responsibility so that technological development equitably complements human ability and does not undermine it. This twin focus on innovation and human-centered design will ensure HMI remains a beneficial and moral aspect of our growing digitized lives.

9.1. HISTORICAL DEVELOPMENT OF HMI

The historical inception of Human-Machine Interaction (HMI) starts during the early 20th century with the inception of mechanization in workplaces. The computer's invention and the subsequent development of the computing technology brought a significant milestone with the need of the equipment for increasingly ease-of-use-supportive user interfaces. Early efforts were at mechanical interfaces in the guise of levers and dials that would ultimately be replaced by electronic interfaces as digital technology developed. More user-centered and intuitive designs were required, sparked by an awareness that enhancing the user experience was essential to achieving optimum productivity within machines, and user-centered design principles have since been given greater emphasis. With added technology, the machines became more complex, demanding collaboration across disciplines to address in an effective way the multidisciplinary problems of HMI. Information came to play a central role in this process, much as oils do in industrial society, as both source material and primary force in the society of knowledge. This development made it possible to describe, represent, and model more complex and detailed descriptions, representations, and models, often multi-parameter, multi-perspective, and multidimensional. As noted, the design of technological artifacts turned more user-centered, incorporating some of the non-functional elements beyond functionality like beauty, user experience, ethics, and sustainability. This developed into designing the need for interdisciplinary work, hence the solidification of HMI through knowledge and practice synthesis from different disciplines, as in hybrid projects blending technology with social and environmental concerns (Almér et al., 2016). The development of HMI has always reflected changes in society's outlook towards the position of technology in everyday existence. Personal computing and internet technology democratized access to high-level machines, greatly changed user bases and usage patterns. As machines evolved from being simple tools into central elements of many aspects of human life, the ethical issues of HMI began to attract attention. This change brought to the forefront basic questions about user agency, responsibility, and technological development sustainability. Furthermore, as is seen in interdisciplinary understanding projects and conferences in HMI, ongoing dialogue and cooperation between designers, researchers, and users is essential. This collaborative effort not only aims to enhance the performance of machines but also the broader societal effects, placing HMI within the context that responds to human and technological factors on how the future landscape of interaction will be determined (Golding et al., 2010) (Figure 9.2).

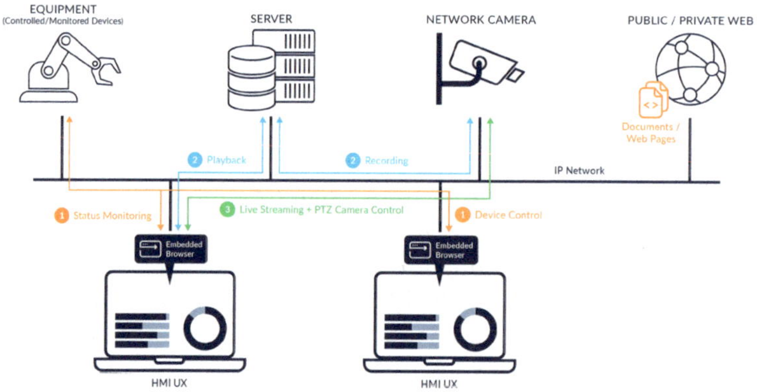

Figure 9.2. *The evolution of human machine interfaces.*

Source: https://www.enghousenetworks.com/wp-content/uploads/2022/08/espial-browser-windriver-diagrams-02.jpg.

9.1.1. Early Examples of Human-Machine Interaction

The origins of human-machine interaction (HMI) date back to early application of automated systems and displays for enhancing user communication and interaction. One such example is the application of intelligent situated displays, such as in workplaces to allow social awareness between users. Panorama is one such HMI initiative where a screen is used to disperse user-generated content in the form of photos and messages in an effort to build community engagement among staff members. From research on Panorama, it was found to successfully mimic curiosity and learning, initiated conversations, and facilitated the relishing of shared memories among employees. This innovation underlines the initial recognition of the importance of social factors in HMI, giving way to later innovations that focused on user interaction and socialization as key aspects of technological advancements (Eliens et al., 2011). Moreover, as various technological innovations emerged, automotive design began to intersect with human-machine interaction principles. The use of ergonomic considerations in vehicle design was not only aimed at functionality but also at the optimization of user experiences and promoting behavioral change.

Ergonomic focus became paramount as private car travel became known for its detrimental impact on the environment. Incorporating ergonomic design in the design of low-carbon automobiles, such as hybrids and electric cars, promotes manufacturers to steer driving habits in the right direction. The relationship between design and user behavior matters; there is evidence that design can significantly affect performance outcomes, including fuel use, and minimize issues such as range anxiety and distraction (Harvey et al., 2013). The relationship is indicative of the general direction of HMI, whereby user interaction is carefully tailored to encourage safer and greener behaviors. The early human-machine interaction studies illustrate the early understanding of technology's ability to enhance social connections and user activity. Panorama is an example of how visionary, interactive screens can trigger user engagement throughout society, using technology for social purposes. Similarly, the ergonomic redesign of automobiles marks the significant paradigm shift towards the inclusion of human-centered principles in technology. These early illustrations underscore the dual role of HMI as being both a social enabler and behavior modifier, and they provide useful lessons for contemporary applications. Progressing toward future advancement in HMI, it is worth remembering the complementarity of technological advancement and human nature and ensuring that designs are empathic and user-oriented and also promoting sustainability and community growth (Arvey et al., 2013; Eliens et al., 2011).

9.1.2. Evolution of Technology and Its Impact on HMI

The progress in technology transformed Human-Machine Interaction (HMI) significantly, from traditional interfaces to sophisticated and natural interfaces. Initially, the interaction was mostly dependent on mechanical input, which created obstacles in usability and accessibility. As technology advanced, especially with the dawn of personal computing and the internet, the model of interaction transformed so that humans began interacting with machines through graphical user interfaces (GUIs). This transition set the stage for subsequent inventions, such as touchscreens and voice recognition systems, to facilitate higher levels of user engagement and operational performance. In greater detail, networked devices, enabled through the Internet of Things (IoT), have reshaped interaction dynamics by empowering systems not only to perform user inputs but also to reply dynamically to context-aware information perceived from their environment with sophisticated sensors and algorithms. The increasing complexity of such interactions challenges users' flexibility while at the same time vowing to make everyday activities easier. In addition, the application of smart technologies has accelerated the development of context-aware systems that highly complement HMI. The incorporation of sensors that provide real-time data has led to machines gaining the capability to understand and respond to users' requirements better. For instance, in automotive technologies, vehicles equipped with HMI systems are no longer passively receiving the commands of the driver; they are now capable of interpreting driving situations, predicting user behavior, and providing vital information, thereby improving safety and convenience. Specifically, the integration of HMI and Semantic Sensor Web technologies provides an intricate ecosystem wherein human interpretations of traffic situations or environmental problems are shared and processed for the common good.

Figure 9.3. *Graphical user interface.*

Source: https://cdn.mos.cms.futurecdn.net/v2/t:0,l:137,cw:946,ch:710,q:80,w:946/Xj58EYWQAj93cby2KK6xNY.png.

This evolution points to the complementary relationship between humans and machines, where technologies become increasingly intelligent and responsive to the nuances of human interaction, enhancing the user experience and operational effectiveness (Bizer et al., 2012). The future for HMI, particularly with advancements in smart grid technologies, promises opportunities but poses challenges. The smart grid is an illustration of the potential of HMI to revolutionize an industry by facilitating a two-way communication flow between consumers and utilities and enabling wiser management of energy. The system allows users to talk to their energy consumption habits in a complex manner while integrating green technologies such as solar and wind power. However, with this increased reliance on electronic communication comes with it the related security concerns, noted by the different types of cyber-attacks directed towards the availability and integrity of information in smart systems. Accountability in such systems is paramount, and this calls for an end-to-end approach towards cybersecurity that does not just incorporate component-based solutions but the entire system plan for its safeguarding. These are issues that must be addressed if the possibility of HMI to transform in smart technologies is to be realized to the fullest (Ghazi et al., 2018) (Figure 9.3).

9.1.3. Key Milestones in HMI Research and Development

The history of Human-Machine Interaction (HMI) has been characterized by various milestones that have revolutionized our conceptualization and use of technology in everyday life. One of the important milestones was the development of graphical user interfaces (GUIs) in the 1980s, which completely transformed the way users communicated with computers. GUIs substituted text-based command lines with icons, thereby making technology more accessible to the masses. This change allowed non-technical individuals to work with computers readily. The expansion of the personal computer during this time period played an integral role in making systems straightforward for users with a clear emphasis on ease of use and intuitive design. The implications of this shift went beyond being convenient, ushering in a new era of humans and machines in which machines began to cater to the needs and desires of individuals, setting the stage for future HMI innovations, including the use of touch screens and responsive designs. As HMI research went on into the 21st century, yet another milestone was achieved with the creation of mobile computing technologies. The proliferation of tablets and smartphones not only changed communication but also opened up a new paradigm in human-computer interaction. Touchscreens became ubiquitous, allowing for more physically engaging and immersive form of interaction, which was significantly different from traditional input mechanisms such as keyboards and mice. This interactive shift led to the creation of applications with user-centric designs, offering instant feedback and tailored experiences by app developers. The impact of mobile technologies on HMI also extends as far as how individuals communicate with not just particular applications but entire ecosystems like smart homes and IoT. These advances underscore the pivotal role that HMI plays in building an increasingly interconnected global world in which human beings and machines need to be able to communicate seamlessly in order to make productivity and interaction smoother. Moreover, recent advancements in HMI research bring to the forefront the capabilities of artificial intelligence (AI) and machine learning in creating adaptive systems that learn from user engagement. Through the integration of these technologies, machines have been capable of understanding and predicting user behavior, resulting in a more personalized experience. For instance, AI-based systems are increasingly becoming competent enough to adapt their functions based on user interests, thus making them more efficient and user-satisfying. This adaptability is a milestone in

HMI, the point where fixed interfaces give way to adaptive systems that evolve with user need. Furthermore, in their pursuit of these advances, researchers start to ask significant questions about the future of privacy, ethics, and human relationships with machines. Knowing these challenges is crucial for proper advancements of HMI, which, according to research on ICTs evolution, can dramatically modify the ways that society interacts and anticipates technology's contribution to our life (Canela et al., 2017; Martí et al., 2019).

9.2. TYPES OF HUMAN-MACHINE INTERFACES

The evolution of Human-Machine Interfaces (HMIs) has given rise to various types, each of which is aimed at supporting user interaction and experience. One of them is the Speech Recognition Interface, where users employ voice commands in interacting with machines. This interface relies on interaction between high-tech and human elements, thus making the process of interaction even more natural. By focusing on how users integrate their speech into dialogue, developers have developed systems such as interactive voice response programs used via telephone connections. Such systems provide an excellent illustration of the benefits of merging technological potential with human behavior knowledge and eventually to speech interfaces' acceptance and success in diverse applications, as described by Kloosterman et al. (1994). Increased improvement of such interfaces highlights the importance of user needs- and context-oriented design implications in order to make speech recognition technology easy as well as effective. Another revolutionary type of HMI is the Tangible User Interface (TUI), which bridges the distance between physical and digital interaction. TUIs take advantage of users' experience with physical objects to enable their interaction with digital systems in order to give a richer experience. This method is founded on cognitive science and psychology, as users are able to physically interact with representations of digital data. Although study on TUIs continues evolving, its potential to transform the interaction between data and users is promising. As Shaer et al. (2009) explains, creating TUIs requires a lot of knowledge of human rules of interaction and integration of technology, necessitating more empirical work to enhance their use. The inherent strengths of TUIs are that they are capable of delivering naturalistic interaction, and their vulnerabilities challenge innovative design methodologies to mitigate present-day shortcomings, laying the foundation for more effective HMIs of the future. As HMIs continue to become increasingly sophisticated, progress in interface design continues to reshape the user interaction landscape. In consideration of the diversity of needs in users, designers now look to hybrid interfaces that combine modes of interaction. For instance, combining touch, voice, and gesture input into one coherent system allows for a more adaptive and accommodating user experience. Such convergence provides seamless transition among interaction modalities, enhancing the overall efficiency of human-machine collaborations. With an emphasis on multi-faceted strategies, developers are seeking to create interfaces that are responsive in the true sense but also flexible to individual users' settings and preferences. This move towards more integrated strategy in Human-Machine Interaction recognizes the importance of user-centered design at its center. Hence, the future of HMIs will likely revolve around the integration of various styles of interaction into one cohesive framework so that sophisticated systems can be effortlessly handled by users (Figure 9.4).

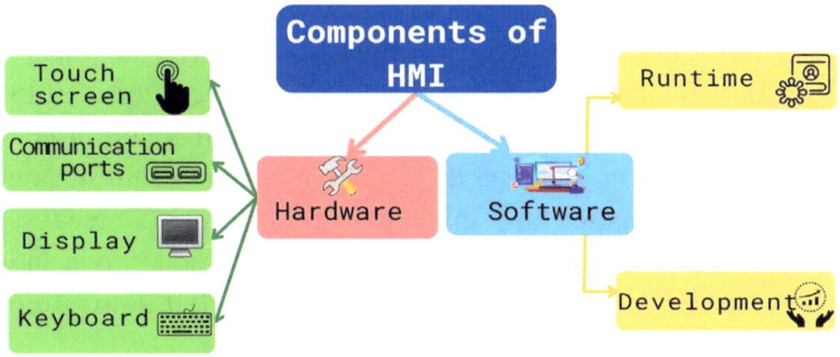

Figure 9.4. *Types of human-machine interfaces.*

Source: https://www.erp-information.com/wp-content/uploads/2022/09/trends-of-hmi.png.

9.2.1. Graphical User Interfaces (GUIs)

The evolution of graphical user interfaces (GUIs) has significantly altered the shape of human-machine interaction, simplifying technology to use and natural in its operation. Through the presentation of information in a visual state, GUIs enable users to communicate with computing systems without extensive technical training. Such a shift away from command-line to graphical interfaces has enabled computing to reach individuals of all spheres of society, facilitating anyone to operate superior technologies with ease. The use of windows, icons, and menus in GUIs not only enhances usability but also facilitates a more interactive user interface. Thus, GUIs form the backbone of modern-day HMI, outlining the way users interact with software applications and devices in commercial and personal life. In addition, the incorporation of Tangible User Interfaces (TUIs) into the field of GUIs is a turning point in HMI, bridging the gap between physical and virtual interaction. TUIs manipulate digital information with the use of physical objects, leveraging users' natural tendencies borrowed from the real world. This supplements the fundamental principles of GUIs by making it more interactive and enticing users on a multisensory level. As in the study of TUIs, extensive research has to be conducted to understand them fully and fine-tune interface design with empirical knowledge (Shaer et al., 2009). The evolution of GUIs thus does not merely reflect technological progress but also signifies a greater understanding of user cognition in interaction design. Aside from structural benefits of GUIs, there is a natural need to implement the cognitive aspects of the tools used in the interfaces. Cognitive artifacts become evident in such a setting, where functionality and design of GUIs influence users' thinking processes and choice-making. As underscored in present studies, understanding the afore-mentioned computer cognitive artifacts critically is essential in perceiving their effects on both user cognition and knowledge creation (Huggett et al., 2017). Examining GUIs from a cognitive influence point of view gives one a better understanding of how usability and general human-machine performance are influenced by design choices. Therefore, analysis of the user experience and cognitive process interaction is an important aspect of HMI studies that can offer insight for future GUI innovation.

9.2.2. Voice User Interfaces (VUIs)

The application of Voice User Interfaces (VUIs) within the context of Human-Machine Interaction (HMI) has transformed user interaction with technology. Leveraging natural language processing, VUIs offer free-hand, natural interaction for users to interact with devices ranging from smartphones to smart homes. The technology makes devices more accessible to disabled individuals so that they can perform tasks that might be challenging for them otherwise. Secondly, the efficiency that VUIs bring to carry out commands can make operations easier, thereby improving productivity in environments that span from workplaces to residences. However, with these advantages comes the problem of interface design that works well with a large variety of accents, dialects, and speaking styles, something that affects the user experience and accessibility. Thirdly, evaluating VUI needs a reconsideration of traditional quality review models. Since the classic measures of software quality may not be applied to the same degree in ubiquitous computing environments where VUIs operate, marking the challenge of measuring their performance (Hong et al., 2006), VUI-specific qualities such as responsiveness, accuracy, and user satisfaction must be determined and validated empirically. Creating a solid foundation serves to not just support the product evaluation but also inform the design process to implement improvements that meet user needs. With greater demand for more intelligent and responsive systems, it is essential to define these measures of quality in assisting in driving the development of VUIs as effective components of HMI. Ultimately, VUIs serve more functions than function; they engage users on an emotional and aesthetic plane with resultant increased levels of human to machine relationships. A focus on automotive HMI where switch design and tactile feedback are of key importance is used to demonstrate the influence that people's perceptions can have on their experience of technology (Pitts et al., 2008). The same principles apply to VUIs, with user satisfaction not only a function of function but also emotional attraction to interaction. Voice tonality, clarity, and even virtual assistant personality can have a strong impact on user preference and adoption. This integrated view of VUIs places them at the heart of the grand story of Human-Machine Interaction, inviting further study and creativity in this rapidly expanding field (Figure 9.5).

Figure 9.5. *Exploring voice user interface (VUIs).*

Source: https://www.calsoftinc.com/blogs/wp-content/uploads/2022/09/Voice-User-Interface.png.

9.2.3. Touch and Gesture-Based Interfaces

The evolution of human-machine interaction (HMI) continues to bring forward the significance of touch and gesture interfaces as a means of offering more intuitive forms of user to digital system communication. Such interfaces leverage the inherent human nature to use touch and gestures and thus make interaction more natural as well as remove the burden of cognitive load normally associated with traditional input devices. For instance, the introduction of devices like 3DTouch illustrates technological advancement that puts emphasis on user convenience without any compromise on efficiency. With touch input for various manipulation tasks, 3DTouch enhances users experience through passive haptic feedback and thus facilitates prolonged usage without excessive fatigue. Therefore, the device illustrates how thoughtful design of gesture interaction can fill significant gaps in today's technologies, and subsequently, will result in more powerful HMI applications across different contexts and environments (Banic et al., 2015). In addition, the touch and gesture-based interface effect propagates even to more advanced systems such as DP systems for maritime shipping. These usability problems with legacy control interfaces can prevent the full exploitation of operational effectiveness and also pose safety risks. Recent research in this field has suggested that blending consumer technology multi-touch and hand gestures can significantly reduce interaction complexity. User studies of prototypes demonstrated that users performed tasks more quickly using direct gestures rather than by using legacy button and menu interfaces. These findings support the potential for gesture manipulation to lead to enhanced user control and lessened cognitive load, paving the way for more efficient and secure operations in case of critical tasks in maritime applications (Bjorneseth et al., 2010). Ultimately, integrating touch and gesture-based interfaces with HMI represents a paradigm shift and encourages creativity in terms of functionality and design. As researchers and developers continue to explore these technologies, the design space expands, offering potential for new interaction methods that address human needs. Continued innovation of tools like 3DTouch and their applications across various platforms can revolutionize how people interact with technology, making interfaces more interactive and responsive. As HMI develops, the ongoing debate regarding usability and user experience will remain equally important to ensure that gesture-based technology development keeps pace with user requirements and operation requirements. This alignment will be crucial to deliver more intimacy and efficiency in human-machine interaction, from common consumer equipment to dedicated industrial applications (Banic et al., 2015; Bjorneseth et al., 2010).

9.3. THE ROLE OF ARTIFICIAL INTELLIGENCE IN HMI

The application of artificial intelligence (AI) in Human-Machine Interaction (HMI) has transformed the world of user experience so that machines can communicate with humans more responsively and intuitively. AI-driven systems employ high-end algorithms that personalize interactions by learning user habits and interests. This adaptability is particularly important in applications such as voice recognition and smart assistants, where the ability to read context and nuance is key to effective communication. For instance, by means of machine learning algorithms, these systems can process enormous volumes of data in an effort to become more responsive and accurate over time. Furthermore, the presence of smart feedback mechanisms creates a sense of control for users and therefore is more satisfied with the technology, a paradigm shifts in the way we interact with machines at home. Furthermore, the application of machine learning (ML) within HMI systems has profound implications for efficiency and capability across a range of areas such as telecommunications and smart devices.

Figure 9.6. *The role of AI in connected automation platforms.*

Source: https://www.plantengineering.com/wp-content/uploads/2025/04/PLE2504_MAG_AI_EMERSON_ASPEN_Fig1-DV_GRF_Cloud-Powered-Ai_1-slider.png.

In next-generation wireless networks, as highlighted in more recent research, ML algorithms are central to enhancing communications via the facilitation of advanced applications and ensuring trustworthy information exchange. For example, the powerful capabilities of ML enable systems to optimally perform adaptively according to real-time data analysis so that they still offer users low-latency and high-rate services. This is particularly worthwhile in heterogeneous networks where unbroken interaction among various devices and platforms is required. With greater levels of adaptability and dependability, AI-driven HMI solutions can better facilitate the demands of increasingly demanding environments, paving the way to innovation in user interaction and machine control. The future with 6G wireless networks on the horizon, the part of AI in HMI is going to become even more important. These next-generation networks will not just need fast connectivity but also the ability to accommodate heterogeneous applications with zero-latency responsiveness requirements. One foundation for such development lies in AIs capability to revolutionize network operation through zero-touch optimization and intelligent decision-making engines, as evident in papers centered around modern telecommunications. Leveraging sophisticated ML models, networks gain the ability to learn resource allocation, manage network traffic, and optimize service delivery. This harmony between HMI and AI not only predicts the needs of users but also adjusts to new trends, so that human interaction with machines keeps improving. Finally, AIs impact on HMI is nothing short of a fundamental transformation toward an integrated and more efficient technological environment wherein people can interact with devices that not only react but also predict their needs in a seamless manner (Figure 9.6).

9.3.1. AI-Driven Personalization in User Experiences

The advent of artificial intelligence (AI) has transformed user experience design in ways that enable personalization on an unimagined scale. This change is most strikingly seen within the domain of Human-Machine Interaction (HMI), where AI facilitates personalized user interfaces and interactions. Depending on user behavior and interests, AI systems can dynamically modify content in real time, thereby enhancing usability and satisfaction. This level of personalization not only fosters more active participation but also signals a shift towards a more user-focused way of design. Notably, however, attention needs to be paid to the interlinkage of personalization and user privacy. As AI systems are gathering significant amounts of data with the intent to make well-informed recommendations, users may be concerned about the effects of their data being utilized, emphasizing the necessity of transparency and user consent to build confidence in AI-driven systems. The merging of sensory stimulation within AI-powered personalization is an emerging area of study. Merging auditory with touch stimuli is one that promises to make user experiences more immersive and tailored to personal preferences. An example would be the addition of combined auditory-tactile stimulation, which could improve user well-being, having a positive impact on mental states by attentive technology use. This innovation aligns with growing evidence that a multi-sensory approach to HMI would entail great therapeutic benefits, eliciting more user engagement with devices. By resolving the research questions involved in such integration, such as the knowledge of the mechanisms of human perception of sensory stimuli, the developers can lay the platform for technologies that not only meet functional specifications but also enable the overall well-being and health, as suggested by the research break-throughs in the sector (Dijk et al., 2010). Moreover, the Quantified Self (QS) movement illustrates the immense potential of AI-based personalization in rendering everyday interactions with technology more rewarding. Using various tracking devices to track personal data, users gain knowledge about their activities and health parameters, enabling them to become more conscious of their lifestyle. However, the problem is that the tools have to be designed to process and display personal data well so that they can assist people to some extent. Next-generation QS systems will focus more on intuitive interfaces that simplify data interpretation, bringing it within reach for the non-expert user. Breakthroughs in this area would include leveraging AI capabilities to convert raw data into usable knowledge, increasing user control over lifestyle choices. By linking data complexity management and human experience, these innovations in HMI could yield a more effective and educated user base while maintaining privacy and ethical concerns, as represented in existing scholarly debates (Cena et al., 2015).

9.3.2. Machine Learning Algorithms Enhancing Interaction

Machine learning algorithm evolution is at the core of enhanced human-machine interaction (HMI). The algorithms facilitate the development of systems that can learn user behavior and preferences, ultimately paving the way for more intuitive interfaces. For instance, intelligent systems can learn user behavior to identify patterns, thus personalizing responses and suggestions to individual needs. This adaptability also minimizes the friction normally experienced by users and therefore increases the overall usability experience. With continuous learning from user input and feedback, these systems can continually evolve and evolve further into the nuances of human behavior. As a result, people are engaging with computers that not only respond but also act pre-emptively to assist with tasks and provide context-specific information, contributing to an environment for more productive and engaging interaction. In the medical context specifically, machine learning algorithms will further enhance interaction, as described by the vision of the Bionic Radiologist. Through methodically aggregating gigantic amounts of medical imaging information, such algorithms allow radiologists to make informed decisions that reflect probabilities of disease and adhere to patient values. Through integration, reliable image analysis conducted at scale can be assured, and treatment recommendations created and dispatched in timely fashion. In addition, data aggregation allows algorithms to discover time-critical patterns that avoid care misses. The Bionic Radiologist is a holistic approach to interpreting information that finally frees professionals to engage more intimately with patients and referring clinicians, hence humanizing the health process without compromising clinical accuracy and effectiveness (Dewey et al., 2019). Another shot in the arm of machine learning algorithm capacity, the inclusion of Knowledge Graphs (KG) introduces complementary, real-world data that can complement training of the systems. By leveraging KGs, machine learning systems such as recommendation systems and community detection can improve in terms of accuracy and interpretability in their outputs. These KGs assist in supplementing inadequate labeled data through giving additional context where user interactions occur. Recommendation systems, for example, can leverage background knowledge from KGs to offer users customized content that is more relevant to them. This not only augments the interaction but also instills faith in the system since users believe that there is improved comprehension of their needs. Lastly, the integration of machine learning and KGs illustrates a stunning enhancement of HMI, crafting interactions that not only work more efficiently but are more personally responsive to user context and preference (Bhatt et al., 2020) (Figure 9.7).

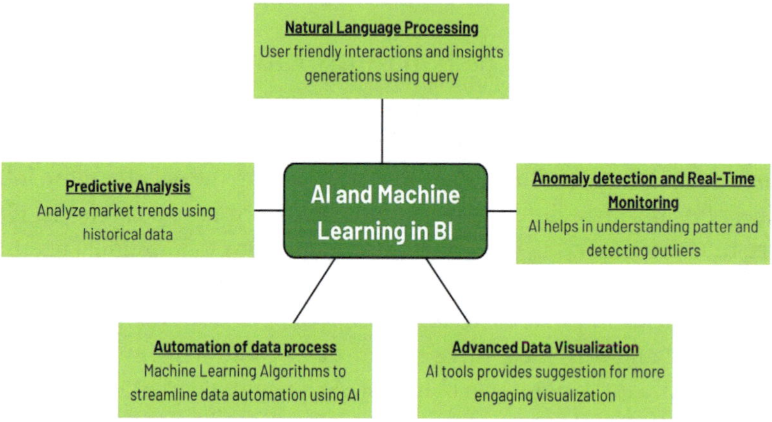

Figure 9.7. *Five key points on how AI and ML enhance business intelligence (BI).*

Source: https://saplingfinancial.com/wp-content/uploads/2024/10/AI-and-Machine-Learning-in-BI.jpg.

9.3.3. Ethical considerations in AI and HMI

In exploring the ethical dimensions of artificial intelligence (AI) in the contexts of Human-Machine Interaction (HMI), it is crucial to see how the intricate dynamics of man and technology are established. With the introduction of AI systems, industries and societies have been transformed, and therefore accountability, decision-making, and the implications of autonomous action have come into question. As machines are taking on major roles, especially in sensitive spheres like healthcare and the food supply chain, implications for ethics cannot be ignored. The importance of embracing reflexive governance models to steer these waters has been stressed by researchers, engaging stakeholders and facilitating openness in data sharing and AI use. This integrated approach aims to ensure that ethical aspects are considered in the context of the entire as a whole, thus bringing about a deeper understanding of the interconnected ethical concerns of human-technology interactions that render our technoscape enormously profound (Brewer et al., 2023). Ethical principles should also be incorporated into AI system design from their inception so as to minimize adverse impacts. Value-Sensitive Design (VSD) is one potential approach to starting this endeavor-by prioritizing the incorporation of stakeholders' values in the earliest phases of AI design. Including ethical issues in the design process will guarantee that risks at development phases such as unintended bias, transparency issues, and user privacy are alleviated. This forward-looking response not only enhances the utility of smart agents, but also induces trust among users, such that technology would be operating in the best interest of humanity. The greater reliance of society on AI makes the implementation of VSD a necessity. Developers, policymakers, and users, amongst stakeholders, must collaborate in an effort to overcome the complexities of ethical design in creating systems that represent shared values and aspirations, as well as addressing the ethical issues that accompany HMI (Bellis et al., 2018). Ethical considerations of HMI are not just limited to technical design but also extend to high-level societal issues. Issues such as job substitution, equality of access to technology, and the potential misuse of AI are serious questions that must be critically analyzed for their moral dimensions. The understanding of these dimensions requires a broad approach, one that entails not just scholarly debate but also public debate and policy formulation. As AI technologies continue to advance, so too must our ways of addressing ethical concerns. Incentives that encourage inclusive dialogue and flexible regulatory approaches can address risks while leveraging the opportunities of emerging technologies. Most importantly, an ethical culture surrounding AI and HMI is necessary to manage the intricacies of human-technology interdependence and ensure fairness for all parties concerned (Brewer et al., 2023).

9.4. USABILITY AND USER EXPERIENCE IN HMI

The intersection of user experience and usability in Human-Machine Interaction (HMI) is crucial for ensuring that users can interact effectively with complex systems. Modern interfaces must not only meet functional requirements but also relate to the user's emotional condition. Awareness of this has resulted in a move towards the new usability, where the case is presented that tried methods may be insufficient in responding to current user demands. As digital products grow more advanced, their user interactions become increasingly complex and require designers to adopt a more advanced usability strategy that considers diverse user environments. Breakthroughs such as adaptive interfaces and individualized user experiences are at the heart of usability enhancement, culminating ultimately in increased participation and enjoyment. By focusing on the emotional and functional aspects of user interaction, designers can engineer systems that meet the needs of the users as well as engage their expectations and maximize overall user experience. In the area of HMI, how successful engagement strategies turn out can be highly dependent on the cultural quality of usability testing and resultant design amendments. For instance, studies indicate that a novel vision-based CAPTCHA system, CAPTCHaStar, is more usable than traditional text-based CAPTCHAs, which are frustrating and difficult for users to fill out. Utilizing user interaction to solve shapes in a dense visual image, the new system is more usable without compromising resistance to automated attacks. The user-facing nature of such innovations reflects a broader movement across HMI, in which user behavior and preference are afforded increased consideration. Consequently, the use of disciplined usability testing—such as that integrated into CAPTCHaStars development—ensures that systems not only protect against exploitation but also provide an uninterrupted experience to users, highlighting the benefit in applying usability principles to design. Effective user experience in HMI also calls for ongoing reconsideration of design approaches, as emerging technologies present new issues and prospects. The concept of the new usability is that the environments of digital interaction are dynamic and change constantly, often making settled-down design ideas obsolete. This paper presumes that the challenges of usability in today's world call for new strategies in usability engineering and assessment. By identifying the distinct needs of digital consumers, researchers and designers can formulate an innovative research agenda to address current gaps in usability practice. This anticipatory approach is imperative in developing HMI systems not only to be user-centric but to enhance overall experience by addressing evolving expectations of technology users. In this view, we can see that the future of HMI is comprised of responsive and adaptive design methodologies with an emphasis on sustained user engagement and satisfaction to create more intuitive and rewarding human-machine interaction (Figure 9.8).

Figure 9.8. *User experience in autonomous cars.*

Source: https://d23rjziej2pu9i.cloudfront.net/wp-content/uploads/2022/06/Blog/Bringing_digital_life_of_the_user_into_the_car.png.

9.4.1. Principles of Effective Interface Design

Successful design of interfaces is vital in enhancing human-machine interaction (HMI), as it lays the foundation for frictionless communication between technology and people. At the forefront of this is the principle of usability, which encourages the creation of intuitive and seamless interfaces. A simple-to-use interface reduces cognitive load, enabling users to communicate with systems readily without needing much training. Therefore, the major aspects like clarity, consistency, and feedback are becoming primary in the process of guiding users through their procedures. Moreover, using popular design conventions can cut the new user learning curve significantly, enabling a healthier environment supporting a wider audience. Here, the inclusion of human-computer interaction (HCI) principles will make interfaces more usable as a whole and reduce the gap between technological sophistication and user friendliness, ultimately resulting in user satisfaction and productivity. The flexibility of interfaces is another fundamental principle that interfaces must adhere to, one that enables them to accommodate a very diverse set of users and situations. This flexibility allows varying interaction modes and preferences to provide customized experience according to needs of individuals. Successful interface design has to implement this principle through offering customization options, allowing the user to customize layouts, color schemes, and functions according to his or her preferences. Further, successful implementation of adaptive technologies, such as gaze tracking or voice control, can increase accessibility for individuals with disabilities. Being a part of ongoing user feedback loops is important in order to continually refine and improve interfaces, so they remain relevant and functional in an ever-evolving ocean of technology. By emphasizing flexibility, designers can create responsive interfaces that adjust as user needs and technological advancements evolve, thus maximizing total HMI. Lastly, the integration of empirical evaluation techniques into the design process is a foundation of effective interface design. User interface testing with real-world user input can determine their usability and efficiency. This perpetual cycle of designing, testing, and revising is important to establish potential usability issues before implementation. With structured evaluation methods, designers can get actionable feedback that informs upcoming iterations with user experience in mind the entire time. In the instance of new applications such as brain-computer interfaces, where stakes are extremely high in terms of user experience, the need for robust evaluation is even more critical. Such methodologies will strive towards making better interactive systems, creating a more seamless integration of human abilities and machine capabilities, ultimately leading to enhanced overall user experience in various HMI applications (Baker et al., 1987; Cincott et al., 2010).

9.4.2. Importance of User Feedback in HMI Development

User input is crucial during the development of Human-Machine Interfaces (HMIs), impacting both user experience and usability. Though designers will always attempt to design to accommodate a broad set of user needs, input from actual users becomes increasingly important. Feedback mechanisms allow developers to get feedback on how the users are interacting with the interface, where users get lost or are less than optimal.

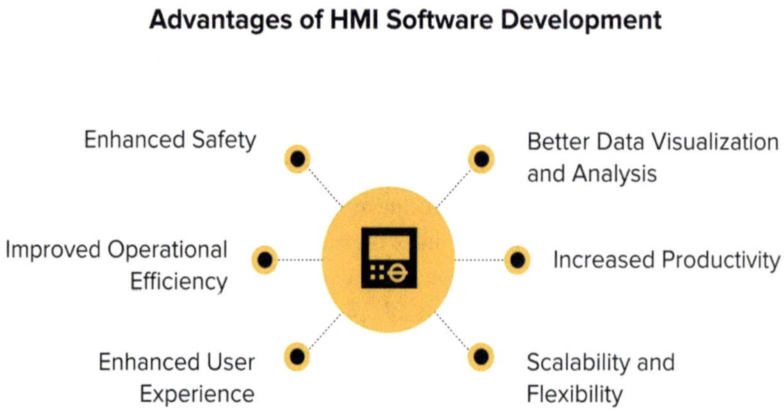

Figure 9.9. *HMI software development.*

Source: https://appinventiv.com/wp-content/uploads/2024/09/HMI-Software-Development-Benefits-Process-Features-Costs-and-More-8.webp.

For instance, the application of ergonomic principles can lead to increases in user comfort and task performance. In the design of low-carbon automobiles, the interaction between user behavior and interface design is critical. Through the utilization of user input, designers can create interfaces that promote ecologically friendly driving practices and decrease cognitive load and distraction and, as a result, yield safer, more efficient driving (Harvey et al., 2013). User input is just as important in dynamic systems, like those used in ship navigation. The character of Dynamic Positioning (DP) systems dictates that the operator experience be untroubled to ensure safety and effectiveness. Because the operators' sole link with the system is through the interface, it is important that the design be different from conventional practice and provide user choice. User feedback sessions will define users' patterns of interaction so that designers can integrate natural motions and intuitive controls. Not only does this enhance usability but also develops more transparent interactions, which are necessary in high-risk situations. Technologies such as multi-touch gestures, fueled by consumer-grade technology, can be tested and refined based on user feedback, ultimately being converted into staggering operation performance gains during typical and critical operations (Bjorneseth et al., 2010). Ultimately, the influence of user input on HMI design transcends functionality; it provides a human-centered design philosophy that endows technology with human capabilities. Engaging the users in the process assures that ensuing interfaces are actually usable and intuitive. Secondly, the iterative nature of the feedback loop allows for the potential for continuous improvement and refinements, which is essential in an ever-evolving technological world. The challenge facing designers—whether in automobile interfaces or complex ship systems—is to better understand human needs in creating engaging, effective, and safe human-machine interaction. As the landscape of HMI development continues to change, prioritizing user feedback will continue to be a mainstay of good design practices (Figure 9.9).

9.4.3. Case Studies of Successful HMI Implementations

Efficient implementation of Human-Machine Interaction (HMI) can be exemplified through case studies that denote robotics and user-oriented technology improvements. A cited example is the use of human demonstrations for robotic system training to carry out object handling tasks. In a study presented at the 2019 ACM/IEEE International Conference on Human-Robot Interaction, a natural input technique known as instrumented tongs was developed. The method extended the input method design space and facilitated effective human-robot demonstration by improving the quality and experience of the user. Unlike the traditional methods of free-hand manipulation and kinesthetic teaching, instrumented tongs articulated a significant achievement in the capability of the demonstrators in conveying complicated procedures, illustrating how feedback from the users can calibrate HMI systems for enhanced effectiveness and naturalness of interaction (Gleicher et al., 2019). Similarly, mobile innovation in technology is another instance of successful HMI implementation that has brought smartphones to intelligent status where they can understand and predict user needs. The latest trends in mobile sensing have equipped phones with sophisticated sensors and machine learning capabilities, which enable them to make inferences such as location and social context. This evolution from mere communication devices to anticipatory devices is a success story in implementing HMI. With the support of anticipation-based mobile computing, not just are these machines capable of predicting user intention but also responding to it within a feedback loop that is mutually rewarding for user experience. This integration of context awareness into mobile applications underlines the importance of HMI in anticipation of the user's needs and giving seamless interaction between man and machine (Musolesi et al., 2015). The implications of these successful HMI case studies extend well beyond mere operational efficiency; they revolutionize our understanding of user interaction with technology. In robotics, human-robot interaction success leads to collaboration, enabling robots to be efficacious co-workers in production, healthcare, and education industries. The ability to represent complex tasks in terms of advanced input modes can significantly eliminate the skill gap that normally occurs in human-robot collaboration. Concurrently, in mobile technology, the shift towards anticipatory systems has redefined consumer expectations and thereby spawned a demand for smarter and more responsive devices. Collectively, these case studies demonstrate the transformative power of HMI and assert its central role in technology and its uses in everyday human practice.

9.5. CHALLENGES IN HUMAN-MACHINE INTERACTION

Expansion of technology into daily life emphasizes resolving Human-Machine Interaction (HMI) problems, particularly in interface design that supports diverse user needs. One such field is the needs of ageing users, who might experience monolithic challenges while interacting with modern computing technology. Designing interfaces that support cognitive, visual, hearing, and motor impairments is key in designing accessible technology. User-specific adjustment facilitates individual adaptations to enhance usability and quality of life in the elderly. With the use of adaptive systems that cater to these specific challenges, the HMI can become more efficient and beneficial to all, especially those who have historically struggled to integrate new technology (Ahamed et al., 2013). Without such adaptations, many elderly individuals would be likely to be frustrated and excluded from technologies that could otherwise make everyday life easier. One of the central challenges in the field of HMI is to make non-designers into capable interaction designers. As design schools come to realize the value of design thinking, challenges are issued to enable students with no training at all in design to learn about principles of interaction design. Challenges such as a lack of awareness of design principles, psychological resistance to adopting a designerly mindset, and practice-based difficulties in application can hinder this transformative journey. Problem-orientated educational models that focus on these can reduce these challenges and provide a conducive learning environment. By embracing methods that emphasize practical exposure and teamwork, institutions are in a position to optimize opportunities of success for future interaction designers (Siegel et al., 2009). Solving these issues not only facilitates individuals to acquire essential skills but also helps in the broader discipline of HMI in making methods and perspectives richer in diversity in the practice of designing. Moreover, the interaction between technological development and flexibility of use provides another layer of complexity in HMI. When computers become more intelligent, the challenge is to ensure that users can communicate and take advantage of the abilities of such machines. Miscommunication between user intention and machine capability can result in annoyance, erosion of trust, and rejection of technology altogether. This is best illustrated in fields such as virtual assistants and autonomous systems, where fine nuances of human communication may be misinterpreted or imperfectly addressed by machines. In order to bridge this gap, designers must focus more on intuitive interfaces that accommodate multiple communication styles and incorporate robust feedback mechanisms to allow users to feel more in control and listened to. Through breaking down these communication barriers, HMI development can lead to smoother human-machine interaction, eventually contributing to greater user satisfaction and engagement in technology-rich spaces (Figure 9.10).

Figure 9.10. *Challenges in human-robot interaction.*

Source: https://media.licdn.com/dms/image/v2/D5612AQHY0mGKY2JgUw/article-cover_image-shrink_720_1280/B56ZcpD5w_HgAI-/0/1748740583160?e=2147483647&v=beta&t=UeYw0PBwo0dlmaZHpOIZVh395kdLC4vtWaxCd472xCg.

9.5.1. Technical Limitations and Barriers

Human and machine interaction (HMI) is often hindered by significant technical barriers and hurdles sourced from the regulatory and infrastructural policies in use. One of the primary hurdles is in the complexity of adhering to evolving rules, particularly where medical devices are concerned. The dominant mixture of state-led and autonomous traditions influencing the implementation of European Union directives to medical technology is pivotal. As has been observed by Altenstetter et al. (1995), the introduction of new rules for governing the operation of medical devices does not always introduce institutional capacities and routines. That is, local medical customs will continue to play a role in the man-machine interface such that interactions are uneven and do not adhere to universal standards. This disparity can nullify the global success of HMI systems since they are unable to accommodate diverse practices in different regional settings. Another significant hurdle is the technological limitations built into the machines and platforms that facilitate human-machine interaction. Current manufacturing environments are shifting towards sustainable practices, but this process is detrimental. As outlined by Casta et al. (2019), while the industry is working hard to find solutions for increased energy efficiency and reducing resource consumption, existing control systems and management strategies may not be ready to seamlessly incorporate these innovations. The result is a disconnection whereby new technologies are rendered useless by inadequate infrastructure or incompatibility. This conflict can delay the evolution of HMI technologies, forcing actors to remain in a universe of old-fashioned systems which cannot cope with the demands of contemporary operational needs. Furthermore, technological barriers to effective HMI extend to the realms of user interface design as well as information processing capability. Human users are likely to be challenged when interacting with complex machines, which may not have been designed with user-friendliness and natural simplicity in their conception. Restricted interoperability among heterogeneous machines imposes a greater cognitive load upon users, increasing the complexity of their decision-making at pivotal instances. This is then added to by limited empirical evidence regarding the performance of various HMI methods due to the fact that the field of HMI continues to evolve without a substantial amount of research to guide development. As the example of regulation and technological adjustment indicates, the institutional setup and working habits need to align closely in order to overcome these barriers and enhance the human-machine interface in a more automated world.

9.5.2. User Resistance to New Technologies

User resistance to change through new technology is a key challenge in the human-machine interaction (HMI) field. It most often stems from deeply rooted distrust regarding the capabilities and performance of new systems by users with knowledge of existing workflows. When confronted with innovation in the shape of advanced gesture-based interfaces or automated check-in procedures, users can react with suspicion based on usability, effectiveness, and technological breakdown issues. The challenge is even more pronounced in terminals like airports, where work pressure and user stress can accelerate resistance, ultimately impacting user experience and satisfaction in general. It is necessary to grasp these dynamics in a way that they guarantee the proper adoption of new technologies, as developers can modify solutions that solve people's issues and enable simpler transitions to new systems. Technology development aimed at enhancing interaction, like gesture interfaces, exhibits possibilities and limitations within users' acceptance. For instance, Carmein et al. (2014) states that gesture and natural language interfaces reduce cyber-sickness when used in immersive settings, but users will be resistant to using such interfaces since they are new and are felt to have a learning curve. As the study states, the interface can indeed provide a more comfortable and more efficient experience to users, but this is achieved after some time and some brain power in the initial stages of adjustment. The reasons for resistance are typically founded in fear of the new and uncertainty regarding changes of tried patterns. It is thus necessary to have good training and stakeholder involvement to overcome resistance, inducing a culture of technology acceptance while enhancing HMI. Resistance is most evident in cases where convenience forms a critical aspect, an example being airport operations. As Grone et al. (2019) points out, the increasing volume of travelers facing long check-in processes and security checks reveals the necessity for technical realignment. Still, users tend to react to such innovations in disbelief, fearing that new systems will exacerbate existing inefficiencies rather than alleviate them. By understanding these points of pain and the necessity for intuitive HMI design, developers can better anticipate users' needs and wants, creating a smoother integration of new technology. On top of that, overcoming resistance through purposeful communication and demonstrating effectiveness can be an important consideration in driving adoption and creating smoother interaction with new technological solutions in high-stress environments (Figure 9.11).

Figure 9.11. *HMI improving user interaction in control systems.*

Source: https://static.blikai.com/prod/20240709/Human-machine-interface-(HMI)%20(1).webp.

9.5.3. Privacy and Security Concerns

The development of human-machine interaction (HMI) has necessarily put issues of privacy and security at the forefront of contemporary debate. With machines increasingly employed in daily life through vehicles such as smartphones and smart home automation, data acquisition and application have grown exponentially while user consent and control issues remain unresolved. This results in the users unknowingly granting consent for the harvesting of their information, which the companies may abuse or let the cybercrime group do so also. Technology further compounds the problem by advancing quicker than law enforcement ever can. Machines are collecting information by engaging with the users, so transparent privacy policies become a huge problem. The interplay between user trust and technology needs to be re-examined in the context of how privacy is managed in HMI to ensure that a safe and respectful experience is provided to the user. Furthermore, the emergence of Cyber-Physical Social Machines highlights the complexities in the human-machine relationship when it comes to privacy and security. These systems are the intersection of human aggregations and networked digital infrastructures, leading to new socio-technical environments and real-world practices. Because such machines have IoT features, they present greater risks due to their expansive sensing and actuating capabilities. Such complexity surpasses conventional comprehension of security due to difficulty in ascertaining the amount of data gathered and protection measures for such data. In a part of such a system, vulnerabilities can ascertain the integrity of the network, and therefore there is a need for advanced security measures. An understanding of these systems' development is essential for the administration of the privacy and security threats they pose to users in HMI environments (Creese et al., 2018). The proliferation of Ambient Intelligence (AmI) and ubiquitous computing makes privacy and security issues in HMI even more challenging. The technologies foretell a future in which computation seeps naturally into all facets of daily social interaction, making the analysis of data handling practice even more necessary. Recent studies on users' attitudes towards AmI scenarios illustrate elevated concern with the management of personal information, particularly among sectors such as healthcare and finance. The ability to trust is essential for users to embrace such systems, highlighting the need for robust data breach defense mechanisms. As individuals interact with AmI systems, they often have to make complex privacy trade-offs, and this leads to an overall concern about what becomes of their information and how it gets used and distributed. These issues need to be addressed in order to establish trust and ensure the users can securely interact with the possible benefits of HMI without compromising their privacy (Briggs et al., 2006).

9.6. FUTURE TRENDS IN HMI

The evolution of Human-Machine Interaction (HMI) is becoming increasingly involved with the issue of design ethics and responsibility. As technology advances, good practice in responsible design is being given greater importance in the HMI field. A workshop that explored future education environments identified the importance of instilling moral responsibilities in designers, particularly at formative stages of the design process. This is reflective of a wider trend which puts ethical considerations at the forefront of HMI innovations. Practitioners are seeking to incorporate technology into everyday life, and consequently, they have to come to terms with the effects of their design on society and users. By reflecting on the ethical dimensions of technology, designers can design more user-focused and inclusive machine interactions, ensuring that future HMI developments enhance user experience while promoting social responsibility and design accountability (Abbey et al., 2019). The other significant trend of HMI is rising reliance on science fiction as an inspiration for pioneering research and design. The appeal of sci-fi narratives comes from their visionary paradigms that challenge current technological paradigms, pushing researchers to explore uncharted territories in HMI. A recent review of the CHI proceedings showed how science fiction metaphors have populated some of HCI research areas, such as Human-Robot Interaction and Human-Body Modification. In these endeavors, scientists are not merely taking inspiration from science fiction narratives; they are using them to speculate about future human interfaces and experiences with technology. This cross-relevance of HCI research and science fiction means that as technology advances, the boundary between imagination and fact can begin to grow indistinct, encouraging the development of new systems of HMI that consider innovative concepts from sci-fi tales and films (Hemminger et al., 2018). Finally, AR/VR technologies are on the verge of transforming HMI, moving towards more intelligent and richer user interaction. Both these technologies allow for improved interactivity and immersion and change the very nature of user interaction with machines. The transition to incorporating AR and VR into ordinary applications is a sign of the direction toward leaving behind conventional interfaces and migrating to environments in which users can navigate and engage with digital elements naturally. With these technologies advancing, they have the potential to shape more adaptive and responsive interfaces for users. This will likely promote deeper emotional interactions between humans and machines, in effect altering how people feel, experience, and relate to technology in their daily lives. The future of HMI thus promises not only enhanced efficiency but also richer human experiences through paradigm-shifting technological engagements (Figure 9.12).

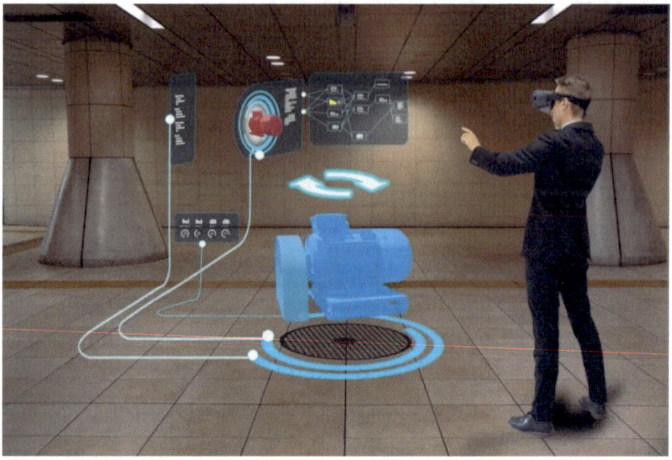

Figure 9.12. *HMI trends engineers.*

Source: https://res.cloudinary.com/rs-designspark-live/image/upload/c_limit,w_829/f_auto/v1/article/shutterstock_1240268662_e8f48f-be99a84a19da41aa6519800d5b8d7c7f7d.

9.6.1. Advancements in Virtual and Augmented Reality

The recent advances in virtual reality (VR) and augmented reality (AR) are transforming the landscape of human-machine interaction (HMI). With increasing computational power and top-level software development, VR and AR applications are now shattering traditional bounds, delivering innovative experiences across multiple industries. In educational environments, for instance, the use of AR technology has revolutionized the way students engage with content, uniting virtual and real-world attributes in a way that amplifies learning gain. Such an interaction between learners and technology demonstrates how such technologies can handle many forms of learning, increasing student engagement and information retention. Thus, the evolution of VR and AR not only improves user experience but also pinpoints the importance of intuitive and flexible interfaces that address human capabilities and limitations. In addition, the evolution of BCIs is a major aspect that facilitates maximum HMI via VR and AR technology. BCIs seek to address the human cognition-technology gap and facilitate a new way of interaction. These interfaces, which are being developed to interface with neural signals, have the potential to enable users to control machines and software with thoughts. As offered by Kerick et al. (2012), the applications of BCIs vary from assisting individuals with severe disabilities to enhancing cognitive training for healthy participants. By providing entry into brain activities, such technologies have the ability to create more natural and potent interactions, transforming how humans engage and experience the world around them. The future of HMI will thus be dominated by the seamless incorporation of brain activities in VR and AR environments. Additionally, empirical findings establish the efficacy of AR in enhancing user experiences in learning environments, with evidence of clear effects on interactions. A newer study equated the effectiveness of AR and video presentation modes in learning settings, displaying complex differences in learner engagement as a function of levels of understanding among subjects. No significant difference in performance between lesson modes was demonstrated by the findings, although they did stress that students with varying reading comprehension levels had varying participation results, particularly for the AR condition (Mart et al., 2020). This is to underscore the significance of personalized interactions in learning settings, where individual user experience may be significantly enhanced by leveraging AR. As emerging VR and AR technologies become ever more sophisticated, their implications for HMI extend beyond mere function to facilitate personalized and adaptive user experiences that dynamically respond to the individual skills and learning needs (Figure 9.13).

Figure 9.13. *The future of augmented reality.*

Source: https://www.reydar.com/wp-content/uploads/2023/02/AdobeStock_159799700-scaled.jpeg.

9.6.2. The Rise of Brain-Computer Interfaces

The emergence of brain-computer interfaces (BCIs) is a significant step in human-machine interaction (HMI), an era that has brought about a never-before-seen array of opportunities for direct communication between the brain and external devices. This revolution is underpinned by technological and neuroscience advances, which enabled the development of systems to interpret neural signals and translate them into implementable commands. These interfaces hold the potential to advance numerous fields, including medicine, gaming, and assistive technology, and consequently, blur the gap between human intellectual potential and machine effectiveness. With researchers probing the potential of BCIs, it is evident that these technologies hold the ability to redefine the mode of user interaction with machines entirely, allowing a more profound and intuitive experience with the potential to redefine the HMI potential. In addition, the application of BCIs lends itself to important issues about the naturalness of human-machine interaction. Studies on interaction salience that have been prioritized in recent research suggest that developing an understanding of user engagement's qualitative aspects is central to creating effective BCI systems. Studies of driver-car interaction suggest that more physical and analogue forms of control allow for a perception of naturalness to be transferred into BCI contexts as well. By focusing on human-like behaviors and mental abstractions, the development community stands to create BCIs that not only respond to commands but also anticipate users' needs, for a more empathetic union between machine response and human thinking. User experience and overall design of the interface have critical implications that require close examination. In addition to the technical challenges posed by BCIs, their cognitive and epistemic dimensions also need to be explored. To grasp how BCIs modify cognition, one must adopt situated and ecological conceptions of cognition, which emphasize the dynamism between individuals and their worlds. This perspective highlights the interconnected nature of human thinking and machine behavior, envisioning our cognitive processes as considerably relied on by the interactive nature provided through BCIs. With the increasing integration into everyday life, these technologies not only bring enhancements in communication but also a deeper reworking of our cognitive processes. By expanding the scope of BCI research, we are able to enhance our understanding of its probable implications for group cognition, and the human-machine cooperation need to decide on the direction of HMI.

9.6.3. Predictions for the Next Decade in HMI

As the decade goes on, one of the overriding trends for the HMI market will be the integration of input from more than one technology to enhance the way that machines understand and respond to human input. With the application of advanced computational methods, including machine learning and artificial intelligence, systems will increasingly be able to interpret rich behavior and biological signals. This capability would lead to richer user experiences in health settings, where technology is tailored to individual patients' needs through combined data processing on health and lifestyle. As argued in recent debates regarding the future of technology, the operation of such systems will be determined by their ability to process and manage large volumes of heterogeneous data, generating more personalized and efficient human-machine interactions (Goldenberg et al., 2018). Another sector that will undergo a highly important alteration is the Internet of Things (IoT), whose contribution to the development of HMI is of significant importance. As more and more devices become connected, the devices will have to talk to each other seamlessly and autonomously, remapping the landscape of interaction from human-initiated to automatic decision-making. Not only will this make quotidian tasks even simpler but also raise some essential questions regarding privacy, safety, and ethical application of data gathered through these ubiquitous networks. Experts predict that with the expansion of IoT, data governance challenges will continue to become increasingly pressing, and new policies and frameworks are needed to secure the rights of users while stimulating innovation (Barroso et al., 2010). The next decade will, thus, see a revolutionary interface between technology and regulatory techniques within the HMI paradigm. Furthermore, advances in natural language processing and machine perception will probably ripen HMI interfaces in the next 10 years. The developments in conversational interfaces and voice recognition programs will make people's interactions more natural, allowing them to interact with machines in the same way they interact with humans. These developments would minimize the learning curve typically associated with new technologies such that they are made more accessible to a broad cross-section of the population, including the less technology savvy individuals. As these technologies mature, they will also help in creating more compassionate machines that can sense emotional cues and reciprocate accordingly. Such empathetic involvement has the promise of enhancing user satisfaction across various applications, from customer support to emotional guidance, and ultimately shift the attitude of humans towards embracing technology in their everyday lives.

SUMMARY

The topic of Human-Machine Interaction (HMI), it is then imperative to value the profound meaning with which advancing technology has an impact on the efficiency and efficacy of human interaction with machines. The addition of new interfaces, like gesture-based systems and natural language processing, represents a stark deviation from the traditional device-based interactions. Research suggests that these newer modalities not only enhance the user experience but also reduce such effects as cyber-sickness to a significant degree, thus enabling more immersive and intuitive usage of technology (Carmein et al., 2014). This stands in contrast to the latter, as usage of older mechanical interfaces results in frustration and disengagement, suggesting the imperative need for a total paradigm shift in the interaction method in HMI. Furthermore, the worth of interdisciplinary approaches for HMI cannot be underestimated. As discussed in the Handbook of Human-Machine Interaction, a well-rounded understanding of physical, cognitive, social, and emotional factors is central to the successful design and application of interactive systems (Lahlou et al., 2011). This overall strategy allows developers and researchers to address user needs completely, leading to more adaptive and intuitive interfaces. Through combining knowledge from fields such as psychology, ergonomics, and artificial intelligence, HMI may go beyond functionality to deliver constructive user experience that enhances productivity and satisfaction. Finally, the future of Human-Machine Interaction is in continuous research and innovation. With the increased pace of technological advancement, the shape of HMI will surely change, necessitating continuous adaptation and adjustment of interaction paradigms. As emerging technologies like virtual and augmented reality become more prevalent, their integration into everyday interactions also brings risks and opportunities for both user and designer alike. Stakeholders must remain vigilant in considering ethical factors and user accessibility in every step forward, making HMI systems accessible and beneficial to every user. The results obtained through current research and field implementations will certainly give a foundation for the next effort in this rapidly developing field.

Human-machine interaction (HMI) research reveals some prominent issues that emphasize the evolution of human-technology relationship. One of the central topics of debate is the application of interface design for enhanced user experience and safety. Evolution in in-vehicle information systems (IVIS) does not aim merely to facilitate greener driving but also assist in mitigating distractions that threaten road safety. Foot-LITE and related work demonstrate the success of ecological interface design (EID) techniques grounded in user-centered development, applied with prototyping acceleration. Results from evaluations presented by potential users and domain experts confirm the relevance of such designs for future generation interface design, illustrating that thoughtful interaction design can contribute to safety in addition to environmental goals. This collaboration of function and user engagement is key to propelling HMI technologies toward greater adoption and effectiveness, a key shift in existing discussion. The role of communications technology in the evolution of smart grids offers another iteration of the discussion regarding HMI. The marriage of information technology with mainstream energy infrastructure offers forward-thinking answers to energy delivery management and environmental issues. However, this unification has its own unique set of security problems that cannot be overlooked. The complexity of two-way communication creates vulnerabilities that expose these systems to cyberattacks, necessitating an integrated approach to security. The majority of current research focuses on deficiencies in addressing all sides of security because it approaches classifying the threats in very specific terms and detailing countermeasures, thereby excluding the broader context of system-level vulnerabilities. The need for a unified model of HMI that addresses both user functionality and security is required to build a healthy technological environment in the future. Finally, the creation of HMI is more than just the basic interface between man and machine; it is a paradigm shift in the general way these interactions are being conceived and executed. Emerging technologies, powered by machine learning and artificial intelligence, increasingly are dictating the pace and nature of human engagement with sophisticated systems. The controversies underscore the need to react to potential ethics concerns and responsibility in these interactions. For instance, while smart grid technology embraces clean energy and greater efficiency, it must also look ahead to likely accountability issues in terms of cyber-attacks that can undermine user data and system integrity. As the HMI environment continues to evolve, recognizing and solving these complex problems will become essential to making intelligent systems a part of everyday life in a sustainable way. The relationship between convenience, security, and ethics will revolutionize the trajectory of human-machine interaction in profound ways.

REFERENCES

1. Abbas, R., Ahmad, I., Ali, S., Amuru, S., et al. (2020). 6G white paper on machine learning in wireless communication networks. http://arxiv.org/abs/2004.13875 (accessed on 10th September 2025).
2. Abbey, B., Brown, B., Buijs, G., Candy, L., Csikszentmihalyi, M., Feland, J., et al. (2019). The 'responsibility' factor in imagining the future of education in China. https://core.ac.uk/download/237010983.pdf (accessed on 10th September 2025).
3. Ahamed, S. I., Alam, M. A. U., Chu, W., Williams, J., et al. (2013). Considerations in designing human-computer interfaces for elderly people. https://core.ac.uk/download/213078692.pdf (accessed on 10th September 2025).
4. Almér, A., Dodig-Crnkovic, G., Holstein, T., Kade, R., et al. (2016). Trans disciplinarity seen through information, communication, computation, (inter-)action and cognition. http://arxiv.org/abs/1604.04711 (accessed on 10th September 2025).
5. Altenstetter, C. (1995). Regulating healthcare technologies and medical supplies: A comparative overview. https://core.ac.uk/download/5079662.pdf (accessed on 10th September 2025).
6. Baber, C., & Stanton, N. A. (1997). Rewritable routines in human interaction with public technology. https://core.ac.uk/download/334503.pdf (accessed on 10th September 2025).
7. Banic, A., & Nguyen, A. (2015). *3DTouch: A Wearable 3D Input Device with an Optical Sensor and a 9-DOF Inertial Measurement Unit*. http://arxiv.org/abs/1406.5581 (accessed on 10th September 2025).
8. Barroso, L. A., Cerf, V. G., Chandy, K. M., Clark, D., et al. (2010). Internet predictions. https://core.ac.uk/download/4880707.pdf (accessed on 10th September 2025).
9. Bhatt, S., Shalin, V., Sheth, A., Zhao, J., et al. (2020). Knowledge graph semantic enhancement of input data for improving AI. http://arxiv.org/abs/2005.04726 (accessed on 10th September 2025).
10. Bizer, C., Bröring, A., Conejero, D., Díaz-Pardo, D., Goodchild, M., Harel, A., Hervás, R., et al. (2012). Sharing human-generated observations by integrating HMI and the semantic sensor web. https://core.ac.uk/download/148660144.pdf (accessed on 10th September 2025).
11. Bjorneseth, F. B., Dunlop, M. D., & Hornecker, E. (2010). Assessing the effectiveness of multi-touch interfaces for DP operation. https://core.ac.uk/download/9033680.pdf (accessed on 10th September 2025).

12. Brewer, S., Craigon, P., Frey, J., Gutierrez, J., et al. (2023). Reflexive governance architectures: Considering the ethical implications of autonomous technology adoption in food supply chains. https://core.ac.uk/download/554842529.pdf (accessed on 10th September 2025).
13. Briggs, P., & Little, L. (2006). Using ubicomp systems for exchanging health information: Considering trust and privacy issues. https://core.ac.uk/download/17298610.pdf (accessed on 10th September 2025).
14. Brooke, J., Burns, C. B., Gibson, J. J., Hellier, E. J., Jones, S. D., Young, M. S., Rasmussen, J., et al. (2012). Ecological IVIS design: Using EID to develop a novel in-vehicle information system. https://core.ac.uk/download/9846649.pdf (accessed on 10th September 2025).
15. Carmein, D. E. E., Chang, E., Hand, C., Mirzaei, M. A., & Oinuma, M. (2014). Navigation and interaction in a real-scale digital mock-up using natural language and user gesture. https://core.ac.uk/download/29137374.pdf (accessed on 10th September 2025).
16. Cena, F., Hopfgartner, F., Kay, J., Kummerfeld, B., et al. (2015). New frontiers of quantified self: Finding new ways for engaging users in collecting and using personal data. https://core.ac.uk/download/42365401.pdf (accessed on 10th September 2025).
17. Chen, K. C., Hanzo, L., Jiang, C., Ren, A., et al. (2019). Thirty years of machine learning: The road to Pareto-optimal wireless networks. http://arxiv.org/abs/1902.01946 (accessed on 10th September 2025).
18. Cincott, F., del R. Millan, J., Giugliemma, C., Kubler, B., et al. (2010). Combining brain-computer interfaces and assistive technologies: State-of-the-art and challenges. https://core.ac.uk/download/80918.pdf (accessed on 10th September 2025).
19. Clowes, R. W., Heersmink, R., & Smart, P. (2017). Minds online: The interface between web science, cognitive science, and the philosophy of mind. https://core.ac.uk/download/131214931.pdf (accessed on 10th September 2025).
20. Creese, S., De Roure, D., Hall, W., Madaan, A., et al. (2018). A storm in an IoT cup: The emergence of cyber-physical social machines. http://arxiv.org/abs/1809.05904 (accessed on 10th September 2025).
21. De Bellis, A. F., & Umbrello, S. (2018). A value-sensitive design approach to intelligent agents. https://core.ac.uk/download/146502298.pdf (accessed on 10th September 2025).
22. Dewey, M., & Wilkens, U. (2019). The bionic radiologist: Avoiding blurry pictures and providing greater insights. https://core.ac.uk/download/228064788.pdf (accessed on 10th September 2025).

23. Diaz Castañeda, J. L., & Ocampo-Martínez, C. (2019). Energy efficiency in discrete-manufacturing systems: Insights, trends, and control strategies. https://core.ac.uk/download/288625543.pdf (accessed on 10th September 2025).

24. Dijk, E. O., van Erp, J. B. F., Kuyper, E., Nijholt, A., et al. (2010). Audio-tactile stimuli to improve health and well-being: A preliminary position paper. https://core.ac.uk/download/pdf/11475236.pdf (accessed on 10th September 2025).

25. Eliens, A. P. W., Makedon, F., Nijholt, A., Poelman, R., et al. (2011). Exploring community building with an awareness display. https://core.ac.uk/download/pdf/11482223.pdf (accessed on 10th September 2025).

26. Garry, T., & Harwood, T. (2019). Cyborgs as frontline service employees: A research agenda. https://core.ac.uk/download/228181962.pdf (accessed on 10th September 2025).

27. Garza Martínez, A. E., López Chao, V. A., Martin Gutierrez, J., del Río Guerra, F., et al. (2020). The limited effect of graphic elements in video and augmented reality on children's listening comprehension. https://core.ac.uk/download/286590736.pdf (accessed on 10th September 2025).

28. Ghazi, H. E., Ghazi, H. E., Kaabouch, N., Mrabet, H., et al. (2018). Cyber-security in smart grid: Survey and challenges. http://arxiv.org/abs/1809.02609 (accessed on 10th September 2025).

29. Giacomin, J., Malizia, A., Ramm, S., Robertson, J., et al. (2014). A first approach to understanding and measuring naturalness in driver-car interaction. https://core.ac.uk/download/30339378.pdf (accessed on 10th September 2025).

30. Gleicher, M., Mutlu, B., Praveena, P., Subramani, R., et al. (2019). Characterizing input methods for human-to-robot demonstrations. http://arxiv.org/abs/1902.00084 (accessed on 10th September 2025).

31. Goldenberg, A., Hoffman, M. M., Leskovec, J., Nguyen, T., et al. (2018). Machine learning for integrating data in biology and medicine: Principles, practice, and opportunities. http://arxiv.org/abs/1807.00123 (accessed on 10th September 2025).

32. Golding, S., et al. (2010). *The Allergic Bodies Conference: Postgraduate/Postdoctoral Research Conference, 2010.* https://core.ac.uk/download/16744969.pdf (accessed on 10th September 2025).

33. Grone, E. J. (2019). Disrupting complex systems with emerging technologies: A study on United States airport operations. https://core.ac.uk/download/215541134.pdf (accessed on 10th September 2025).

34. Harvey, C., McIlroy, R., Robertson, D., Stanton, N., et al. (2013). Sustainability, transport and design: Reviewing the prospects for safely encouraging eco-driving. *(No link provided.)*

35. Harwood, T., Maltby, J., & Mukaetova-Ladinska, E. (2019). Role of artificial intelligence (AI) art in care of ageing society: Focus on dementia. https://core.ac.uk/download/228191193.pdf (accessed on 10th September 2025).

36. Hemminger, B. M., Kirby, D., Asimov, I., Beall, J., Larson, J., Lin, J., Mingers, J., et al. (2018). Exploring the referral and usage of science fiction in HCI literature. http://arxiv.org/abs/1803.08395 (accessed on 10th September 2025).

37. Hong, G. Y., James, H., & Ryu, H. (2006). Quality assessment technique for ubiquitous software and middleware. https://core.ac.uk/download/148639200.pdf (accessed on 10th September 2025).

38. Huggett, J. (2017). The apparatus of digital archaeology. https://core.ac.uk/download/82965040.pdf (accessed on 10th September 2025).

39. Kerick, S. E., Lance, B. J., McDowell, K., Oie, K., et al. (2012). Brain computer interface technologies in the coming decades. http://arxiv.org/abs/1211.0886 (accessed on 10th September 2025).

40. Kloosterman, S. H. (1994). Design and implementation of a user-oriented speech recognition interface: The synergy of technology and human factors. https://core.ac.uk/download/pdf/11461320.pdf (accessed on 10th September 2025).

41. Lahlou, S. (2011). Socio-cognitive issues in human-centered design for the real world. https://core.ac.uk/download/286249.pdf (accessed on 10th September 2025).

42. Lowe, A., Nielsen, J., Thomas, P., Macredie, R. D., & Sharpe, B. (2002). Introduction to the new usability. https://core.ac.uk/download/333163.pdf (accessed on 10th September 2025).

43. Martí Janda, J. G. (2019). Automation of the continuous integration (CI)–continuous delivery/deployment (CD) software development. https://core.ac.uk/download/286456226.pdf (accessed on 10th September 2025).

44. Minguella Canela, J., & Muguruza Blanco, A. (2017). Digital manufacturing: What are we able to print? https://core.ac.uk/download/157811997.pdf (accessed on 10th September 2025).

45. Musolesi, M., & Pejovic, V. (2015). Anticipatory mobile computing: A survey of the state of the art and research challenges. https://core.ac.uk/download/304118168.pdf (accessed on 10th September 2025).

46. Norman, D. A., Lopresti, D. P., Kanizsa, G., Hinton, G. E., Baird, H. S., Ahn, L., Shirali-Shahreza, M. H., et al. (2015). CAPTCHaStar! A novel CAPTCHA based on interactive shape discovery. http://arxiv.org/abs/1503.00561 (accessed on 10th September 2025).
47. O'Brien, S. (2013). The Borrowers: Researching the cognitive aspects of translation. https://core.ac.uk/download/11311754.pdf (accessed on 10th September 2025).
48. Pitts, M. J., Wellings, T., & Williams, M. A. (2008). Customer perception of switch-feel in luxury sports utility vehicles. https://core.ac.uk/download/48924.pdf (accessed on 10th September 2025).
49. Shaer, O., & Hornecker, E. (2009). Tangible user interfaces: Past, present and future directions. https://core.ac.uk/download/9033674.pdf (accessed on 10th September 2025).
50. Siegel, M. A., & Stolterman, E. (2009). Metamorphosis: Transforming non-designers into designers. https://core.ac.uk/download/99962.pdf (accessed on 10th September 2025).
51. Vega, E. S. (2019). Designing women: Essentializing femininity in AI linguistics. https://core.ac.uk/download/322983206.pdf (accessed on 10th September 2025).

CHAPTER 10

SAFETY, SECURITY, AND ETHICS OF AUTONOMOUS SYSTEMS

Contents

Introduction 373
10.1. Safety Concerns in Autonomous Systems 377
10.2. Security Challenges in Autonomous Systems 383
10.3. Ethical Implications of Autonomous Systems 388
10.4. Regulatory Frameworks for Autonomous Systems 393
10.5. Public Perception and Acceptance 398
10.6. Technological Advancements and Innovations 403
Summary 408
References 410

INTRODUCTION

The advent of autonomous systems, particularly in transportation, marks a revolutionary era with promise for precedent-shaping innovations as well as complex challenges. With self-driving automobiles becoming a reality, various stakeholders are compelled to tap into a multifaceted argument over their implications. This argument raises engineering advancements, social impacts, and ethical concerns, and all point to the necessity of overarching frameworks for safety and security protocols. Engineers and developers must contend with complex issues that go beyond technical details, requiring a good understanding of the way these systems work out in human situations. As Dodig-Crnkovic et al. (2018) observes, sanitized representations of ethical dilemmas featured in arguments, such as the trolley problem, assume too much simplicity in the choice involved in applying these technologies.

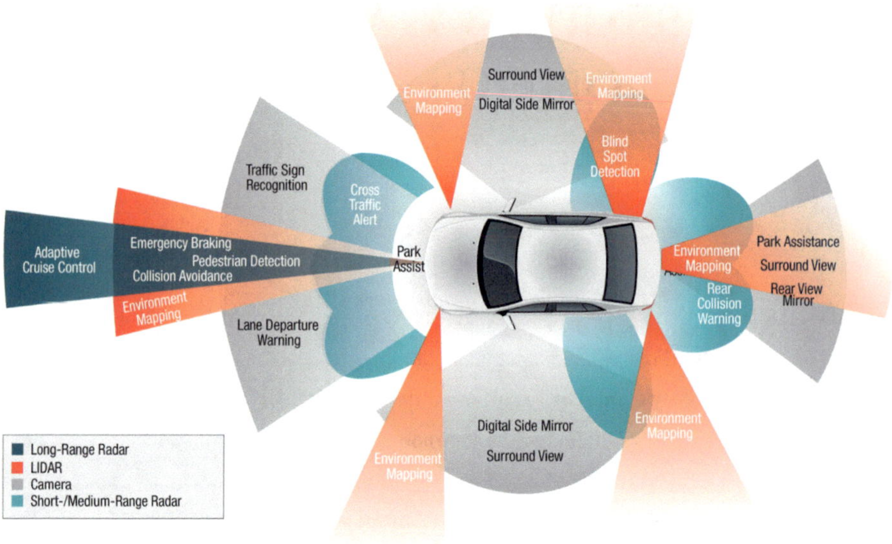

Figure 10.1. *The ethics of autonomous vehicles: Balancing safety and control.*

*Source: https://miro.medium.com/v2/resize:fit:1248/1*YXHgiTuND7Ej9jcM40KzRA.png.*

The need for strong safety and security requirements becomes imperative, especially as society begins to embed autonomous systems into daily existence. Current frameworks have to adapt to encompass not just mechanical operations but the psychosocial aspects of safety within autonomous systems. This adaptation implies intensive testing and verification procedures to provide reliability across various conditions. The findings of the Tigar's project emphasize the importance of assurance methods and resilience strategies, which are critical in establishing public confidence in autonomous technology. As Bloomfield et al. (2025) elaborates, full understanding of concerns in simulation, dynamic testing, and security-informed safety analysis will provide secure integration of the systems into existing transport networks. So, a sound strategy is needed to counter the inherent risks provided by such cunning devices. Ethical matters also have to become a part of the discussion regarding autonomous systems, since they bear profound consequences for human agency and responsibility. Autonomous system choices normally reflect larger societal values and must contend with moral issues capable of impacting human life. A practical engineering ethical position, as enunciated by Dodig-Crnkovic et al. (2018), addresses solutions to real-world complexities rather than abstract disputes. Both policymakers and engineers are faced with the challenge of coming up with rules that not only regulate the technology of autonomous systems but also ensure that ethical factors are the impelling motives behind the development and deployment activities. Through promoting transparency and accountability, stakeholders are able to handle the uncertainties and complexities associated with autonomous systems, ultimately contributing to their safe and ethical use in society (Figure 10.1).

Importance of Safety and Security in Technology

The implication of autonomous systems into the day-to-day existence significantly strengthens the inseparable link between technological progress, security, and safety. With autonomous vehicles being one of the most significant illustrations of this technology, there is a need to research their engineering misfortunes alongside ethical concerns. As the technologies continue to develop, the role of software becomes paramount; it not only governs operational efficiency but also public safety. The development of secure and safe autonomous systems requires tight emphasis on software engineering solutions that adequately capture ethical concerns, particularly decision-making in high-risk scenarios, such as those similar to the trolley problem. Engineering standards compatibility and ethical frameworks are essential for successful integration of such technologies within society, indicating the necessity of responsibility in design and deployment (Dodig-Crnkovic et al., 2018). Borrowing from the concept of ethical considerations, the topic of Value-Sensitive Design (VSD) comes up as a major methodology in the incorporation of human values into autonomous system development. This method appreciates that design must extend beyond technical specifications to embrace the larger social impact of technology. Through the incorporation of stakeholder values at the inception of artificial intelligence development, designers will actively mitigate possible security and safety risks from autonomous systems. The emphasis on VSD marks a paradigm shift in the way technology is developed and implemented, one that seeks not only to enhance functionality but also to provide ethical precedence. Thus, anticipatory VSD integration in the development of autonomous systems is a central response to the sophisticated challenges of safety and security concerns (Bellis et al., 2018). Moreover, as there is continued controversy surrounding autonomous systems, the establishment of an effective regulatory framework is a necessary imperative. The technology safety and security sector cannot thrive without a full set of standards and regulations to lead developers as well as consumers. Regulators have to account for some factors, including the evolving nature of technological advancements, the possibility of vulnerabilities in software, and morality of such breakthroughs. Transparent regulatory tools not only guarantee compliance with established safety measures but also promote public confidence in such mechanisms. Therefore, the consultation between engineers, ethicists, and policymakers becomes critical to guaranteeing that autonomous technologies are developed with security and safety prioritized at the core. This collaborative approach aims to create an environment by which optimal social returns are realized with minimized risks, finally in line with the ethical needs of the modern technology development (Dodig-Crnkovic et al., 2018).

Overview of Ethical Considerations

As the rapid development of autonomous systems is unfolding, ethics is becoming a focal point with increased intertwinement of these technologies into daily life. Ethical issues usually arise from the decision-making process in these systems, particularly when human intervention could be delayed during the instance. Autonomous vehicles, for example, need to decide at once during times of high-risk moment and wonder about how these kinds of systems would respond to approaching danger. Clearly defined ethical algorithms become a requirement so that these systems operate on a level that emphasizes human safety and ethical alignment with societal norms. Also, as technologies like AI and machine learning become more dominant, concerns of transparency and accountability in decision-making are of utmost relevance, as highlighted by the intricacy of algorithms being deployed in autonomous systems (Singh et al., 2024). Without robust ethical guidelines, the likelihood of unintended consequences grows, requiring an active policy of ethical management. Privacy is also an important ethical issue in the use and deployment of autonomous systems. Data collection practices employed by AI-based systems can lead to intrusive surveillance and a violation of the rights of individuals to privacy. The combination of the couple with fog/cloud computing and the Internet of Things makes it even more challenging since data gathered can be gigantic and often personal in character (Singh et al., 2024). The stewardship of data's ethics thus has become imperative to study. Organizations must place priority not only on data protection but also on data usage and its implications for user privacy. Tipping the balance between innovation and privacy rights needs robust guidelines to ensure that data is not used or misused, thereby allowing for a trusting relationship between technology and society. Balance is particularly necessary in fields where sensitive information is processed continuously, such as medicine and banking. Further, ethical dimensions of prejudices inherent in autonomous systems pose a great challenge. These biases typically come about as a consequence of the training data employed to develop machine learning algorithms, which may be unrepresentative of disparate groups or not an equitable representation of decision-making. As autonomous systems apply their presence in industries ranging from manufacturing to healthcare, the impact of such biases has serious consequences, including that of sustaining discrimination and creating unjust outcomes for marginalized groups. Such biases need to be managed through ongoing monitoring and refinement of data feeds supplemented by a commitment to creating more representative datasets. There is a need to create frameworks that emphasize ethical authorizations and auditing of AI systems to identify and prevent instances of bias. Apart from technologies, an organizational culture of ethical awareness is necessary to deal with the complexities of implementing autonomous systems ethically.

10.1. SAFETY CONCERNS IN AUTONOMOUS SYSTEMS

With the introduction of autonomous systems, particularly in the shape of autonomous vehicles, there has been a need for extensive debate on safety issues. These issues of safety are more to do with the solution of how these systems can be designed to operate under varying conditions without causing undue harm to human life and property. In their response to these challenges, researchers urge an integrated solution that is both by means of sound design methods and real-time testing processes. For instance, formulation of testing standards and validating the reliability of autonomous systems is necessary when preventing grand failures. Through concentration of research on developing technological systems that are resilient and secure, the industry can attempt to optimize the utility of autonomous systems while minimizing inherent risks to customers and society in general (Dewey et al., 2015). Such comprehensive safety systems are essential to encouraging public confidence in the deployment of such technologies. Moreover, the complexity of integrating artificial intelligence into autonomous systems makes matters even more complicated from the perspective of safety. The interplay between machine learning algorithms and real-time scenario assessment produces new vulnerabilities since algorithms may misinterpret data or remain slow to respond in critical situations. In response, experts in the field contend that robust validation through simulation and dynamic testing is critical to enhance system resilience. Diversification of the testing environments allows scientists to experiment with how the systems react to unforeseen circumstances and, therefore, can identify the potential weaknesses before subjecting them to real-world applications. Additionally, research works indicate towards the significance of establishing standards involving security-aware safety analysis, which ensures that the systems are not only secure but also safe against possible cyber-attacks (Bloomfield et al., 2025). It is by considering safety and security simultaneously that one can formulate an enhanced approach for autonomous technologies. Finally, ethical considerations pertaining to safety concerns in autonomous systems must be appropriately accounted for.

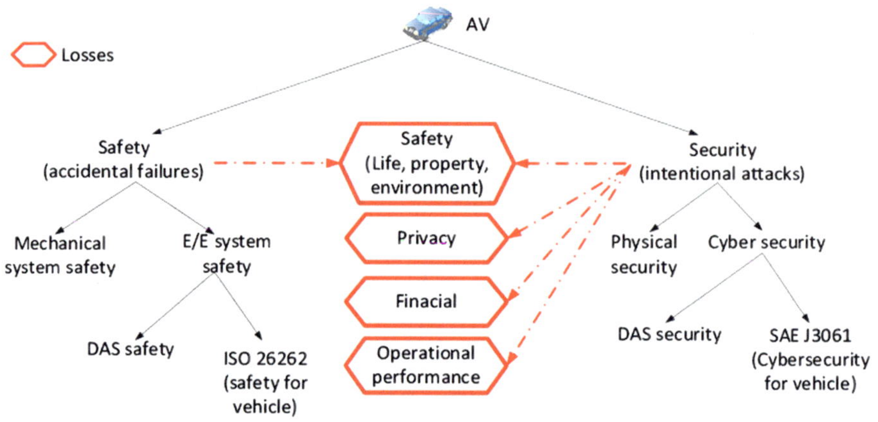

Figure 10.2. *Safety and security in autonomous vehicles.*

Source: https://www.researchgate.net/profile/Jin-Cui-20/publication/324006032/figure/fig1/AS:608269311688705@1522034113566/Safety-and-security-in-Autonomous-Vehicles.png.

As the operation burden is transferred from human operators to algorithms, ethical questions arise regarding accountability in the event of accidents or malfunction. Questions regarding whose moral values govern these systems—i.e., of manufacturers, developers, or regulatory bodies—are still highly debated. Moreover, biases embedded in machine learning algorithms can automatically lead to discriminatory outcomes, particularly in high-risk applications like autonomous vehicles. In a bid to navigate these moral challenges, there is a requirement for multidisciplinary collaboration between policymakers, ethicists, and technologists with the aim of crafting harmonious standards that prioritize equity and justice on the agenda of design processes for autonomous systems. Addressing these multidisciplinary viewpoints not only solves safety problems but also promotes a responsible mindset towards innovation that brings advantage to society in general (Figure 10.2).

10.1.1. Risk Assessment and Management

Risk analysis and management are key factors in planning and implementing autonomous systems, particularly ensuring safety and security. With advancing technologies, the complexity of their operations brings on new risks that necessitate a scientific risk analysis. An appropriate risk analysis involves cataloging potential dangers, evaluating the risks associated with them, and having safeguards to reduce adverse effects. This is especially so for autonomous vehicles, where the dynamic nature of road environments will lead to unexpected situations.

The observations provided under the Technical Topic Notes reflect the need to ensure not only intensive early tests but also continuous monitoring and adaptation of risk management approach throughout the life of the vehicles (Bloomfield et al., 2025). This forward-thinking approach allows the identification of future risks as the technology and deployment environments evolve. Mandating security-conscious safety analysis in the risk management process is the most important in autonomous systems. The traditional boundaries between safety and cybersecurity continue to erode, requiring converged assessment methodology. Cyber-attacks may disrupt security, resulting in catastrophic impacts when self-managed systems rely on real-time data and communications networks. System designers can create systems that not only satisfy performance specifications but are also immune to hostile interference using fortified safety analyses to test for possible cyber vulnerabilities.

This aligns with present discourse on resilience within autonomous systems, encouraging diversity and defense-in-depth practices that stack layers of safeguarding against various kinds of threats (Bloomfield et al., 2025). By the inclusion of these concepts, general stability and dependability of autonomous systems can be significantly enhanced. The ethical aspects of risk management and assessment in autonomous systems should also be given careful consideration. As such technologies become more and more entrenched in daily life, responsibility for ensuring they are used safely and ethically falls to both developers and society as a whole.

Participation in the risk management field with human factors is particularly critical as younger and more susceptible individuals are targeted for brain manipulation in digital environments. The potential of cognitive vulnerabilities requires a delicate understanding of human interaction with such systems and the nature of bringing ethical thought to risk analyses so that adopted management strategies do not aim merely to minimize technical errors but also to protect individuals and populations from the broader psychological and social risks of autonomous technology rollout.

10.1.2. Fail-Safe Mechanisms and Redundancy

Deployment of fail-safe mechanisms and redundancy within autonomous systems is required in order to enhance operational reliability and safety. Fail-safe mechanisms are deployed to ensure systems revert to a safe condition when faulty or showing anomalies. This reduces threats likely to lead to catastrophes, thus enhancing the degree of trust in autonomous operations. Redundancy, on the other hand, serves as a backup approach where essential parts of the system are replicated so that if one fails, the other can replace it without loss of function. This double-layered strategy for guaranteeing continuity and security is just one example of the proactive steps developers have to implement in autonomous systems, thus adhering to ethical imperatives regarding user safety and open operation. By applying these principles in the development and operation of autonomous technology, developers move closer to establishing more robust frameworks that prioritize safety in uncertain environments.

Additionally, the multi-layered application of fail-safe methods and redundancy provide more flexibility in autonomous systems, especially in dynamic environments where conditions around them can significantly impact operation. For example, in the case of the Netherlands' climate change adaptive capacity, institutions have begun incorporating innovative technologies that utilize predictive modeling to analyze environmental impacts (Bergsma et al., 2010). The implementation not only protects systems from unforeseen circumstances, but it also prioritizes preparedness and resilience to emerging vulnerabilities. The same principle applies to autonomous systems, whose constant review of performance in varied settings enables the immediate updating of fail-safe protocols. In addition to enhancing the operational capabilities of these systems, such a process works to put users at ease regarding their reliability, an imperative element in the management of ethical concerns in autonomous technology adoption.

In space travel, the application of high-tech fail-safe methodologies and redundancy enhances the safety and efficiency of automated systems utilized in cases where human intervention is minimal or not feasible. Importantly, research in the area of the Space Station determines ways in which automation with AI possesses the potential to enhance crew safety by minimizing extravehicular activities (EVA) and overall efficiency through automating routine tasks (Firschein et al., 2025). With the use of multiple redundant systems, the risk of single-point failure is significantly reduced, thereby ensuring constant operation in critical circumstances. This approach not only supports the technical safety of autonomous systems but also raises profound ethical issues regarding transparency and accountability in such technology. As autonomous systems become ubiquitous in high-hazard environments, foundations laid by redundancy and fail-safe modes are essential to managing the intertwined complex interaction of safety, security, and ethics (Figure 10.3).

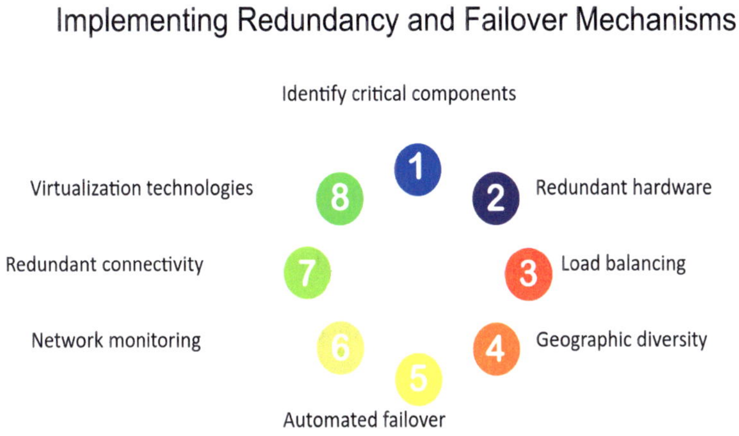

Figure 10.3. *Implementing redundancy and failover mechanisms.*

Source: https://fastercapital.co/i/Strengthening-Network-Resilience-with-Best-Efforts-Techniques--Implementing-Redundancy-and-Failover-Mechanisms.webp.

10.1.3. Case Studies of Safety Failures

An examination of safety deficiencies in autonomous technology involves analytical study of some case studies illustrating potential threats to human health and well-being from the application of such technologies. Problems in Autonomous Weapons Systems (AWS) are a prime example, which bring unparalleled military achievements but serious safety and ethical concerns. Delaying action to address the inherent risks in using AWS can lead to unexpected interactions and calamitous outcomes in a war situation. An emerging study identifies socio-technical risks of AWS, including system crashes and vulnerability to cyber-attacks. Addressing these challenges is essential to guaranteeing that AWS is ethical and secure to operate as it has to embed robust control processes to offset risks without impairing efficiency of operation. This placement of safety and effectiveness highlights the complexities of the application of autonomous military technology, requiring close watching and orderly approaches to regulating it (Maathuis et al., 2025). In civilian application, that of Autonomous Vehicles (AVs) provides the context for broad failure in safety that has placed big ethical and security concerns. AVs, which are autonomous and lack human intervention, are founded on advanced technologies like sensors and artificial intelligence. The rate of accidents involving AVs has put their reliability into question and created concerns regarding the effects of failures. AVs are said to be susceptible to cyber-attacks, and their operation integrity is impaired, posing risks to public safety. These weaknesses not only challenge the engineering skills that go into these systems but also raise huge ethical issues surrounding accountability in the event of an accident. It is a joint initiative among the stakeholders—be it manufacturers, researchers, or policymakers—to develop preventive measures that can provide assurance on the reliability of AVs, thus engendering greater confidence towards their adoption as well as maintaining public safety (Ogunlade et al., 2025). The complexity of safety failure extends beyond technical issues; it represents the imperative need for ongoing discussion of the ethical implications of deploying autonomous systems. As the AWS and AV case studies demonstrate, ensuring safety is not so much a matter of functionality as it is one of adhering to ethical principles where human safety is a priority. The danger of unintended injury emphasizes the need for ubiquitous regulation of the way autonomous technologies are designed and deployed. The policies also need to ensure transparency and accountability so that engineers, designers, and policymakers can have genuine conversation about ethical behavior. As society continues to deploy autonomous systems in daily life, proactiveness in safety must go hand in hand with an unbending commitment to ethics in a bid to instill public trust in the technologies that hold the promise to change our world (Maathuis et al., 2025; Ogunlade et al., 2025) (Figure 10.4).

Figure 10.4. *Autonomous weapons systems (AWS).*

Source: https://i0.wp.com/opiniojuris.org/wp-content/uploads/technology-g44e877d9b_1920.jpg?fit=1200%2C656.

10.2. SECURITY CHALLENGES IN AUTONOMOUS SYSTEMS

The introduction of autonomous systems in the various industries, particularly transport, raises challenges of security that must be met to ensure safety and dependability. Among the key challenges is the vulnerability of systems to hacking, which compromises their operational integrity. Considering that autonomous automobiles are very reliant on software and connectivity, they are susceptible to being tampered with by hackers intent on creating malfunction. This calls for the setting up of robust cyber security measures capable of detecting and mitigating against the potential threats. Underpinning this is the understanding of the environment of potential threats; hence, research must be ongoing in the identification of vulnerabilities in hardware and software components of autonomous vehicles so that safety and resilience are ingrained in their development processes. In addition to this, the ethical aspects of security in autonomous systems cannot be downplayed.

Figure 10.5. *Autonomous vehicles be protected against cyber security threats.*

Source: https://innovationatwork.ieee.org/wp-content/uploads/2020/05/bigstock-196026349_1024X684.png.

The discipline is generally confronted with complex ethical issues that are similar to thought experiments, for instance, the trolley problem. Framing such issues in strictly moralistic terms may not result in solutions applicable in real life. Instead, emphasis on an engineering ethics of application is to be encouraged because it offers means of dealing with the complexity of real-world decision-making on autonomous systems. Since Dodig-Crnkovic et al. (2018) asserts, in addressing engineering problems, ethical issues must always be at the forefront of software engineering practice. This way, implementers are able to bring their technological developments in line with societal values, something that will enhance trust and acceptance by users of autonomous systems. Finally, the setting of standards and regulations concerning security within autonomous systems is vital. According to recent research, there is a need for comprehensive frameworks that address assurance, resilience, and security-informed safety analysis (Bloomfield et al., 2025). Such frameworks can offer performance and accountability standards that guarantee autonomous cars have high safety requirements. Through the implementation of a defense-in-depth strategy, stakeholders are able to implement diverse layers of security that enhance system resiliency against a wide range of attacks. Coordination of engineers, policymakers, and ethicists is critical in the development of these standards because it guarantees that there is an integrated effort in responding to the intricate security issues posed by autonomous systems. Through tireless work in developing robust protocols, the industry can guarantee the safety of the users alongside guaranteeing public confidence in technology (Figure 10.5).

10.2.1. Vulnerabilities to Cyber Attacks

In today's rapidly evolving autonomous systems environment, the cyber-attack vulnerabilities present an increasingly critical problem. As those systems become increasingly reliant upon artificial intelligence (AI) and machine learning programs, the structure becomes increasingly complicated, inadvertently exposing several lines of assault for malicious parties. They exploit these vulnerabilities to gain system access, potentially with catastrophic implications, such as the hijacking of autonomous vehicles or unauthorized access to sensitive information. Moreover, their integration into infrastructures critical to areas like transportation or healthcare intensifies stakes and demands an in-depth understanding of such vulnerabilities. Effectively responding to such cyber threats hinges not just on technical measures but also on governance frameworks and planning processes that take into account the peculiar nature of concerns presented by AI-enabled systems, as highlighted by the discussions in the Task Force on Artificial Intelligence and Cybersecurity (Fantin et al., 2020). Further, the intersection of AI and cybersecurity has far-reaching ethical concerns about managing vulnerabilities. One of the chief concerns is that efficiency in operations comes at the expense of safety, particularly when AI systems are optimized to perform well in real-time environments. However, in doing so, such optimizations sacrifice security resilience, leaving systems highly vulnerable to exploitation. For example, automation reliance triggers complacency in human monitoring and neglects vigilance against possible cyber threats. The High-Level Expert Group on AI Ethics lays great emphasis on the fact that guidelines must be established that ensure AI systems are trustworthy and secure (Fantin et al., 2020). Therefore, vulnerability mitigation constitutes a collective effort that traverses the entire gamut of stakeholders, ranging from developers to policymakers who must prioritize security culture across every phase in the lifecycle of AI. Lastly, to effectively safeguard autonomous systems from cyber threats, there is a need for robust standards and guidelines to be in place. The Technical Topic Notes formulated in the Tigar's project emphasize the requirement of resilience and assurance in the cybersecurity strategy of autonomous vehicles. The guidelines will help establish an overall strategy that includes dynamic testing, formal verification, and defense-in-depth solutions to minimize risks. Through the application of these methodologies, the autonomous systems industry can progress toward the creation of more secure systems that can resist and recover from cyber-attacks. This calls for continuous cooperation between engineers, researchers, and regulatory bodies to ensure that these standards keep up with the rapidly changing technological landscape. Passing of such precautionary measures will be vital in ensuring the ethical application of autonomous systems and safeguarding against threats that would be usurped by malicious actors (Figure 10.6).

Figure 10.6. *Vulnerabilities to cyber-attacks.*

Source: https://www.adv-polymer.com/hs-fs/hubfs/common-cyber-vulnerabilities-MED-1.png?width=600&name=common-cyber-vulnerabilities-MED-1.png.

10.2.2. Data Privacy and Protection Issues

The evolution of autonomous systems, particularly in transport, introduces tremendous progress but also tremendous data protection and privacy issues. Autonomous cars collecting tremendous amounts of data to move around and make operations efficient, the sensitive information that they are capable of collecting introduce ethical concerns on consent, ownership, and utilization. For instance, data produced by employing such data can indicate individual habits and preferences, which, if improperly managed, can be utilized for illegal exploitation or discrimination. This unity of engineering solutions with social responsibility necessitates a holistic knowledge of both the technological potentialities and the implications of collecting data. An ethical emphasis in software development is critical to ascertain that such systems possess robust countermeasures to potential data breach and misuse, thereby sustaining the privacy rights of the individual while benefiting from technological advancements (Dodig-Crnkovic et al., 2018).

Figure 10.7. *Data privacy and protection issues.*

Source: https://www.datagrail.io/wp-content/uploads/2022/05/Social_6_Common_Data_Privacy_Issues_D1-01.jpg.

Additionally, as AI technologies become central to autonomous systems, they pose fundamental questions about privacy and related ethical concerns. Digitization of human behavior patterns allows for the possibilities of unintended bias and manipulation, raising questions regarding the users' privacy. Ethical considerations must address how autonomous systems process personal data and how transparent the processes are. For example, if the users are not aware of the quantity of data gathered or how data is used, then informed decision-making cannot take place. Such transparency can lead to a lack of trust in autonomous systems, and therefore, transparent policy that guides data protection and privacy is of significance. Narrowing the gap between ethics and technological advancement can be met by the construction of interfaces that focus on user control and awareness of the data they have (Müller et al., 2021). Finally, the data protection and privacy regulatory environment is transforming with the intricate interaction between technology and ethics in autonomous systems. The legislatures are increasingly being tasked with establishing a balance between innovation and consumer protection. Such regulations need to mandate open standards on data collection, consent mechanisms, and the use of personal data in machine learning algorithms. Furthermore, industry stakeholders must actively engage in dialogues with policymakers to develop pragmatic solutions that not only safeguard individual privacy but also promote technological progress. By fostering collaboration between engineers, ethicists, and lawmakers, it becomes possible to construct a more resilient data protection ecosystem. The general goal remains creating autonomous systems that are not only efficient and reliable but also respect the natural rights of human individuals in a rapidly data-driven world (Dodig-Crnkovic et al., 2018; Müller et al., 2021) (Figure 10.7).

10.2.3. Strategies for Enhancing Security

Enhancing security in autonomous systems calls for introducing high-performance work practices (HPWPs) that support both innovation and safety. Making collaboration a priority high on the agenda that encourages open exchange of possible security risks, organizations can leverage the collective knowledge of their workforce to bridge gaps. For instance, empirical studies indicate that innovative work practices can have a significant effect in making production processes more streamlined, leading not only to greater efficiency but also to an improved employee experience. During a period when random cyber-attacks can cause operations to be undermined, integrating security training and fostering a culture of awareness can allow employees to detect and report vulnerabilities. Besides technical controls, the emphasis on workplace innovation aligns with the dual benefit of enhancing organizational resilience combined with spearheading a proactive security posture, setting the foundation for an effective defense mechanism in autonomous systems (Cox et al., 2012). Moreover, the confluence of artificial intelligence (AI) and cybersecurity both presents challenge and opportunity for security enhancement. The Centre for European Policy Studies has initiated a Task Force on the AI future of cybersecurity to manage these intricacies efficiently. The initiative encompasses diverse stakeholders assessing the dynamics between cyber attackers and defenders and the ever-evolving AI application landscape. In particular, the Task Force points out the need for rigorous information-sharing practices regarding emerging threats and vulnerabilities specific to AI systems only. Implementing periodic threat assessments in developing AI technologies is crucial in a bid to restrict exposure to cyberattacks. Jointly, the Task Force aims to build standards that ensure both the effectiveness of AI in cybersecurity and its ethical application, guaranteeing security integrity across sectors (Fantin et al., 2020). Lastly, addressing security in autonomous systems calls for organizations to not only develop effective technology solutions but also imbue them with ethical considerations. Establishing frameworks that place emphasis on trustworthy AI is crucial in mapping the ethical landscape of autonomous technologies. Initiatives like the High-Level Expert Group (HLEG) on AI Ethics Guidelines for Trustworthy AI illustrate the necessity of a non-exhaustive checklist of assessment and the framework to guide AI development. With developers integrating ethical principles with technological solutions, security measures are then framed in relation to broader societal values. Since autonomous technologies inherently affect central decision-making processes, it becomes a necessity that they include ethical consideration throughout their design and deployment. Ultimately, creating an integrated relationship between security practices and ethics will foster public trust as well as allow for greater societal acceptance of autonomous technology (Figure 10.8).

Figure 10.8. *High performance work system.*

Source: https://3.bp.blogspot.com/-QJWWlrkQyxA/W62rLF8fW2I/AAAAAAAAADQ/N5aDaN8l2fs7rzwUqq2bIFJSUbZ8kl_zgCEwYBhgL/s1600/Shurlly1.jpg.

10.3. ETHICAL IMPLICATIONS OF AUTONOMOUS SYSTEMS

The emergence of autonomous systems holds revolutionary potential across sectors including healthcare, transport, and security. But the ethical implications of developing such technologies are acute and complicated. One of these is accountability for autonomous decisions, particularly in life-or-death matters. As these systems take increasingly autonomous decisions, the question is who will be accountable for them. The challenge lies in developing frameworks which grant human control with respect for the complexities of automated systems. Given that success with artificial intelligence can be extremely rewarding necessitates ongoing research on enhancing the robustness and ethical integrity of AI systems. It is important to carry out such research to render assurances that integrating autonomous technologies truly fosters societal welfare without compromising safety and moral standards (Dewey et al., 2015). One of the most pressing ethical concerns is the notion of trust in autonomous systems, particularly when implemented in caregiving roles.

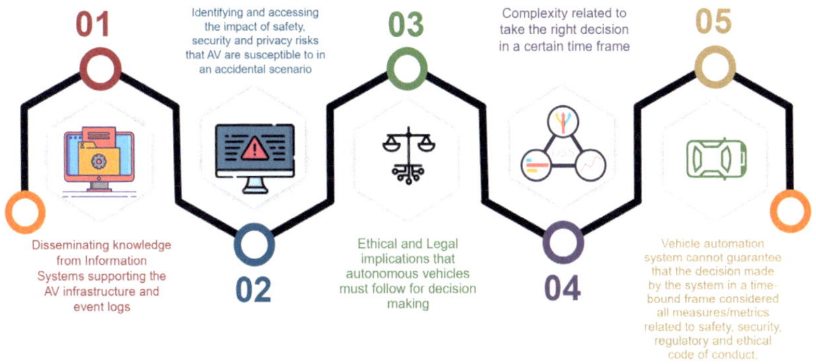

Figure 10.9. *Ethical dilemmas and privacy issues in emerging technologies.*

Source: https://pub.mdpi-res.com/sensors/sensors-23-01151/article_deploy/html/images/sensors-23-01151-g003.png?1674121521.

Dynamics between human users and robots must be studied to gain insights on how expectations of dependency on availability and responsiveness influence user behaviors. Researchers have found that robots can serve as supplementary caregivers but can never equate to the emotional depth provided by human caregivers. This limitation raises ethical issues about the morality of employing robots in delicate circumstances where attachment and emotional support are necessary. Although robots can provide a longer length of parental care and amplify developmental exploration, their inability to develop true emotional relationships may complicate their effectiveness as caregivers. It is imperative to examine these ethical dimensions in order to avoid the unintended consequences of replacing human interaction with machine support (Petters et al., 2010). Also, intersection of safety, security, and ethics should be more than the operationality of autonomous systems. Development of these technologies has implications related to surveillance, data privacy, and abuse. With autonomous systems gathering and processing more and more personal data, the ethical dimensions of consent and user control must be considered. Users should have the ability to keep their data in control in order to establish trust and foster ethical use in technology application. As autonomous systems keep expanding in their applications, policymakers should also offer guidelines to govern their usage effectively, including the ethical dimensions based on social inequality and access. The overall examination of such factors is crucial for establishing a future where autonomous systems have a positive role to play in society while keeping their ethical values intact and ensuring security (Petters et al., 2010) (Figure 10.9).

10.3.1. Moral Responsibility and Accountability

The creation of autonomous systems, particularly in the case of autonomous vehicles, introduces hard problems of moral responsibility and accountability. As technology becomes more advanced, it presents us with ethical dilemmas framed in terms of hypothetical scenarios, such as the infamous trolley problem. Even though the dilemma addresses the potential moral nuances of decisional processes, it is flawed to thc extent that it simplifies the question at hand. More precisely, it requires an applied engineering ethical approach that focuses on the study of actual problems in the social and ethical aspects of autonomous systems. Thus, software plays a fundamental role in ensuring that these technologies are guided by societal values, and as such, it is imperative that developers embed moral responsibility into their design model (Dodig-Crnkovic et al., 2018). Moral accountability is more than basic algorithmic choice; it also encompasses the broader ramifications of how autonomous systems are integrated into society. The question is asked: who is accountable for the actions of an autonomous car when it makes choices autonomously? Accountability may lie with various stakeholders, including manufacturers, software developers, and lawmakers. Liability concerns are very controversial in accidents involving self-driving vehicles, such that the allocation of blame is not clearly defined. Giving accountability mechanisms thus becomes important in fomenting confidence in public trust with respect to such systems. Stakeholder debates on moral responsibility will assist in formulating rules that not only govern the use of autonomous vehicles but also emphasize ethical considerations in their implementation. In order to effectively address the problems of moral responsibility and accountability in autonomous systems, the incorporation of methods like Value-Sensitive Design (VSD) becomes significant. VSD promotes the incorporation of ethical values within the design process from the initial phase so that the technologies developed are sensitive to the interests of stakeholders as well as the values of society. This forward-thinking approach supports a comprehensive examination of the implications related to intelligent agent usage, ultimately addressing the ethical stakes involved. With stakeholder values, developers can create systems that not only meet safety and security standards but also adhere to ethical concerns that resonate with the public. With more developments in autonomous systems, employing methods like VSD can substantially enhance the ethical responsibility embedded in their designs (Bellis et al., 2018).

10.3.2. Impact on Employment and Labor Markets

The advent of autonomous systems through artificial intelligence (AI) has initiated a shift in paradigms of labor markets and work that demands keen examination. Traditional work, particularly in sectors such as manufacturing, transport, and service industries, faces severe threats from automation. As autonomous technologies advance, the cost benefits and efficiency of such systems make them increasingly appealing to employers seeking competitive advantage. Consequently, this transition is projected to displace a substantial number of workers, with estimates suggesting up to 30% of jobs being affected by automation within the next two decades. This disruption highlights a growing divide between low-skill occupations vulnerable to replacement and high-skill jobs that are less likely to face immediate automation, emphasizing the urgent need for workforce retraining and upskilling initiatives to mitigate job losses. Besides, since the displacement caused by autonomous systems poses a threat to employment, it also opens doors for the creation of new work. As advancement in AI and machine learning technology expands, there is a growing requirement for employment concerned with their maintenance, development, and ethics monitoring. Those fields of data analytics, robotics, and AI ethics are bound to see significant growth as organizations attempt to harness the potential of these technologies in a sustainable manner. Yet, the challenge lies in how the affected workforce is able to transition into such new positions effectively. In the absence of suitable educational systems and skill-enhancement programs, the employment generated benefit may not be realized when displaced workers are not well prepared to take up new labor markets that require a different set of skills. Hence, both the disturbances and opportunities introduced by autonomous systems have to be tackled so that a sustainable transformation in the labor market can take place. Furthermore, integration of autonomous systems into the workforce comes along with ethical concerns beyond levels of employment. Management of equitable job distribution, as well as ethical applications of AI technologies, needs to become a forefront priority to avoid further exacerbating socioeconomic divides. As pointed out by present controversy in the field, strategic policies are necessary in order to ensure that the benefits of autonomous systems are shared across a wide base rather than being possessed by a reduced base. For example, initiatives that are aimed at investing in educational capital among marginalized populations can contribute to the development of a diversified workforce to thrive in a more automated society. Additionally, as AI continues to evolve—leveraging advancements in machine learning and intelligent agent paradigms as noted in studies (Dewey et al., 2015) and shifting toward probabilistic decision-making (Dewey et al., 2015)—care must be taken regarding the impact these systems will have on labor. Balancing ethics with innovation is key to building a sustainable future for the economy and its workforces.

10.3.3. Bias and Fairness in Algorithmic Decision-Making

The emergence of autonomous systems created consternation regarding bias and fairness in algorithmic decision-making. With autonomous systems becoming more intermediaries for human-human interaction and access to resources, it is essential to examine the algorithms upon which they operate. Algorithms mirror their creators' biases, leading to undesirable consequences favoring some and harming others. This reality also necessitates models for assessing fairness in such systems, where the moral implications of biased algorithms build entrenching biases. The criticality of algorithmic fairness emerges from stakeholders' calls for end-to-end assessment mechanisms to handle such biases, challenging a more nuanced comprehension of how independent systems work in multicultural environments, and pushing for an ongoing dialogue of their effects on society (Fantin et al., 2020). Frameworks used to ensure algorithmic fairness tend to adopt a limited definition, addressing only the most obvious categories of bias in input and output data. They may overlook the more nuanced systemic problems that arise from the broader context within which algorithms operate. Thus, algorithmic fairness will in most cases respond only to accusations of discrimination, ignoring the wider social and economic harms that these technologies cause to populations and ecosystems. Reliance upon the service providers alone to ensure accountability worsens the situation, particularly if these providers have conflicting interests or are unclear. In order to counter such drawbacks, the development of Protective Optimization Technologies (POTs) can provide innovative alternatives. By allowing affected groups to contest harms brought about by algorithms beyond algorithms themselves, POTs provide a way of questioning embedded biases in classical forms and pushing for more inclusivity in design (Muhammad et al., 2020). The quest for fairness in algorithmic decision-making raises some basic ethics and governance challenges requiring collaboration between academia, business, and government. Multi-stakeholder collaboration is essential in formulating inclusive guidelines which counteract algorithmic accountability in accordance with a commitment to safety and security in autonomous systems. Initiatives such as the High-Level Expert Group (HLEG) on AI Ethics Guidelines for Trustworthy AI illustrate this multi-stakeholder collaboration and its demand for cooperative responsibility in the deployment of AI technologies. When these issues are being debated by stakeholders, it becomes increasingly necessary to intertwine technological advancement with ethical values for the promotion of equality. Such an integrated approach can bring about the responsible use of autonomous systems and cause them to enhance, rather than reduce, social equity and security (Fantin et al., 2020) (Figure 10.10).

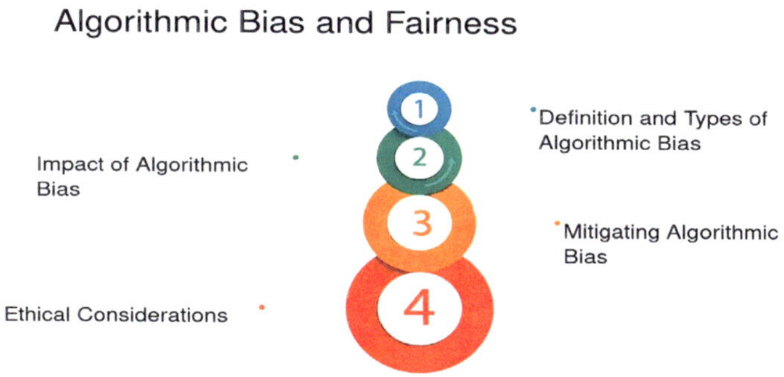

Figure 10.10. *Algorithmic bias and fairness.*

Source: https://fastercapital.co/i/Social-Media-Challenges--How-to-Overcome-the-Common-and-Emerging-Challenges-of-Social-Media-Analysis--Algorithmic-Bias-and-Fairness.webp.

10.4. REGULATORY FRAMEWORKS FOR AUTONOMOUS SYSTEMS

The design and incorporation of autonomous systems, namely the transport through self-driven vehicles, need a robust regulatory framework that addresses technical as well as ethical concerns. Such frameworks need to be developed in such a way that they not only ensure safety and efficiency of operations but also adhere to ethical benchmarks suitable for social norms. Current controversies over autonomous cars illustrate this complexity where technological advancement must work alongside ethical aspects, such as the issues raised by the trolley problem. The dilemma emphasizes the necessity of using an applied engineering ethical framework where regulation addresses practical engineering problems and the influence of software engineering on social ethics. Hence, rules have to be responsive as these systems are tested and rolled out, emphasizing ongoing evaluation alongside technological innovation in order to enable smooth integration into public existence (Dodig-Crnkovic et al., 2018). Within the context of national security, application of artificial intelligence (AI) in autonomous systems adds additional challenges that make regulation more complex. As identified in extensive research, AI has the capability to enhance the efficiency of intelligence activities; however, incorporation into existing security mechanisms also poses significant concerns of privacy and human rights.

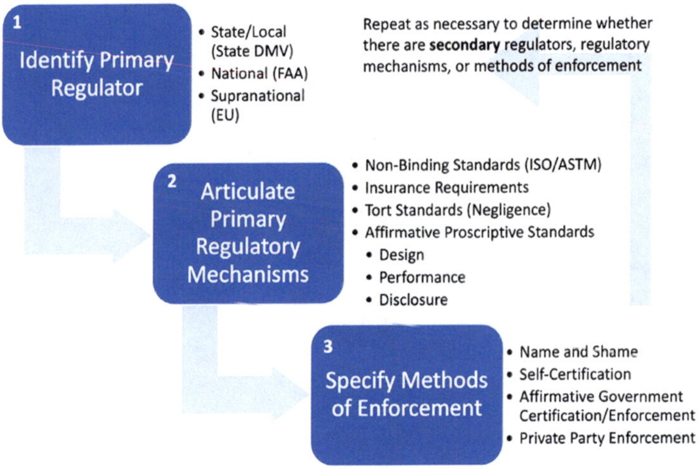

Figure 10.11. *Work on the future of regulating autonomous systems.*

Source: https://mnemosyne.umd.edu/tomcat/newsengine/articleImg/article15254.large.jpg.

Regulatory frameworks must therefore be adapted to contain these risks while balancing the benefits AI technologies have to offer. In particular, continuous policymaking is required to respond to the evolving threats posed by AI, making the use of such technologies open and accountable to human rights norms. This requires foresight in the creation of policies that continually assess the implications of using AI in national security operations (Babuta et al., 2020). Moreover, there should be an integrated regulatory framework for autonomous systems that addresses not only technical and ethical considerations but also incorporates stakeholder participation. Stakeholder participation from several stakeholders—industry experts, researchers from academia, and members of civil society—is likely to lead to more comprehensive rules that anticipate and address various issues concerning safety, security, and ethics. This process facilitates openness and enables standards to be developed to govern the production and implementation of such technologies. The multi-dimensional approach laid out above ensures that the regulations are not just reactive to problems but are anticipatory and hence, so increase innovation and decrease risks. When evolving, regulations need to be adaptive in that they can still accommodate the rapid innovations in autonomous technology, as with the fluidity in technological innovation and society values (Babuta et al., 2020; Dodig-Crnkovic et al., 2018) (Figure 10.11).

10.4.1. Current Regulations and Standards

The development of autonomous technology, particularly in the transportation industry, has created a pressing necessity for strong regulations and guidelines. These systems are meant to ensure security and safety of such systems and address ethical concerns regarding their use. The regulators are starting to realize the complexity of autonomous technologies, which are accompanied by not only intricate engineering challenges but also intricate social implications. To negotiate these difficulties, what is required is an expansive body of standards that focuses on the operational parameters that characterize autonomous vehicles. This means implementing rigorous testing protocols designed to determine safety and robustness in a variety of scenarios. It is necessary to tackle these layers because they collectively have the function of implanting public trust in autonomous systems, which is crucial for their widespread acceptability and integration into existing transport infrastructures (Dodig-Crnkovic et al., 2018). Further, the latest advancements have emphasized harmonization of technical standards with ethics in autonomous technology. Due to progress in machine learning and automated decision-making, a greater need is felt to bring the underlying algorithms and their effects on societal values under an examination lens. The governing frameworks are not required to merely incorporate safety and performance criteria but also address the ethical dimensions of decision-making processes intrinsic in the technologies. For instance, the so-called trolley problem is typically argued philosophically; however, laws need to make these philosophical puzzles into operative norms that developers can carry out. The need for an applied engineering ethics exists since it fosters a richer understanding of how autonomous systems interact with the subtleties of the real world (Bloomfield et al., 2020). This convergence of regulation, ethics, and engineering is critical to leading-edge governance of future technologies in our society. Lastly, standard-setting on a global basis is becoming increasingly necessary as autonomous technologies extend beyond geographical boundaries. Collaboration between nations and industry stakeholders is critical to developing harmonized standards that support equitable and safe deployment of autonomous technologies globally. Such standards must be capable of adjusting to the ever-changing nature of technological advancements, tools, and methods supporting autonomous systems, like in software engineering and vehicle assurance. Multiple different regulatory models can offer opportunities for the identification of gaps in current frameworks, offering prospects for development. As evident from the efforts taken in projects that are aimed at bolstering the trust of self-driving cars, an approach based on multidisciplinary research incorporating resilience, safety requirements, and safety analysis grounded in security is critical (Bloomfield et al., 2020). These coordinated efforts will not only augment the efficiency of regulations but will also foster innovation coupled with concern for public safety and ethical issues during the age of autonomous systems.

10.4.2. International Cooperation and Agreements

International cooperation has more and more become essential in the case of autonomous systems, particularly when safety, security, and ethical concerns are involved. Nation-states are faced with escalated challenges regarding the application of such technology, necessitating cooperative regimes in addressing collective issues. The creation of lethal autonomous robotics, for example, is representative of the necessity for regulation of their design, deployment, and utilization. Global standards can establish common legal and moral requirements, promoting accountability and transparency among nations. The debate for such guidelines needs to be inclusive with a wide variety of stakeholders ranging from governments to technologists, ethicists, and civil society organizations so that diverse perspectives are incorporated into policymaking. Finally, the cooperation of nations on an international scale can lead to consensus where not just the advancement of technology but its effect on human rights, security, and ethical leadership on a global platform takes the forefront. Regulation of emerging technologies like autonomous weapons systems is a significant concern area for international cooperation. The complexities involved with such systems necessitate an overall approach that includes harmonization of policies and ethics, particularly as it pertains to averting misuse. Current discourse touches upon the need for standards capable of addressing these issues in an effective manner, invoking the establishment of international agreements or treaties as an attempt to avert threats and ensure proper utilization. As dictated by present policy debates, such pacts can range from ethical standards to real legal treaties controlling the blueprint and implementation of hazardous autonomous weapons. Secondly, the diversity of national interests can make it challenging to achieve common agreement. However, with constant dialogue and collaboration, it is possible to devise a cohesive global approach that promotes security and safety, enabling compliance with critical ethical standards among states. The role of modern EU systems for the extension of international cooperation once again proves the nexus between security, safety, and ethical governance of autonomous systems. The EU has been actively engaging in the formulation of policies that affect a wide range of sectors, including food safety, which involves intricate value chains spanning multiple nations. The strict regulations established within the EU, relating to such matters as fair trading practices and food integrity, provide the example to follow in relation to autonomous systems globally. The examples of the EU illustrate the manner in which tougher regulation will lead to improved monitoring and adherence and thus greater transparency and eradication of fraud in advanced technological environments. Such structures not only ease national policy coordination but also allow broader discourse on equity and accountability for the governance of future technologies around the world, leading to safer and more ethically sound utilization of autonomous systems across industries (Allenby et al., 2011; Barling et al., 2019) (Figure 10.12).

Figure 10.12. *Emergence of lethal autonomous robotics.*

Source: https://static01.nyt.com/images/2018/11/15/magazine/15atwar-robots-2/15atwar-robots-2-superJumbo.jpg.

10.4.3. Future Directions for Policy Development

Increasing deployment of autonomous systems across sectors highlights the need to develop comprehensive policy for their safe and responsible use. Emerging policy directions should be on the basis of a collaborative way forward among governments, industry, and academia. Multistakeholder involvement is essential to foster multi-dimensional discourse for addressing the complexity of safety and security concerns. Policymakers must present a priority for establishing standards governing the interoperability and performance of autonomous systems because these standards will directly influence determining trust and responsibility. Policymakers must simultaneously ensure that ethical considerations are not set aside while pursuing technological advancement. As noted by the need for strong and effective AI, policies must develop an environment for innovation while ensuring public interests (Dewey et al., 2015). An important aspect of developing future policy is comprehending the peculiar vulnerabilities imposed by autonomous systems in various contexts. As autonomous technologies expand, they also open doors to cyberattacks and duplicitous practices that take advantage of human users, particularly vulnerable sections such as students and young people. Having learned from the hybrid war in Ukraine, which demonstrated the paradigm shift of military engagement from targeting infrastructures in isolation to targeting people, policy platforms must prioritize educational programs to increase awareness on cybersecurity and digital literacy. This will help create a robust citizenry that can navigate the complexities that accompany autonomous systems, thus reducing exposure to social engineering and disinformation campaigns (Burov et al., 2020). Finally, policy design should be foresighted, with adaptive mechanisms capable of addressing emerging challenges and rapid technological developments. Policymaking should consider implications of artificial intelligence, machine learning, and other autonomous technology on societal values and norms. This will require an evolving policy climate that has the ability to respond to new findings and shifting mores in technology. Continuous research and advice from technology experts, ethicists, and legal scholars will be important to help craft these nimble policies. Besides, with further deliberations on the safety, security, and ethics of autonomous systems, it will also be necessary that policies are not rigid in nature to allow for timely adjustments reflective of real-time societal needs and ethical compulsions (Dewey et al., 2015).

10.5. PUBLIC PERCEPTION AND ACCEPTANCE

Public acceptance and perception of autonomous systems are crucial to ascertain their integration into everyday life. Arguments favoring or opposing these technologies have been prevailed by concerns at a societal level. Societal aspects such as safety, security, and ethics play substantial roles in determining society's attitude towards autonomous systems. A systematic literature review suggests that while economic gains related to autonomous systems are expected, societal concerns have remained profound. The analysis points to regulatory issues and issues of environmental impact, showing the need for inclusive communication practices to detail the benefits and mitigation strategies for any potential risk due to such technology (Biehle et al., 2020). Finally, gaining public trust should involve open discussion that focuses on the potentialities and potential weaknesses of such technology so that people can easily embrace it with more information. Following the pillar of public trust, the social implications of autonomous systems should not be overlooked, particularly with the unintended consequences outlined by expert measurement. The problem of digital transformation affirms that technology develops along with accompanying problems that may exacerbate social divisions. For example, key challenges such as the digital divide continue to hinder equal access to innovations that could potentially enhance safety and security in autonomous systems (Garcia et al., 2020). Because public distrust is most likely to stem from experiences of inequality and exclusion, stakeholders must prioritize making inclusiveness central in creating and rolling out these technologies. These include involving interesting diverse societies in the conversation regarding ethical action and making sure that technology serves to connect, not isolate, social cleavages. Through active involvement in addressing such underlying problems in society, public acceptance is greatly increased. Finally, establishing an active conversation between technology producers and impacted societies makes sure that collaboration is promoted. Public perception is most influenced by the degree of participation and transparency offered by stakeholders in the design phase. Projects that include the public in the discussion, solicit feedback, and incorporate citizen issues into the design can lower apprehension over safety and ethics. An emphasis on empirical evidence over speculative benefits, particularly within traffic and environmental effectiveness, also serves to substantiate the claims of the autonomous system (Biehle et al., 2020). Secondly, reconciling the potential rebound effects of digital transformations is a signal of developers' duty to anticipate and defend against unforeseen effects (Garcia et al., 2020). More often than not, the development of a spirit of shared ownership and responsibility promotes an environment for the adoption of autonomous systems, molding them to meet communal dreams and demands.

10.5.1. Factors Influencing Public Trust

The subtleties of public trust, particularly that in autonomous systems, are shaped by several critical factors that must be taken with seriousness. Foremost among them is the transparency of operation of the technology, involving not only how decisions are made but also the algorithms that influence autonomous behaviors. An open system fosters a culture whereby users can understand and anticipate outcomes, thereby elevating their comfort with the technology. Apart from that, the regulatory frameworks also enable confidence that standards of safety are upheld, thus ensuring confidence. Without regulations and guidelines, potential users would doubt whether or not such systems can be relied upon, thus their reluctance to use such technologies. Therefore, understanding how transparency is balanced against regulation becomes important in establishing a confidence foundation that enhances more people accepting and using autonomous systems. Furthermore, public opinion regarding the moral concerns of autonomous systems has a significant influence on trust levels. Moral concerns vary from privacy of data through to whether machines have the moral obligation to take decisions that determine life and death.

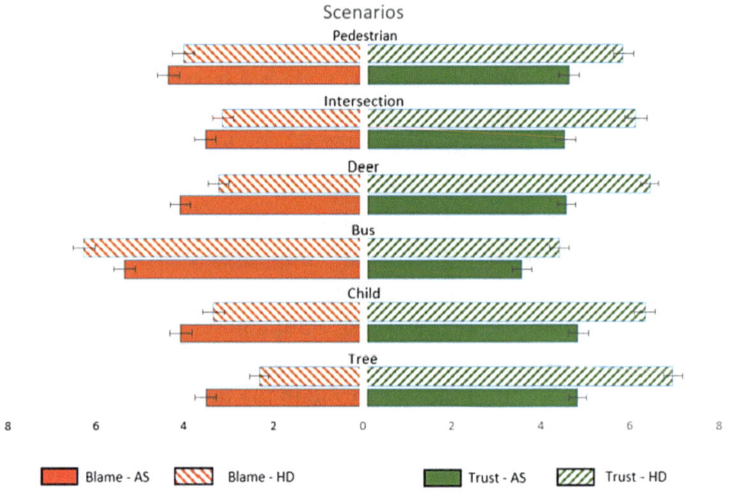

Figure 10.13. *Mean ratings of blame on the target vehicle and post-incident trust in both the human driver and autonomous system conditions across all scenarios.*

Source: https://www.sciencedirect.com/science/article/pii/S0965856423003075.

As autonomous systems become increasingly present in everyday life, public discourse on these moral concerns takes center stage. Involving the community through open forums is capable of demystifying these technologies, making room for thorough discussions that cover concerns and emphasize responsible practices. The inflow of opinion also ensures that developers keep in touch with societal values and expectations, hence building trust. As noted in the discourse on digital transformation, making use of a diversity of stakeholders in the conversation can open up understanding of potential vulnerabilities and ethical threats, which enhances public confidence in the technology (Garcia et al., 2020). Last, but most certainly not least, the role played by empirical proof in shaping public trust cannot be overstated. Displaying tangible benefits through case studies of successful implementations of the technology adds credibility and ensures the public of the technology's capability to improve safety and efficiency. For instance, advancements in work organization have shown that high performance work practices (HPWPs) can be used to achieve positive results for employees and organizations, therefore providing an example of where trust can be earned on the basis of proven results (Cox et al., 2012). With the evolution of autonomous technologies, showing their reliability and efficiency when put into action benefits not only in eliminating fears about these technologies but also in highlighting the advantages they can bring to society. A clear, evidence-based deliverance assists developers in minimizing disbelief and establishing a faith-based relationship between the public and autonomous technologies (Figure 10.13).

10.5.2. Role of Education and Awareness

The significance of education and awareness in autonomous systems cannot be stressed too much, particularly when it comes to safety and security. As technology advances, it becomes important that students and professionals learn the ethical implications and threats surrounding autonomous systems. Educational institutions play a vital part in this, as it is their responsibility to educate about new technologies, the good they do, and their potential downsides. By integrating cybersecurity education, artificial intelligence ethics, and governance into the curriculum, institutions of learning prepare individuals with skills to question and interact with autonomous technologies. This proactive approach fosters a culture of awareness in the field of cybersecurity needed to maintain public confidence in such systems, as reflected in reports that highlight the need for leveraging education to build public resilience to emerging threats from technological leaps (Burov et al., 2020). Awareness campaigns targeted at the masses and most susceptible categories of people, including children, are part of educating the public on cyber security and safety. Special training programs promoting digital literacy and familiarizing users with a sophisticated understanding of social engineering methods prove to be useful in preparing individuals for identification and proper response to threats. By educating their students as well-informed users and guardians of responsible technology practice, campaigns can accomplish a lot towards minimizing risks associated with autonomous systems. The cognitive vaccination concept, presented in formal literature, synthesizes the idea of equipping individuals with critical thinking skills to counteract false data and manipulation (Burov et al., 2020). Good education practices and awareness campaigns can ultimately provide a safer and responsible context for the use of autonomous systems. Furthermore, comprehensive training courses are needed for professionals in charge of the development and operation of autonomous systems. This type of training would have to cover not just technical proficiency but also the ethical dimensions of application of such systems. A collaborative effort between schools, industry stakeholders, and policymakers, as in initiatives such as the Centre for

European Policy Studies Task Force, reflects the common responsibility in having a solid understanding of AI and cybersecurity interfaces. By integrating learning structures that reflect actual conditions, the participants are better placed to cope with the complexities of cyber-attacks and ethical challenges surrounding autonomous systems. This emphasis on continuous learning and inter-disciplinary debate is critical in the establishment of a secure and secure world in terms of technology since it broadens participants' horizons and makes them better placed to cope with ethical challenges.

10.5.3. Case Studies of Public Response to Autonomous Technologies

The adoption of autonomous technology in everyday life has elicited varied public responses, largely influenced by factors such as perceptions of safety, morality, and technological literacy. Case studies reveal a rich brocade of responses, with municipalities grappling with the implications of such innovations. For instance, autonomous car usage in urban centers has provoked concerns on liability and accountability in the event of an accident. When citizens witness these technologies in operation, they lose confidence in the systems, bringing to light the importance of robust safety measures and clear regulatory principles. In addition, the different reactions point to the importance of engagement with public issues by way of frank communication and education to secure approval of autonomous systems and prevent hostility based on fear and misunderstanding. An integral part of public response to autonomous technology is the ethical issues raised by its arrival. Case studies illustrate that as some parts of society perceive options for increased efficiency and availability; others protest against privacy violations and loss of control. Such scenarios call for the need for a Value-Sensitive Design paradigm, wherein one emphasizes the embedding of stakeholder values into the design process of autonomous systems. These strategies require developers to heed the broader social context and ethical implications of their technologies, thus creating trust and responsibility. By progressive evaluation of ethical concerns and inclusive design principles, stakeholders manage to better cope with public skepticism issues, ensuring autonomous technology development evolves according to societal expectations and norms, hence facilitating overall public acceptance and support (Bellis et al., 2018). Furthermore, public reactions are largely determined by perceived safety assurances for autonomous systems, like in the reception of autonomous cars. Technical notes and reports, such as written in the Tigar's project, highlight the very need for strict evaluations that fully take into consideration safety and security concerns. The assurance measures like simulations, dynamic testing, and formal verification of machine learning systems are not just technical necessities but also a public expectation. If people observe thorough testing and transparent safety procedures, they will trust autonomous systems more. By incorporating safety standards and regulatory norms in a way that makes them publicly acknowledged and disseminated, developers can construct public fear so that there is more informed debate on the social aspects of such technologies (Bloomfield et al., 2020). Incorporation of safety, ethics, and security is thus essential in designing public opinion around the future of autonomous systems (Figure 10.14).

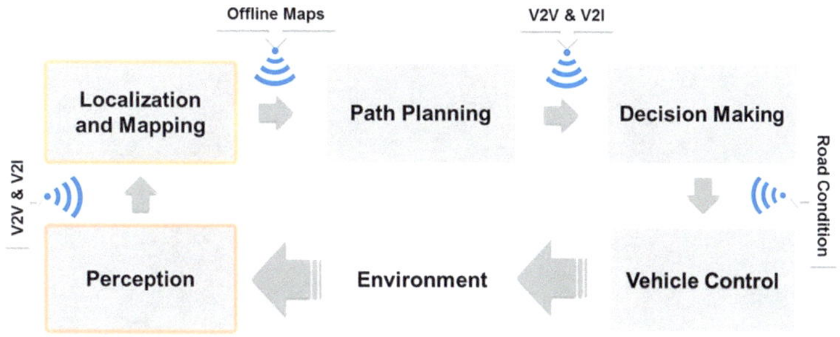

Figure 10.14. *Autonomous vehicle perception.*

Source: https://ars.els-cdn.com/content/image/1-s2.0-S0968090X18302134-fx1.jpg.

10.6. TECHNOLOGICAL ADVANCEMENTS AND INNOVATIONS

Autonomous systems represent a significant shift in technology innovation that reshapes how tasks are performed in sectors. Innovations in artificial intelligence (AI) and machine learning have enabled the creation of autonomous agents with the capacity for real-time decision-making from large sets of data inputs. Merging the technologies raises serious ethical issues around design strategies. Value-Sensitive Design (VSD) is the concept which encapsulates the imperative to incorporate human values in the design of AI systems. Value-sensitive design promotes proactive ethics, citing that early design reflections play a significant role in assuring safety and reliability of autonomous systems in real-world contexts, and emphasizing the threats arising from neglect of ethical aspects in the process of technology development (Bellis et al., 2018). With self-driving cars coming to the market, the effects of these technologies extend beyond the boundaries of technological potential to brush against social organization and labor markets. The future arrival of autonomous cars has prompted much debate regarding their broader implications, particularly in terms of work. There are reports that a maximum of 30 firms is near to massive deployment of such vehicles, leading to large-scale job displacement, especially in driving jobs (Austin et al., 2017).

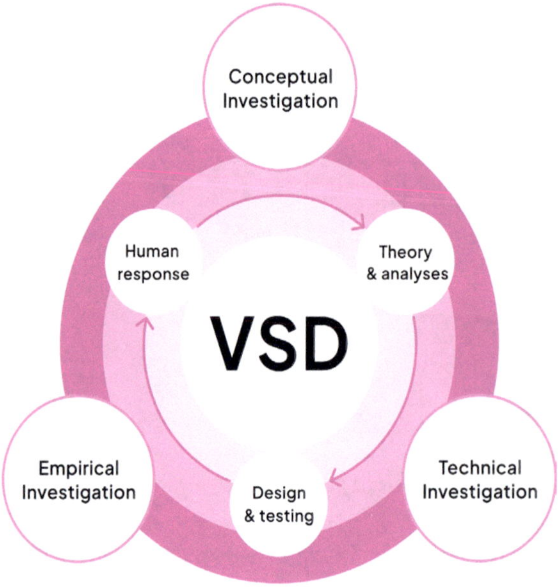

Figure 10.15. *Value sensitive design.*

Source: https://muteo.co/wp-content/uploads/2022/09/muteo-value-design-diagram-min.svg.

A transition of this nature will require a robust safety net to support workers affected by such changes, which implies that ethical considerations of autonomous technologies will have to deal with socio-economic effects as well. Protection of livelihoods of displaced workers and promoting technological development present a strange challenge to maintain innovation and social responsibility. Complying with the challenge of addressing safety and security of autonomous systems demands diverse approaches taking into consideration both technological advancement and ethical factors. Multidisciplinary testing and regulatory architectures must be developed for safe deployment of such systems in everyday life. The prioritization of accountability in AI development and deployment speaks to the essence of VSD principles in prioritizing stakeholder engagement throughout the development process. In accepting that the pace of innovation today requires regulatory paradigms to be rethought to address emerging risks, academics and business professionals accept that regulatory evolution is necessary. Lastly, constructing a safety and ethical culture in the innovation ecosystem of autonomous systems not only protects the users but also sustains public trust, which is vital for the long-term success of such disruptive technologies (Bellis et al., 2018) (Figure 10.15).

10.6.1. Emerging Technologies in Autonomous Systems

The rapidly improving pace of technology has brought with it an era in which autonomous systems, driven by sophisticated algorithms and machine learning, are revolutionizing industries and everyday life. The new technologies can potentially streamline processes, enhance accuracy, and reduce the requirement of human intervention in countless sectors, including transportation, healthcare, and manufacturing. The interface of these systems, however, brings with it an intense examination of safety protocols, particularly in high-stakes environments. As autonomous systems become more prevalent, safety features have to be incorporated during design time to reduce the risks that arise from failures. In this context, the Value-Sensitive Design method appears to be a progressive method, encouraging the integration of stakeholder values into the design process. By integrating safety aspects from the outset, designers have the ability to address issues surrounding the dependability and predictability of autonomous systems and ultimately cultivate public trust in such technologies (Bellis et al., 2018). Additionally, the ethical aspects of the application of cutting-edge technology in autonomous systems cannot be overlooked. Because the systems have a tendency to operate in environments that are filled with moral ambiguities, algorithms that govern them must be questioned based on their ethical underpinnings. The challenge is to remain efficient but also be morally accountable, particularly when autonomous systems are faced with decisions that have significant effects on ethics. In the instance of autonomous vehicles, decisions regarding avoiding collisions involve tough ethical computation. It is important to have the ethical systems determine how autonomous technologies are designed and used to keep societal values intact. This is particularly significant in light of the potential for biases to be unintentionally embedded into algorithmic decision-making systems. For this reason, it is imperative to develop inclusive design practices to prevent reinforcing existing inequalities and advancing fairness and transparency in autonomous systems (Apesoa-Varano et al., 2020). Moreover, the intersection of emerging technologies and allied ethical concerns raises immense security and privacy concerns. As autonomous systems increasingly rely on data-driven knowledge, the risk for unauthorized access and exploitation of sensitive information increases. The merging of technologies such as robotics, wireless sensors, and data analytics necessitates robust security frameworks to protect against vulnerabilities that could be exploited by malicious actors. Furthermore, the significance of equal access to these new solutions should not be overlooked, as differences in access to technology can further reinforce social disparities. Addressing these issues through public policy mechanisms and ethical standards will enable stakeholders to strive for an environment within which autonomous systems can thrive while keeping safety, security, and ethical concerns at the forefront. This strategy of balancing is crucial in realizing the potential of the new technologies to establish a world for society better off (Apesoa-Varano et al., 2020).

10.6.2. Integration of AI and Machine Learning

The union of artificial intelligence (AI) and machine learning with autonomous systems is changing industries in a fundamental manner, enhancing operations and decision-making effectiveness. By applying machine learning algorithms, autonomous systems can process enormous data in real-time, enabling adaptive behaviors essential for high-volatility or uncertainty environments. For instance, AI can be used to improve route navigation in autonomous vehicles to optimize safety and efficiency by learning to respond to dynamic elements such as traffic patterns and weather conditions. While these developments, therefore, require a keen examination of the ethical dimensions accompanying their application. The goal is to maximize the benefits of AI and safeguard against harm so as to ensure that the technologies involved are resilient and of positive character towards society, as stated by ongoing research efforts dedicated to defining this complex issue (Dewey et al., 2015). Additionally, the ethics of integrating AI are significantly dependent on safety and security measures. Autonomous systems must adhere to a framework that includes resilience and safety standards, as described in the Technical Topic Notes developed in the Tigar's project. The notes provide relevant issues such as formal verification and static analysis of machine learning systems, which are vital in guaranteeing the validity of AI applications in autonomous systems. Such assessments are intended to determine if such systems, in addition to functioning as intended, also continue to be secure from potential penetrations or manipulations that could compromise their operating integrity. The integration of AI with assurance practices emphasizes the requirement for a multidisciplinary system design incorporating technical expertise together with ethical control to adequately address security concerns (Bloomfield et al., 2025). Finally, as AI and machine learning extend their reach in the field of autonomous systems, guidelines and standards become a necessity. Overcoming the various challenges emanating from these technologies is about having an entire framework encompassing simulation, dynamic testing, and defense in depth concepts. The plan includes the deployment of many layers of security controls whereby the autonomous systems are made increasingly resilient to the attacks of adversaries. Also, there is a necessity to foster an open discussion of the ethical application of AI to guide regulatory efforts and stakeholder engagement. Such a procedure can ensure ethical issues are kept at the forefront of development, ensuring responsible integration of AI and machine learning into autonomous systems to maximize their benefits while mitigating risks involved (Bloomfield et al., 2025; Dewey et al., 2015).

10.6.3. Future Trends and Potential Developments

The future of autonomous technologies is increasingly interwoven with the philosophical method known as Value-Sensitive Design (VSD), which promotes the integration of human values into technological advancement from the outset. With autonomous technologies becoming commonplace in most industries, including transportation and healthcare, an anticipatory design approach is essential to steer clear of moral traps. VSD suggests that integrating stakeholders' values at the outset can encourage the harmonization of autonomous agent development with societal values and ethical principles. The methodology addresses not only safety and security but also ensures that these systems do not reinforce existing societal wrongs. With technology evolving at lightning pace, application of VSD principles will be integral to defining more responsible development and deployment of autonomous systems, ultimately leading to acceptance and trust by users and the wider public (Bellis et al., 2018). In the geopolitical sphere, artificial intelligence (AI) will redefine international relations and military power. The application of AI technologies among military forces is a paradigm shift capable of altering the power balance among nations. Nations that can effectively employ AI can potentially enjoy significant superiority in defense and attack capability, leading to an arms race of autonomous weapons. Moreover, with commercial sector innovation driving improvement in AI, its military applications can quickly become widespread. Such a reality makes for a double-edged sword: while innovators can transform warfare, the unpredictability of the technological path of AI can deliver difficulties to current military hegemons. Therefore, it is important to understand how this will unfold in order to inform national security policy in an escalating competition environment (Horowitz et al., 2018). Given the intricacies and the rapid rate of progress in the autonomous system technology, inter-disciplinary collaboration between policymakers, ethicists, and technologists will be important in addressing the myriads of safety and security challenges. Inter-mixing various perspectives will help ensure that new technologies align with moral standards as well as meet safety standards. The more independent systems get, the more likely they are to have unforeseen side effects, which raise the need for balanced regulatory frameworks. Anticipating potential dangers and ethics surrounding AI, stakeholders are able to work together in creating a conciliatory strategy that protects human interests without stifling innovation. Future developments in autonomous technology will demand coordinated effort to align technology and ethics in a symphony so that safety is always the driving factor as autonomous systems increasingly pervade everyday life.

SUMMARY

An inquiry into the safety, security, and ethics of autonomous systems, there is a need to appreciate the dual-edged aspect of such technologies. While on one front they provide revolutionary opportunities in numerous fields, including healthcare, transport, and finance, and can potentially enhance productivity and decision-making capability. For instance, the creation of artificial intelligence can make processes automatic and improve the user experience in complex sectors such as banking and tourism, as seen in the observations derived from professionals across different domains who highlight both the potential applications and intrinsic challenges of such systems (Yogesh et al., 2023). However, these advancements are being brought along with significant ethical and security challenges that need constant monitoring, most prominently privacy, bias, and the possibility of exploitation. The moral implications for autonomous systems call for a rigorous regulative system in order to ensure that such technologies are adopted responsibly. As explained in recent debates about their position in society, topics such as algorithmic prejudice and openness about decision-making processes are paramount issues requiring urgent remedy. This is particularly relevant given that the potential for autonomy misuse can have adverse effects on vulnerable members and foster suspicion over technological advancement. In addition, deployment of these systems necessitates a multidisciplinary approach, utilizing multiple areas of expertise to counter threats of privacy infringement and disinformation, a realization expressed in recent debates around the ethics of deploying AI technology (Yogesh et al., 2023). Lastly, the creation of autonomous systems will continue to influence societal norms and ethical guidelines, particularly with the advent of such environments as the metaverse. The convergence of virtual and augmented reality technologies will break down the boundaries between real and virtual interactions, profoundly affecting users' ways of interacting and relating to one another. Such technological innovations will necessitate fresh reevaluation of existing ethical frameworks to manage the complexities of trust, user privacy, and the psychological implications of interactive immersion. As the future of autonomous technologies unfolds, there will need to be continuing research and discussion to guide a future that prioritizes ethical factors, security measures, and the protection of personal rights. Thus, stakeholders have an obligation to work together to address the challenges and opportunities of these revolutionary systems. The development of autonomous systems, particularly in transport, necessitates a multi-disciplinary understanding of safety, security, and ethical issues. Recent controversies over self-driving cars, for instance, illustrate how such cars pose numerous engineering problems alongside social and ethical issues. Traditional paradigms, including the famous trolley problem, oversimplify too complicated decision-making scenarios that are real for these systems. Instead, an engineering ethical approach with a greater emphasis on practical applications and the integrative role of software in control of autonomous vehicles is needed. Confronting these intricacies requires a determination to not only technologically develop, but to integrate ethics into design and implementation and thereby close the gap between engineering and ethics in this rapidly evolving sector (Dodig-Crnkovic et al., 2018). With autonomous vehicle assurance and safety evolving, solid evaluation frameworks are paramount. Tigar's project research emphasizes how safety requirements and resilience in autonomous systems must be taken into account. This entails a thorough examination of simulation and dynamic testing and the application of formal verification and static analysis methods on machine learning systems. The concerns that arise through the incorporation of security-informed safety analysis in the assurance program reflect existing weaknesses that require improvement in order to increase the credibility and usefulness of autonomous vehicles. Thus, creating robust standards and guidelines is critical to ensure that these systems are secure and safe to use in our transportation infrastructure (Bloomfield et al., 2025). Ethics are a major consideration in shaping the debate for autonomous systems. As engineers and designers wrestle with integrating ethical norms into their work, they need to think through the social consequences of their technological practice. In particular, the potential for bias in algorithmic decision-making necessitates attention to inclusivity and fairness. Developing transparency and accountability, both essential to public trust, can be achieved by collaborating with communities and stakeholders. Secondly, responsible use of autonomous vehicles must include ongoing engagement with the governing bodies so as to ensure continued compliance as well as evolving in response to the fast-paced environment of technology. Ultimately, resolution of these ethical considerations will not only enhance public safety and confidence but also be part of accountable development of autonomous systems (Dodig-Crnkovic et al., 2018).

REFERENCES

1. Ah Muhammad, A. A., Angwin, J., Benner, K., Biggio, B., Bird, S., Browne, S., et al. (2020). POTs: Protective optimization technologies. http://arxiv.org/abs/1806.02711 (accessed on 10th September 2025).
2. Allenby, B., Arkin, R., Barrett, E. T., Borenstein, J., et al. (2011). International governance of autonomous military robots. https://core.ac.uk/download/161437872.pdf (accessed on 10th September 2025).
3. Apesoa-Varano, E. C., Gladstone, C., Kim, K. K., Lindeman, D., et al. (2020). Technology and caregiving: Emerging interventions and directions for research. https://core.ac.uk/download/323070528.pdf (accessed on 10th September 2025).
4. Austin, A., Bucknor, C., Cashman, K., & Rockeymoore, M. (2017). Stick shift: Autonomous vehicles, driving jobs, and the future of work. https://core.ac.uk/download/80511025.pdf (accessed on 10th September 2025).
5. Babuta, A., Janjeva, A., & Oswald, M. (2020). Artificial intelligence and UK national security: Policy considerations. https://core.ac.uk/download/305121521.pdf (accessed on 10th September 2025).
6. Barling, D., Gresham, J., Mylona, K., Sharpe, R., et al. (2019). Characterization framework of key policy, regulatory and governance dynamics and impacts upon European food value chains: Fairer trading practices, food integrity, and sustainability collaborations. *VALUMICS project "Understanding Food Value Chains and Network Dynamics" funded by EU Horizon 2020 G.A. No 727243. Deliverable D3.3.* https://core.ac.uk/download/287582350.pdf (accessed on 10th September 2025).
7. Bergsma, E., Biesbroek, G. R., Brink, M. H., van den, A., et al. (2010). How Dutch institutions enhance the adaptive capacity of society. https://core.ac.uk/download/pdf/29237721.pdf (accessed on 10th September 2025).
8. Biehle, T., Fischer, L., & Kellermann, R. (2020). Drones for parcel and passenger transportation: A literature review. https://core.ac.uk/download/323212148.pdf (accessed on 10th September 2025).
9. Bloomfield, R. E., Fletcher, G., Imai, K., Khlaaf, H., et al. (2020). Towards identifying and closing gaps in assurance of autonomous road vehicles–A collection of technical notes Part 1. https://core.ac.uk/download/305120054.pdf (accessed on 10th September 2025).
10. Bloomfield, R. E., Fletcher, G., Imai, K., Khlaaf, H., et al. (2025). Towards identifying and closing gaps in assurance of autonomous road vehicles–A collection of technical notes Part 2. https://core.ac.uk/download/305120055.pdf (accessed on 10th September 2025).

11. Burov, O., Krylova-Grek, Y., Lavrov, E., Lytvynova, S., et al. (2020). Кибербезопасность в образовательных сетях. https://core.ac.uk/download/322373029.pdf (accessed on 10th September 2025).
12. Cox, A., Rickard, C., & Tamkin, P. (2012). Work organization and innovation. https://core.ac.uk/download/16749552.pdf (accessed on 10th September 2025).
13. De Bellis, A. F., & Umbrello, S. (2018). A value-sensitive design approach to intelligent agents. https://core.ac.uk/download/146502298.pdf (accessed on 10th September 2025).
14. Dewey, D., Russell, S., & Tegmark, M. E. (2015). Research priorities for robust and beneficial artificial intelligence. https://core.ac.uk/download/83234142.pdf (accessed on 10th September 2025).
15. Dodig-Crnkovic, G., Holstein, T., & Pelliccione, P. (2018). Ethical and social aspects of self-driving cars. http://arxiv.org/abs/1802.04103 (accessed on 10th September 2025).
16. Dwivedi, Y. K., Hughes, L., Baabdullah, A. M., Ribeiro-Navarrete, S., Giannakis, M., Al-Debei, M. M., Dennehy, D., et al. (2022). Metaverse beyond the hype: Multidisciplinary perspectives on emerging challenges, opportunities, and agenda for research, practice and policy. *International Journal of Information Management, 66*, 102542. https://doi.org/10.1016/j.ijinfomgt.2022.102542 (accessed on 10th September 2025).
17. Fantin, S., Ferreira, A., & Pupillo, L. (2020). CEPS Task Force on Artificial Intelligence and Cybersecurity Technology, Governance and Policy Challenges Task Force evaluation of the HLEG Trustworthy AI Assessment List (Pilot Version). *CEPS Task Force Report, 22 January 2020.* https://core.ac.uk/download/287647893.pdf (accessed on 10th September 2025).
18. Firschein, O., Georgeff, M. P., Kautz, W. H., Levitt, T., et al. (2025). NASA space station automation: AI-based technology review. https://core.ac.uk/download/pdf/42844682.pdf (accessed on 10th September 2025).
19. Garcia, F. C. O., Cardona, D., Chesñevar, C., Collazzo-Yelpo, G., et al. (2020). South American expert roundtable: Increasing adaptive governance capacity for coping with unintended side effects of digital transformation. https://core.ac.uk/download/286359460.pdf (accessed on 10th September 2025).
20. Horowitz, M. C. (2018). Artificial intelligence, international competition, and the balance of power (May 2018). https://core.ac.uk/download/211333744.pdf (accessed on 10th September 2025).

21. Maathuis, C., & Cools, K. (2025). Risks and control measures for assuring the safety of trustworthy autonomous weapon systems. *European Conference on Cyber Warfare and Security*. https://www.semanticscholar.org/paper/408c416b0497a6e7877cec7ab57df7b31e67488e (accessed on 10th September 2025).

22. Müller, V. C. (2021). Ethics of artificial intelligence. https://core.ac.uk/download/305120702.pdf (accessed on 10th September 2025).

23. Ogunlade, O., Ogunlade, A., & Tenibiaje, M. (2025). Challenges of securing artificial intelligence-powered systems from cyber threats: Case study of autonomous vehicles. *International Journal of Latest Technology in Engineering Management & Applied Science*. https://www.semanticscholar.org/paper/3eae11c4925781e7858cde7265f6dae954ab0e42 (accessed on 10th September 2025).

24. Petters, D., Schönbrodt, F. D., & Waters, E. (2010). Strange carers. https://core.ac.uk/download/12174004.pdf (accessed on 10th September 2025).

25. Singh, K. D., & Singh, P. (2024). Fog cloud computing and IoT integration for AI-enabled autonomous systems in robotics. *EAI Endorsed Transactions on AI and Robotics, 3*. https://doi.org/10.4108/airo.3617 (accessed on 10th September 2025).

INDEX

A

Accurate motor-based robot arm 110
Acquisition phase 95
Actuator technology 108, 109, 110, 112
Adaptable operation 268
Adaptive automatic systems 170
Adaptive filtering methods 103
Adaptive learning 13, 142, 151, 173
Adaptive learning enables 13
Adaptive regulatory frameworks 277
Adaptive systems 38, 71, 133, 212, 341, 342, 356
Adopted management strategies 379
Advanced human-robot interactions 124
Advanced measurement techniques 301
Advanced navigation systems 213
Advanced path planning 33, 207, 221, 238
Advanced sensing techniques 113
Aerial photogrammetry 214
Aerial photography 211
Aerial systems xi, 14, 19, 20
Aggressive measurement and analysis 301
Agricultural technology sector 217
AI-driven frameworks 127
Air traffic controllers (ATCO) 210
Air traffic management 235
Algorithm-based signal processing 100
Ambient Assisted Living (AAL) 16
Ancient Navigation Techniques 212
Anomaly detection 46, 64, 90, 176, 200
Anticipation-based mobile computing 355
Application-specific integrated circuits (ASICs) 55, 196
Artificial Neural Networks (ANN) 63
Automation 2, 5, 6, 7, 8, 9, 16, 17, 21, 22, 23, 24, 26, 30, 33, 36, 41, 44, 45, 46, 47, 48, 49, 51, 52, 54, 59, 62, 63, 65, 67, 69, 72, 73, 74, 75, 76, 83, 85, 86, 89, 90, 91, 93, 97, 102, 104, 107, 109, 111, 112, 113, 115, 116, 117, 118, 130, 134, 144, 146, 161, 165, 166, 167, 171, 172, 179, 180, 193, 199, 216, 217, 230, 235, 240, 246, 253, 254, 256, 265, 266, 267, 275, 278, 318, 324, 347, 360, 380, 384, 391, 411
Automotive embedded systems development 71
Automotive industry 108, 167, 169
Autonomous driving systems 84
Autonomous flight control systems 141
Autonomous system architectures 139
Autonomous system operation decisions 233
Autonomous systems 1, 2, 3, 4, 5, 6, 7, 8, 9, 10, 11, 12, 13, 14, 15, 16, 17, 18, 19, 21, 22, 23, 24, 25, 26, 27,

28, 29, 30, 31, 32, 33, 34, 35, 36, 37, 41, 56, 61, 86, 98, 103, 105, 110, 111, 117, 123, 124, 125, 126, 127, 128, 129, 130, 131, 132, 133, 134, 135, 136, 137, 138, 139, 140, 141, 142, 143, 144, 145, 147, 148, 149, 150, 151, 152, 153, 154, 155, 156, 157, 158, 159, 171, 208, 209, 217, 221, 223, 229, 232, 236, 237, 238, 250, 251, 252, 253, 254, 255, 256, 258, 259, 260, 261, 262, 263, 264, 265, 266, 267, 268, 269, 270, 271, 272, 274, 275, 276, 277, 278, 279, 280, 281, 282, 283, 284, 285, 304, 357, 373, 374, 375, 376, 377, 378, 379, 380, 382, 383, 384, 385, 386, 387, 388, 389, 390, 391, 392, 393, 394, 395, 396, 397, 398, 399, 400, 401, 402, 403, 404, 405, 406, 407, 408, 409, 412
Autonomous system technology 22, 31, 407
Autonomous technology 9, 15, 23, 26, 27, 30, 31, 32, 34, 37, 50, 255, 268, 270, 369, 374, 379, 380, 381, 388, 395, 398, 402, 408
Autonomous vehicles 18, 19, 376, 383, 410
Autonomous Vehicles 182, 227, 381

B

Balanced system 96
Battery Technologies 302
Battery technology 295
Bayesian belief networks 56
Bayesian networks 103
Behavior Trees (BTs) 16
Beliefs, desires, and intentions (BDI) 143
Big Data analytics 180, 188
Big Data processing 167
Boolean Artificial Neural Networks 101
Brain-computer interfaces (BCIs) 363
Business efficiency 77, 276

C

Car-like nonholonomic robots 226
Cases like cooperative robots (COBOTS) 103
Celestial navigation 212, 218
Celestial Navigation 217
Cellular Vehicle-to-Everything (C-V2X) 258
Centralized decision-making 126
Civil society organizations 396
Closed-loop control 123, 126, 127, 132, 133, 134, 135, 136, 150, 152, 155, 156, 158
Closed-loop control systems 123, 125, 132, 133, 134, 136, 148, 150
Cloud of Things (CoT) 271
Cognitive 6, 7, 8, 9, 10, 11, 14, 15, 22, 29, 41, 43, 53, 55, 56, 80, 98, 107, 109, 111, 119, 125, 133, 155, 158, 161, 172, 173, 191, 199, 202, 207, 212, 238, 241, 242, 246, 277, 278, 333, 334, 335, 336, 342, 344, 345, 346, 353, 354, 356, 358, 362, 364, 366, 369, 371, 372, 379, 401
cohesive framework 343
Cohesive interpretation 99
Collision avoidance 116, 208, 216, 225, 227, 230, 236, 245, 246, 254, 286
Color-coded flight strips 210
Combat operations 96
Combined auditory-tactile stimulation 348
Combining conventional 207
Commercial vessels 230
Compiler optimization 170, 176
Complex environments 5, 9, 15, 57, 85, 89, 98, 101, 107, 143, 144, 145, 148, 158, 208, 210, 221, 237, 241, 243
Complex machines 358
Component-Based Simplex Architecture (CBSA) 208
Comprehensive energy management 293
Compromising computational 185

Computational complexity 233, 234
Computational intelligence (CI) 164
Computational paradigm 12
Computational power 34, 48, 71, 75, 77, 95, 196, 234, 362
Computer science 6
Conserving energy 178
Consumer electronics 53, 69, 73, 294, 302, 314, 335
Contemporary embedded systems 43, 297
Contemporary systems 85
Context-aware applications 293, 317
Control architectures 125, 126, 161
Convolutional neural networks 89, 202, 214, 231
Convolutional Neural Networks (CNNs) 189
Cost-effective local motion planner 240
COTS device-based projects 305
Cross-industry production efficiency 8
Cultural diffusion 212
Curbing energy wastage 314
Cutting-edge computer vision technology 172
Cyber-attacks xv, 199, 340, 367, 378, 379, 381, 382, 385, 387, 402
Cyber-physical systems (CPSs) 3, 208
Cybersecurity strategy 385
Cyborg technologies 336

D

Data acquisition 89, 93, 94, 95, 96, 99, 100, 103, 104, 110, 112, 114, 115, 117, 118, 192, 215, 360
Data analysis 9, 37, 62, 69, 85, 97, 166, 174, 196, 200, 233, 265, 319, 320, 347
Data and decision-making 165
Data communication 200, 261, 264, 315, 316
Data-driven intelligent 194
Data fusion techniques 116
Data services 255
Deep artificial neural networks 146
Deep learning-based methods 106
Deep learning techniques 71, 106, 214
Deep neural networks (DNNs) 58, 61, 67, 77
Deep Space Network (DSN) 54
Delivering naturalistic interaction 342
Delivery as a Service (DaaS) 31
Digital products 351
Digital signal processors (DSPs) 55
Digital transformation 21, 251, 399, 400, 411
Dijkstra's Algorithm 223, 224
Distributed agent architecture (DAA) 53
Distributed behavior 145
Distributed Control Architectures (DCA) 49
Distributed control system 147
Dominant hardware 298
Driver-assistance systems 51
Drone technology 20, 75
Dual-layer control algorithm 226, 238
Dynamic adaptation 107
Dynamic environmental conditions 154
Dynamic Positioning (DP) 354
Dynamic systems 85, 88, 133, 154, 354
Dynamic voltage and frequency scaling (DVFS) 296, 314, 324

E

Early navigation techniques 212
Earth navigation 220
Ecological interface design (EID) 366
Edge Impulse 49, 192
Edge Impulses architecture 194
Edge Learning Machine 48, 62, 68, 74, 76
Effective actuator responses 99
Effective communication 12, 148, 252, 280, 293, 335, 346
Effective energy management methods 293
Effective human-machine interaction 335
Effective signal processing 102

Efficient path planning 207, 214
Efficient resource management 136
Electric connected autonomous and shared (ECAS) 108
Electromagnetic energy 292, 308
Electronic navigation 215, 220, 221
Embedded intelligence 43, 44, 45, 46, 47, 48, 49, 51, 52, 53, 55, 56, 57, 58, 59, 61, 62, 65, 68, 69, 70, 72, 73, 74, 75, 76, 78, 79, 164, 165, 166, 167, 169, 170, 171, 172, 173, 174, 175, 176, 177, 178, 180, 185, 186, 187, 189, 192, 194, 195, 196, 197, 198, 199, 200
Embedded processor design 55
Embedded processors 45, 48, 53, 54, 57
Embedded systems 52, 63, 67, 71, 167, 171, 179, 184, 185, 291, 293, 294
Emotional well-being 336
Energy-aware communication protocols 311
Energy-aware design 316
Energy-centric optimization 310, 315
Energy consumption 46, 48, 49, 65, 69, 71, 72, 75, 179, 181, 185, 194, 255, 263, 277, 279, 282, 292, 293, 294, 297, 299, 301, 307, 310, 315, 316, 317, 318, 319, 320, 324, 325, 326, 327, 329, 340
Energy density problem 306
Energy efficiency 11, 36, 44, 46, 48, 51, 55, 58, 63, 69, 77, 107, 186, 268, 281, 292, 293, 294, 295, 296, 297, 298, 300, 301, 302, 304, 306, 310, 312, 313, 315, 316, 317, 319, 322, 323, 325, 326, 327, 328, 331, 358
Energy harvesting 292, 294, 296, 297, 299, 300, 303, 304, 305, 306, 307, 308, 309, 310, 311, 312, 314, 315, 320, 321, 322, 323, 324, 325, 326, 327, 328, 329, 331
Energy harvesting methods 296, 308, 309, 325, 326, 327
Energy management 264, 265, 267, 270, 291, 292, 293, 294, 295, 296, 297, 298, 300, 301, 302, 303, 304, 305, 307, 308, 309, 310, 311, 312, 313, 314, 315, 316, 317, 318, 319, 320, 321, 322, 323, 324, 325, 326, 327, 328, 329
Energy management methods 295, 303, 305
Energy optimization 313, 321
enhance packet routing performance 253
Ensure user satisfaction 297
Environmental change 115, 138, 148
Environmental factors 232
Environmental stimuli 227
Environmental variability 231
Ethical systems 405
Explicit programming 13
Extraterrestrial environments 105
Extravehicular activities (EVA) 144, 380

F

Facilitate decision-making 276
Fail-safe mechanisms 380
Fail-safe operation 236
Fault tolerance 140, 142, 145, 228
Feature Selection Library (FSLib) 176
Feedback-based approaches 150
Feedback mechanisms 124, 125, 126, 127, 128, 129, 130, 134, 135, 136, 137, 142, 145, 147, 148, 149, 151, 152, 157, 159, 324, 346, 354, 357
Feedback processes 136, 138, 149, 152
Field-programmable gate arrays (FPGAs) 196
Field Programmable Gate Arrays (FPGAs) 74, 77, 79, 189
Food supply chain 350
Forward-looking response 351
Free-flying autonomous robot 237
Full-resource management techniques 186
Fuzzy motion planning systems 206

G

General Systems Theory (GST) 125

General traffic regulations 140
Geographical knowledge 212
Geo-referencing networks 179
Gesture interfaces 345, 359
Global positioning systems (GPS) 94, 103, 110, 118
Graphical user interfaces (GUIs) 339, 341, 343
Graphics processing units (GPUs) 317
Greater perception 270
Ground Robotic-Hand Applications for the Space Program (GRASP) 111
Guarantee system-wide safety 208

H

Hardware innovation 189
Health assessments 178
Healthcare monitoring systems 65
Heating, ventilation, and air conditioning (HVAC) 319
Heat management 48
Heavy traffic environments 227
Hierarchical control designs 125
Hierarchical control structures 126, 136, 141
Hierarchical control systems 124, 139, 140, 159
Hierarchical graph (H-graph) 208
Hierarchical reinforcement learning 232
Hierarchical Spatial-Temporal State Machine (HSTSM) 165, 188, 197
High-critical applications 184
High-Level Expert Group (HLEG) 388, 393
High-level path planning 236
High-level sensors 238
High-level strategic planning 138, 139
High performance work practices (HP-WPs) 400
High-performance work practices (HP-WPs) 24, 387
High-power applications 311
High-speed civil transports (HSCTs) 236
Huge data analysis processes 62
Human-based methodologies 22
Human cognition-technology 362
Human-controlled domains 21
Human error 4, 9, 11, 22, 23, 111, 335
Human involvement 5, 163
Human knowledge (HHK) 208
Human-machine interaction (HMI) 335, 338, 345, 349, 353, 358, 360, 362, 363
Human-Machine Interaction (HMI) 333, 336, 337, 339, 341, 344, 346, 348, 350, 351, 355, 356, 361, 365
Human-machine interfaces 10, 85, 334, 343
Human-robot interaction , 16, 99, 111, 116, 121, 132, 147, 203, 356, 357
Human-robot interaction (HRI) 16, 116
Hybrid algorithms 135
Hybrid intelligent systems 206
Hybrid model 156
Hybrid-open-closed-loop (HOCL) 22

I

Industrial Internet of Things (IIoT) 49, 195
Industrial sensor networks 265
Inertial measurement units (IMUs) 83, 85, 86, 88, 94, 95, 103, 110, 115, 116
Innovative hybrid frameworks 206
Integrated system 138, 267
Intelligence-based decision-making 84
Intelligent communicative agents 126
Intelligent devices 44, 68, 165, 256, 310
Intelligent embedded devices 46
Intelligent navigation systems 236, 240
Intelligent system design 111, 158
Intelligent Transport Systems (ITS) 18
International cooperation 396
Internet of Robotic Things (IoRT) 103
Internet of Things (IoT) 33, 38, 46, 59, 69, 72, 74, 76, 103, 116, 145, 186, 250, 251, 252, 254, 255, 259, 262, 263, 264, 265, 266, 267, 269, 270, 271, 273, 274, 276, 277, 278, 279, 281, 283, 284, 286, 293, 297, 299,

312, 315, 318, 321, 339, 365
Irrigation systems 108
ITHACA (Information Technology for Humanitarian Assistance, Cooperation and Action) 210

J

Job substitution 351

K

Kennedy Space Center 97, 112
Kernel cross-correlator (KCC) 20
Kinetic power 307
Knotted ecosystem 192
Knowledge creation 196, 344
Knowledge Graphs (KG) 349

L

Leading-edge governance 395
LIDAR systems 85, 90, 93, 98, 102
Low-barrier technology 130
Low-energy supply 68
Low power consumption 45, 47, 56, 257, 277, 283, 324
Low-power design 291, 292, 294, 295, 296, 297, 299, 305, 306, 309, 312, 314, 315, 317, 318, 320, 322, 324, 325, 326, 327, 328
Low power design techniques 299
Low-rank decomposition (LRD) 166

M

Machine learning 2, 5, 11, 13, 16, 27, 29, 31, 32, 33, 37, 44, 45, 46, 47, 48, 49, 51, 58, 59, 62, 65, 66, 68, 70, 71, 72, 74, 76, 79, 80, 81, 89, 114, 118, 125, 137, 138, 163, 164, 165, 166, 167, 168, 169, 170, 171, 172, 173, 174, 175, 176, 177, 178, 179, 180, 181, 182, 183, 184, 185, 186, 187, 188, 189, 190, 191, 192, 193, 194, 195, 196, 197, 198, 199, 200, 201, 202, 203, 204, 208, 226, 236, 237, 240, 241, 255, 274, 286, 321, 341, 346, 347, 349, 350, 355, 364, 367, 368, 369, 376, 377, 378, 384, 387, 391, 392, 395, 398, 402, 403, 405, 406, 407, 409
Machine-to-human interaction 87
Machine-to-Machine (M2M) 250, 254, 255, 262, 265, 268, 270, 273, 278
Machine-Type Communication (MTC) 293, 328
Maritime navigation 230
Mastered navigation methods 212
Mechanical systems 169
Media Access Control (MAC) 322
Medical Data Architecture (MDA) 28
Medical monitoring systems 50, 63
Memory capacity 185
Metal-air batteries 302
Microcontrollers 43, 45, 46, 47, 48, 49, 51, 55, 58, 59, 60, 61, 62, 63, 65, 68, 69, 70, 71, 72, 73, 75, 76, 78, 79, 181, 192, 297, 298, 315
Microcontroller technology 63
Micro Electro-Mechanical Systems (MEMS) 280
Mobile innovation 355
Model-Based Predictive Control (MBPC) 297, 318, 319
Model-free control methods 115
Modern computing technology 356
Modern maritime navigation 230
Moon surface navigation 218
Multi-agent systems (MAS) 143
Multidimensional data 86
Multi-disciplinary approach 21
Multidisciplinary approach 334, 408
Municipal energy systems 27

N

Narrowband IoT (NB-IoT) 76
Natural phenomena 212
Navigable design 207
Navigate dynamic environments 225
Navigate semi-structured environments 226
Navigation and path planning 205, 206, 226

Navigation performance 215
Navigation route 210
Network operation 347
Next-generation cyber-physical systems (NG-ICPSs) 186
Next-generation technologies 71, 276
Next generation wireless communication 59
Non-controlled environments 4
Non-Gaussian noise 106
Novel sensors 11

O

Obstacle avoidance 115, 221, 229
Obstacle detection 102, 225, 236
Octree grid model 220, 230
Ongoing synthesis 124
Open-loop 123, 126, 127, 128, 129, 130, 131, 133, 134, 143, 147, 148, 151, 152, 154, 155, 156, 158
Open systems standards 54
Operating functions 87
Operational efficiency 9, 11, 12, 26, 44, 63, 72, 102, 112, 116, 125, 141, 147, 158, 165, 168, 193, 209, 228, 230, 251, 265, 266, 279, 282, 284, 295, 298, 307, 314, 321, 326, 328, 336, 356, 375
Operational environments 139, 225
Operational integrity 68, 129, 267, 276, 383
Optical wireless communication (OWC) 282
Optimal navigation systems 205
Optimal productivity 112
Optimize communication channels 273, 281
Optimize sensor 90, 100

P

Pacific Russia 23
Parallel processing 12, 53, 75
Partial differential equations (PDEs) 146
Path integral control methodology 225
Path planning 208, 209, 222, 227, 241, 244, 246, 247
Path Planning Algorithms 221
Patient care 18, 53, 191, 305
Pedestrians 183, 226
Peer-to-peer data transfer 281
Perception systems 14, 15, 84, 85, 86, 87, 89, 91, 92, 93, 94, 95, 97, 99, 102, 106, 108, 109, 110, 111, 112, 113, 114, 117, 118
Perceptual Control Theory (PCT) 124, 132, 138, 149, 152, 159
Perform real-time processing 68
Personal computer 341
Personal preferences 348
Phase-ordering 170, 176
Photovoltaic arrays 302
Physical space 53, 61
Pioneering projects 10
Power consumption 44, 46, 51, 57, 59, 62, 66, 68, 189, 190, 196, 251, 272, 274, 292, 297, 298, 299, 300, 301, 304, 305, 307, 308, 310, 311, 312, 314, 315, 317, 318, 322, 324, 325, 327, 328
Power distribution 64
Power Harvesting (PH) 324
Precise signal manipulation 99
Predictive maintenance (PdM) 49
Principal Component Analysis (PCA) 174
Probabilistic Roadmap (PRM) 229
Problem-solving flexibility 126
Proportional navigation-based fuzzy controller 221
Protective Optimization Technologies (POTs) 392
Proximal-to-source data processing 65
Prudent adjustments 134
Pseudospectral optimization methods 221
Public health practice 182
Public trust 6, 198, 382, 388, 390, 395, 398, 399, 400, 405, 409
Pushing existing technologies 239

PyTorch Mobile 191

Q

Quantified Self (QS) 348

R

Radio frequency (RF) 282, 309, 323
Radiologist 349
Rapidly exploring Random Tree (RRT) 234
Rapid technological development 249
Real-time analysis 88, 103, 107, 229
Real-time applications 110, 183, 221, 225, 234, 269, 272, 282
Real-time biosignals 3
Real-time communication 52, 265, 275, 277
Real-time comprehension 104
Real-time control 105
Real-time data 13, 18, 21, 29, 33, 62, 63, 71, 72, 73, 74, 77, 78, 97, 108, 109, 115, 125, 127, 138, 142, 152, 170, 173, 200, 210, 220, 226, 233, 237, 250, 251, 255, 256, 259, 265, 268, 270, 272, 280, 284, 326, 340, 347, 379
Real-time data processing 62, 71, 72, 74, 97, 170, 220
Real-time monitoring 76, 90, 116, 134, 167, 269, 282, 299, 301, 321
Real-time operation 57, 263
Real-time processing 58
Real-time scenario assessment 377
Real-time systems 93, 100
Real-world applications 57, 78, 84, 87, 107, 136, 151, 178, 192, 203, 221, 271, 318, 323, 378
Real-world environments 168, 189, 241, 252
Recommendation systems 349
Reduce human labor 226
Reducing resource consumption 358
Reinforcement Learning 177
Replacing human interaction 389
Resilient operation 252
Resource-Constrained Project Scheduling (RCPSP) 143
RF spectrum crisis 282
Risk analysis and management 379
Road environments 379
Road traffic management 279, 287
Robotics and telecommunications 145
Robotic systems 3, 5, 16, 18, 21, 60, 101, 102, 104, 105, 112, 118, 136, 137, 138, 147, 152, 154, 226, 229, 238
Robotic Wireless Sensor Networks (RWSN) 253, 273, 279, 281
Robots operating 98
Robust information flow 253
Rotorcraft-Aircrew System Concepts Airborne Laboratory (RASCAL) 94

S

Safe human-machine interaction 355
Secure data sharing methods 191
Secure Device Pairing (SDP) 254
Self-adaptive systems 27
Self-correction 133
Self-driving systems 257
Self-organize 145
Sensitive information 68, 70, 167, 267, 270, 377, 384, 385, 405
Sensor data enrichment 99
Sensor data processing architectures 100
Sensor fusion 10, 74, 84, 85, 89, 91, 92, 94, 95, 97, 98, 99, 100, 101, 102, 103, 104, 105, 106, 107, 109, 110, 112, 116, 117, 118, 239
Sensor fusion processes 84
Sensor fusion techniques 86, 88, 98, 100, 103, 105, 113, 114
Sensor integration data 107
Sensor readings 99
Sensors extract data 85
Sensor technology 14, 15, 87, 90, 93, 100, 109, 237, 238, 239, 298
Sensor technology developments 238
Share information 250, 259, 263, 264,

272
Share knowledge 146
Signal processing algorithms xii, 94, 100, 101
Signal processing techniques 93, 118
Simultaneous localization and mapping (SLAM) 229
Single devices 44
Smart actuator 109
Smart grid technology 147, 367
Smart home devices 66, 70, 179, 194
Smart home systems 71, 75
Smart home technology 69, 115, 116, 180, 181
Social implications 9, 14, 20, 180, 395, 399
Social learning 11
Social responsibility 181, 361, 386, 404
Society values 395
Socio-ethical implications 169
Sodium-ion 302
Soft computing techniques 206
Soft real-time 53
Software development xv, 317, 354, 362, 371, 386
Solar 295, 299, 300, 303, 305, 307, 308, 309, 310, 311, 313, 319, 322, 323, 325, 326, 340
Solid-state batteries 304
Sophisticated architecture 71, 114
Sophisticated autonomous system 251
Sophisticated computing capacity 166
Sophisticated navigation systems 216
Sophisticated systems 2, 28, 75, 83, 95, 102, 144, 150, 187, 343, 367
Space-based systems 215
Space exploration 27, 97, 102, 105, 112, 133, 144, 189, 217
Spatial environments 211
Streamlined communication 155
Supervised learning 174, 175
Supply chain digitization 269, 289
Surgical intervention 3
Surveillance tools 169
Systematic manner 60

T

Tangible User Interfaces (TUIs) 89, 343
Telecommunication sector 141
Telephone connections 342
TensorFlow Lite 190
Terrestrial navigation 218, 219
Thermal 300, 304, 307, 308, 310, 311, 314, 318, 322, 323
Time-Constrained Project Scheduling (TCPSP) 143
Tiny Machine Learning (TinyML) 55, 59
Topomap framework 234
Total ionospheric electron content (TEC) 219
Traditional hybridization 234
Traditional input mechanisms 341
Traditional power supplies 307, 324
Traffic efficiency 141, 252, 253, 282
Traversed actual distance 222
Trolley problem 374, 375, 383, 390, 393, 395, 409

U

Ultralow-power (ULP) 55, 76
Ultra-reliable low-latency communication (URLLC) 255
Unmanned aerial vehicles (UAVs) 141, 210
Unmanned Surface Vehicles (USVs) 220, 230
Unsupervised learning 174, 176, 177
Upper-level decision processes 101
User cognition 344
User experience 16, 21, 46, 53, 69, 70, 77, 89, 115, 116, 141, 164, 167, 180, 194, 196, 199, 242, 307, 310, 324, 333, 335, 336, 337, 340, 343, 344, 346, 348, 351, 352, 353, 354, 355, 359, 361, 362, 363, 365, 366, 408
User privacy 179, 180, 348, 351, 376, 408
User security 86
Users' mental models 335

V

V2V communication 250, 258, 259, 261, 262, 263, 266, 268, 269, 270, 272, 273, 274, 276, 278, 279, 282
V2V (Vehicle-to-Vehicle) 259
Value-Sensitive Design (VSD) 33, 351, 375, 390, 403, 407
Variable vector countermeasure suit (V2Suit) 92
Vehicle to Infrastructure (V2I) 12
Vehicle-to-vehicle communication 261, 267
Vehicle-to-Vehicle (V2V) 250, 254, 255, 257, 266, 268, 270, 273, 278
Very Large-Scale Integration (VLSI) 280
Virtual reality (VR) 209, 362
Visual sensor frailty 98

W

Wastage and resource consumption 26
Wearable sensors 181
Wearable Technology 181
Wi-Fi operates 254
Wireless communication protocols 249, 252, 255, 257, 264, 265, 269, 271, 275, 276, 279, 280, 283, 284
Wireless communication technologies 59, 252, 260, 262, 284
Wireless energy transfer (WET) 319
Wireless powered communication (WPC) 319, 323
Wireless sensor networks 63, 80, 145, 175, 177, 178, 201, 253, 273, 280, 283, 286, 288, 308, 310, 315, 323, 327, 329, 331
Wireless vehicular cooperative systems 252, 262, 279

Z

ZigBee 249, 251, 252, 254, 255, 256, 257, 259, 260, 261, 262, 264, 265, 266, 267, 269, 270, 271, 272, 274, 275, 276, 277, 279, 280, 282, 283, 284